The Vacuum Cleaner

ALSO BY CARROLL GANTZ

*The Industrialization of Design:
A History from the Steam Age to Today*
(McFarland, 2011)

The Vacuum Cleaner

A History

Carroll Gantz

McFarland & Company, Inc., Publishers
Jefferson, North Carolina, and London

Unless otherwise indicated, illustrations came from the following sources:
Hoover Historical Center/Walsh University (abbreviated HHC);
International Housewares Association, Lifshey Collection (IHA/Lifshey);
Tacony Corporation (Tacony); U.S. Patent Office (U.S. Pat. Off.);
Milwaukee Institute of Art & Design (MIAD);
Dyson Ltd. (Dyson); or author's collection (author).

LIBRARY OF CONGRESS ONLINE CATALOG DATA

Gantz, Carroll, 1931–
The vacuum cleaner : a history / Carroll Gantz.
    p.   cm.
Includes bibliographical references and index.

ISBN 978-0-7864-6552-1
softcover : acid free paper ∞

1. Vacuum cleaners — History.
I. Title.
TX298 .G29 2012   643.6—dc22   2012472486

BRITISH LIBRARY CATALOGUING DATA ARE AVAILABLE

© 2012 Carroll Gantz. All rights reserved

*No part of this book may be reproduced or transmitted in any form
or by any means, electronic or mechanical, including photocopying
or recording, or by any information storage and retrieval system,
without permission in writing from the publisher.*

Front cover images: *clockwise from upper left*—straw broom © 2012 Shutterstock;
1910 Allen manual vacuum; 1908 Electric Suction Sweeper Company Model "O"
(both Hoover Historical Center); 1920 Air-Way Sanitary System;
2011 Riccar "Radiance" (both Tacony Corporation); 1963 Hoover Dial-a-Matic
(Hoover Historical Center); 1995 Dyson DC03 (Dyson Ltd.)

Manufactured in the United States of America

*McFarland & Company, Inc., Publishers
Box 611, Jefferson, North Carolina 28640
www.mcfarlandpub.com*

To my many friends and former professional colleagues
at the Hoover Company and Black & Decker,
whose dedication, innovation and contributions
to the vacuum cleaner industry are timeless

Table of Contents

Acknowledgments viii
Preface 1
Introduction 3

1. Before Vacuum   5
2. Suction Cleaners After 1860   34
3. Electrics 1900–1920   59
4. Consolidation 1920–1940   86
5. Postwar 1940–1970   119
6. Globalization 1970–1990   150
7. 1990 to the Present   176

Postscript 207
Notes 208
Bibliography 217
Index 222

Acknowledgments

A smart author finds people who know more about the subject matter than he or she does, and that's what I believe I have accomplished. It was my good fortune to find some of the best vacuum cleaner historians around. I would like to give them the credit they deserve for their generous support and contributions to accuracy and invaluable information they have provided to this book. Many of them have distinguished themselves in careers dedicated to preserving the heritage of man-made artifacts of our popular culture, as well as the history of the men and women who have made those artifacts possible.

One of these is Ann Haines, who for the past 27 years, has been operations coordinator of the Hoover Historical Center/Walsh University in North Canton, Ohio. She heads a docent staff of retired Hoover employees in the company founder W. H. Hoover's original 1852 boyhood home, often just called the Hoover Vacuum Cleaner Museum. Ann spent 36 years serving the Hoover Company, starting in its headquarters billing department in 1972, and running the Hoover Museum starting in 1984. In 2004, when Walsh University assumed stewardship of the museum, Ann became an employee of the University, which has a Museum Studies Program with the Center as part of its curriculum. Ann provided me with reams of literature regarding the Hoover Company's history, including product information, photos, and background information. She answered dozens of my detailed questions promptly and cheerfully, despite her daily commitments to the Center and the University, and I am eternally grateful to her for her generous and kind collaboration.

Another wonderful historical resource was my good friend and design colleague Victoria Kasuba Matranga, H/IDSA, who edited one of my earlier books on design. Since 1992, Vicki has been design programs coordinator of the International Housewares Association (IHA), for which she authored *America at Home: A Celebration of Twentieth-Century Housewares* in 1997. In her role for IHA, she organizes juries for annual student design competitions and, since 2004, annual houseware design awards for professionals, sponsored by *HomeWorld Business* magazine. She serves as IHA's historian, publicist and researcher. As an independent consultant, Vicki curates many historical houseware exhibits for a number of museums, design firms, and private organizations.

In addition to some of her personal files on vacuum cleaners, Vicki generously loaned me IHA's historical files of vacuum cleaners, assembled from 1968 to 1972 and used by Earl Lifshey (1902–1998) to write and publish his book *The Housewares Story* for IHA in 1973.

Acknowledgments

A priceless historical collection, the Lifshey Collection contains primary information, trade literature, and photos from many manufacturers, as well as historical documents from the Vacuum Cleaners Manufacturers Association (now the Floor Care Division of AHAM, the Association of Home Appliance Manufacturers).

Lifshey became a journalist in 1934 as a reporter for the *Retailing Home Furnishings*, later called the *Home Furnishings Daily* (HFD) in New York. He became managing editor in 1944, and in 1948 began writing a widely read daily column of industry comment called *If You Ask Me*, which appeared twice weekly in the HFD. In 1968, Lifshey retired to Ft. Lauderdale, Florida, and that same year, the board of directors of the National Housewares Manufacturing Association (NHMA), now the IHA, asked him to write a history of the housewares industry, much of which is based on primary source interviews and correspondence. One chapter, *From Brooms to Vacuum Cleaners*, covers 19 pages, and it is his primary materials for this chapter that Vicki provided to me. During retirement, Lifshey became vice president of the Galt Mile Community Association and spearheaded a $3.8 million renovation project. He was named Fort Lauderdale's "Citizen of the Year" in 1996, and in 1999, the city built and named the "Earl Lifshey Ocean Park" in his honor.

In addition to the material Vicki provided me, she also introduced me to Tom Gasko, who was also an incredible resource for this book. Tom, an avid collector of vintage vacuum cleaners, currently is curator of the Vacuum Cleaner Museum, which consists of Tom's collection of over 688 vacuum cleaners, located at the Tacony Corporations' manufacturing plant in St. James, Missouri, where Simplicity and Riccar vacuums are made. In these days when many manufacturers are stressed by mergers, takeovers, or bankruptcy, it is collectors like Tom who are at the leading edge of vintage vacuum cleaner information, images, and analysis.

Interested in vacs since preschool, Tom became fascinated in 1966 at age four, when he was invited to visit the collector Stan Kann (see Chapter 5) to view his collection. Tom began collecting vacs in 1978, at age 16, and for many years, sold vacs door-to-door for a number of manufacturers, including Rexair, Electrolux, Health-Mor, and Kirby. In 1990 he opened his own store, The Vac Shac, in Festus, Missouri. From 1995 to 2001, Tom was president of the Vacuum Cleaner Collectors Club (VCCC) with members worldwide, and initiated the Antique Vacuum Contest and the Antique Vacuum Swap Meet. From 1997 to 2010, Tom wrote articles on vacuum cleaner history published in *Floor Care Professional*, a publication of the Vacuum Dealers Trade Association (VDTA). In 2000, Air-Way asked Tom to help design their Signature Series, of which they sold 8,000; the first hundred featured Tom's signature on the serial number plate, making them collector's items. In 2006 Tom and friends started the Vacuum Cleaners Association, for those who work in the vac industry and have diverse collections of vacs from vintage to modern, and in 2009, Tom joined Tacony with his collection, and currently manages the factory retail outlet as well as the museum.

Tom generously answered many of my historical questions, and provided considerable information to me for this book. His articles for VDTA were extremely helpful, and through his efforts, Adrienne Pierson at Tacony provided me with a number of historical and current vacuum photos, which grace this book. Unfortunately, unexpected family events forced Tom's discontinuance as an active source of consultation. Still, I continued to seek advice from prominent collectors of vintage vacuum cleaners, to insure accuracy and to provide interesting information.

Among these is Fred Stachnik, who enthusiastically volunteered to review my manuscript for accuracy and to add suggestions. A seasoned collector of over 250 vintage vacs, with a particularly outstanding representation of Hoover cleaners, he is also the current webmaster of the Vacuum Cleaner Collector's Club at http://vacuumland.org/. By age ten Fred had collected 20 vacs and was repairing and restoring them for neighbors. After a visit to the Hoover Company, where he was royally entertained, a local news story of the event escalated his precocious knowledge to national TV prominence. In 1994, at age 12, he made a big hit on the *Tonight Show* with Jay Leno by demonstrating his collection of vacuum cleaners. Fred's early experiences on TV appear in Chapter 7. Fred generously offered many invaluable suggestions, as well as numerous corrections and encouragement of my efforts.

Vicki Matranga also connected me to Mark Lawson, director of galleries at the Milwaukee Institute of Art & Design (MIAD), to whom I am indebted for his generous permission to obtain and use vacuum cleaner images from MIAD's outstanding Industrial Design Collection. I am also grateful to Erin Adler and Valerie Silvis at Dyson, for providing and granting permission to use outstanding images from Dyson's files.

This book would be much less informative and complete without the generous assistance of all these wonderful people, and you, the reader, are benefited by their knowledge and kind contributions of time and substance. My thanks cannot possibly equal the enormous value they have added.

Preface

The purpose of this book is to tell the fascinating story of the technological evolution of vacuum cleaners, the men and women who invented or improved upon them, and the changing social, economic, environmental, health, artistic, technical, political, and particularly, competitive contexts within which they evolved. Moreover, most people do not realize that such a history began long before the electric vacuum cleaner with which we are familiar today. Early attempts to create suction in carpet sweepers were made not long after the Civil War, and were operated manually with considerable physical effort until well into the twentieth century when electricity was commonly available. Their predecessors were simple hand tools such as brooms and brushes that were known and used from antiquity.

When my publisher originally suggested I write such a book, I was more than astonished to discover that no comprehensive book on vacuum cleaner history existed. As a retired design educator and historian, and upon the urging of my friends and my innate compulsion to fill a "vacuum," I decided such a book was sorely needed. There should not be a library, public or private, with a "vacuum" of vacuum cleaner books among its shelves. How else, indeed, would the bookshelves be relieved of their ever-accumulating film of dust?

Just because there was no comprehensive book on the history of vacuum cleaners does not mean there was no literature on the subject. Articles, papers, and recently websites, about vacuum cleaners have been available over the last hundred years, to say nothing of the thousands of public advertisements, and hundreds of actual products and photos that survive in private and public collections, including several museums and organizations devoted to the subject. Many individual vacuum cleaner manufacturers have documented their own history on websites, brochures, and in several books. U.S. patents, now available online, describe all the individual mechanical inventions regarding cleaning devices over several centuries, as well as design patents for their external appearance. Many articles and books describe the economic, technological, artistic, social, and legalistic historic context in which vacuum cleaners developed. All printed sources are found in the reference notes as well as the bibliography. It's just that until now, no one has organized this information into a sequential and detailed history that explains how the technology developed as part of the industrial revolution. This is what I have done, with the generous collaboration of historical experts described in the acknowledgments.

I am generally familiar with the industry because of my former professional life as head of Hoover's industrial design department from 1956 to 1972, and during that period, I was awarded dozens of design patents for Hoover products, including its Floor Washer, Dial-a-Matic, Portable, Lark, Pixie, Handivac, Floor-a-Matic, and Swingette. Also, as director of industrial design at Black & Decker from 1972 to 1986, I was awarded the design patent and a mechanical patent for the world's first Dustbuster in 1979. A number of my designs have won national design awards. My professional experience of 30 years has resulted in my lifelong interest in the subject, as well as some inside stories and the many multidisciplinary perspectives that I will share with you in this book, particularly those concerning the important appearance aspects of vacuum cleaner design, my own particular specialty.

Introduction

We are all familiar with vacuum cleaners, having used them since childhood, in one shape or another, to perform the somewhat mundane task of removing dust, dirt and debris from household surfaces, furniture, and corners. To many of us, a vacuum cleaner is simply one of the many electrical/mechanical appliances that surround us in today's highly technological environment.

Nevertheless, historians know that the vacuum cleaner was one of the "machine age" marvels of the early twentieth century. In a recent program on the History Channel, *Popular Science* magazine conducted a survey among leading technology authorities to identify the "101 Gadgets that Changed the World," in order of significance. The vacuum cleaner came in eighteenth, not far behind items such as the Global Positioning Device (GPD), the Blackberry, the radio, television, the sewing machine, the phonograph, and the incandescent light bulb. Not too shabby a showing for an invention that is 143 years old, older than the automobile (which, incidentally, was not classified as a "gadget").

But in the same way that the noisy, gasoline-powered automobile grew to dominate our quiet outdoor landscapes, the equally noisy, electric-powered vacuum cleaner grew to dominate our even quieter home interiors. Just as the roar of the new horseless carriages startled horses, so did the annoying whine of vacuum cleaners frighten cats, dogs, and children. Both these inventions were the result of the rapidly advancing technology that characterized the early 1900s. People got used to them, of course, and today both are considered essential to our middle class lifestyle, but are so ubiquitous that they pass almost unnoticed.

Still, the basic technology of vacuum cleaners developed well before twentieth-century motors and electricity, a few years after the Civil War. In fact, primitive cleaning technology can be traced back to antiquity. The fascinating history of housecleaning began with the fabrication of basic hand tools using everyday natural materials and concludes with the recent computerized technology of robotic, cordless vacuum cleaners that require no human operator. Along the way, brooms progressed from twigs to soft brushes, hard floors to carpets, hand-wielded rug beaters to mechanized agitators, manually operated vacuum devices to those powered by electric motors, electrical cords to rechargeable batteries, and a novelty luxury device to a household necessity. All of this progression was the result of the compelling and innate human desire to collect and dispose of household dust and dirt; a passion for cleanliness, orderliness, and healthfulness that still motivates millions of housekeepers around the globe.

This process of evolution was influenced by social, scientific, technological, artistic, and economic changes over recent centuries, as America came of industrial age. Not the least of these was the growing public awareness in the late nineteenth century that disease was caused by nasty microscopic organisms that inhabit the world all around us. Cleanliness soon was considered next to godliness. In addition, home ownership, house size, and furnishings expanded hand in hand with the rise of the middle class and economic independence, and people needed more effective methods of cleaning.

The history of vacuum cleaners, like the history of any technology, is humanized by stories of individual inventors and business entrepreneurs, both men and women, such as James Murray Spangler or Anna Bissell. Some of them became not only household names, or brands, but became descriptive verbs of action when any product of the type they innovated was operated, such as "bisselling" or "hoovering." (Patent attorneys hate when a trademark becomes part of common language — it tends to negate the unique protection intended!)

Intensive competition among numerous manufacturers inspired technological breakthroughs and significant improvements every year. Small local businesses morphed into international giants through performance, quality, and good management. As foreign economic and global competition increased dramatically in the late twentieth century, some of these companies merged with others and flourished, while others failed. The story of vacuum cleaners, however ordinary their place or function in a multitude of modern technological wonders, parallels and echoes the inspiring twentieth-century saga of American industrialization, technology, and global leadership in mass production and distribution to a growing mass market of consumers.

Almost every household has at least one vacuum cleaner, more often several, which, because they are practically indestructible, tend to last for generations, becoming living antiques over the years. Some of my antiquarian friends tell me the older models are equal to, or surpass, the newest versions. Therefore, almost any reader, regardless of age, will recognize specific vacuums in this book from their own, their parents', or their grandparents' homes, and will have used many of them, occasionally with annoyance, often with enthusiasm, and sometimes with amazement. A number of vacuums were designed by national design celebrities and have won national design awards. Others are in the Smithsonian American History Museum along with other cultural artifacts. The U.S. vacuum cleaner industry alone is an annual $3.5 billion business, with active trade associations and many enthusiastic private collectors. Love them or hate them, vacuum cleaners are an integral part of our modern culture.

Chapter 1

Before Vacuum

After church, Johnny tells his parents he has to go and talk to the minister right away. They agree and the pastor greets the family. "Pastor, "Johnny says, "I heard you say today that our bodies came from the dust."

"That's right, Johnny, I did."

"And I heard you say that when we die, our bodies go back to dust."

"Yes, I'm glad you were listening. Why do you ask?"

"Well you better come over to our house right away and look under my bed 'cause there's someone either comin' or goin'!"

I know, I know. It's an old joke. But it highlights an element of truth that is relevant to the origin of vacuum cleaners. Eighty percent of household dust, in fact, does consist of dead human cells. The rest of this insidious stuff, nearly a ton of which falls on one square mile each day,[1] includes soil, pollen, soot, bacteria, animal dander, feathers, moth eggs, fungi, chemicals, insects, dust mites, lint, hair, and mold spores, to name just a few. Along with the dust there is substantial dirt and litter from normal human activities. Back in earlier days, that included wood chips, saw dust, manure, coal dust, ashes, straw, and farm debris of all sorts. It is not surprising that people wanted to collect and dispose of dust and dirt in their homes, probably back to the dawn of human occupation of caves.

Good housekeepers, bless their hearts, have always strived to keep their houses (or caves) clean, even at a significant expenditure of labor. Aside from the unsightliness to which daily occupants are exposed, the visibility of dust and dirt to guests or visitors could brand the housekeeper as inept, uncaring, or worst of all, lazy. Being labeled as any of these was, and still is, the ultimate embarrassment for many women, the traditional homemakers, at least until just a few years ago. Because of this instinctive, fundamental human propensity for cleanliness and pride of place, removal of dust and dirt from homes has been a high priority since antiquity.

Ladies may notice, with slight annoyance, that I mentioned women in the context of house cleaning, probably one of the more onerous tasks implicit in any household division of responsibilities. This is in no way intended to offend the men of the house who happily volunteer to clean, nor to imply that cleaning should necessarily be "women's work." However, it is essential to accept as a historical fact that most cleaning implements were, and in many cases still are, almost exclusively designed for, promoted to, purchased by, and used by women, just as are many household appliances, furnishings, and related cleaning products today.

Blame this bias on a long tradition of social history. Women had always shared the labors of maintaining a home equally with men in preindustrial ages, up to the mid-nineteenth century, because the work of homesteading was endless and exhausting from dawn until dusk. The division of labor appears to have been originally based on perceptions of physical strength and proximity to the home. Men assumed the heavier duties of cutting down trees, chopping wood, plowing, building fences, shelters, barns or structures, butchering the animals, etc., as well as tasks that required being away from the safety of home, or which posed a threat of some external danger, such as hunting, traveling for provisions, supplies, or delivering produce to market, plowing, planting, and harvesting. To women fell the tasks of cooking, cleaning, sewing, spinning, washing, and the nursing of children — essentially those physically lighter tasks in or near the safety of the home.

As far back as the thirteenth century, these respective roles became known and labeled as "husbandry," an activity we now call "farming," and "housewifery," used until 1841 in England, and until 1871 in America, when it began to be called "housework." Thus was established the traditional and almost doctrinal separation of household tasks into "men's work" and "women's work."[2] This remained until the industrial revolution, when, as we will see, household responsibilities changed dramatically for men, but for most women, only in kind and intensity. Women became master managers and keepers of the home, physically, emotionally, and morally.

Electric vacuum cleaners, with which this book is primarily concerned, are only the most recent weapon in the arsenal of indoor cleanliness. Before them, there was an array of manual vacuum cleaners powered by strenuous human effort, and before them, the rotating brushes of carpet sweepers with handles, also powered by humans. Further back were hand-wielded brushes, carpet beaters, and dustpans, and even further into the dawn of human history were a wide variety of brooms, the basic tools of cleaning, just as the axe and handsaw were the ancient and basic tools of male homesteaders. All are still with us today, despite our sophisticated technological advancement. We will work our way back up though this historical sequence so that we can better understand, step by step, what led to each subsequent innovation over time in cleaning methods, attitudes, and tools.

Nothing could be more basic than a broom, essentially a bunch of flexible natural materials crudely affixed to a stick. It is not hard to imagine the use of such a device fashioned of branches or twigs to sweep debris from a cave, or from the dirt or sand floors of a tent or some other crude living structure. Archeologists have found examples of brooms from 2300 B.C. Ancient civilizations in Babylon, Egypt, and India probably used sorghum, a cereal grass, for

Besom or twig broom. HHC.

brooms. Romans used soft sponges on long handles or brooms of wild myrtle and tamarisk twigs. When floors of hard earth or stone appeared in medieval European houses, it was customary to cover the floors with reeds, rushes, or straw, which was occasionally removed and renewed for sanitary reasons. Stiff, round-style twig brooms, similar to those we now associate with witches or Harry Potter's fictional game of Quidditch, were perfect for this, and had been since biblical times called besoms, or as some say, "bezzums" (Isaiah 14:23: "I will sweep it with the besom of destruction"). When floors of tile, mosaic, wood and carpet became more common, softer brooms or besoms of rushes, broomcorn, sorghum, pig bristles, whisk, or cocoa husk fibers became popular.

Used for centuries as the primary tool for the eternal purpose of cleaning homes, brooms also became symbolic of the home. From the 1500s, "jumping the broom" was, and still is today in some cultures, a traditional and ceremonial part of weddings. Scholars can't agree whether the ceremony originated in Africa, or in Wales by Celts, Druids, and Gypsies. Typically, the ceremony required the bride and groom to signify their leap into a new life and shared domicile. Some placed the broom on the ground to jump over; others placed it at an angle in a doorway for the groom to jump over first, followed by the bride.

Over centuries without central heating, fireplace ashes and soot must have been a primary and constant source of house dirt. The center of American colonial daily life was the fireplace, needed in the winter for heat and year-round for cooking. Wood had to be carried in and ashes out, creating a daily mess that required constant attention. In the seventeenth to eighteenth centuries, craftsmen in Europe made such brooms that were imported to America as an item of colonial commerce. Because these were expensive, colonists often made crude brooms by tying straw, hay, rushes or cornhusks to a stick or handle, but these were short-lived and required repeated replacements.

It was the ingenious Native Americans, well experienced with the use and qualities of native wood, who provided colonists with an inexpensive alternative to costly imports or self-crafted versions. Variously called "splinter," "splint," "split," "birch," or "Indian" brooms, these were made from a yellow birch stick or branch by peeling numerous thin strips a foot or so down-

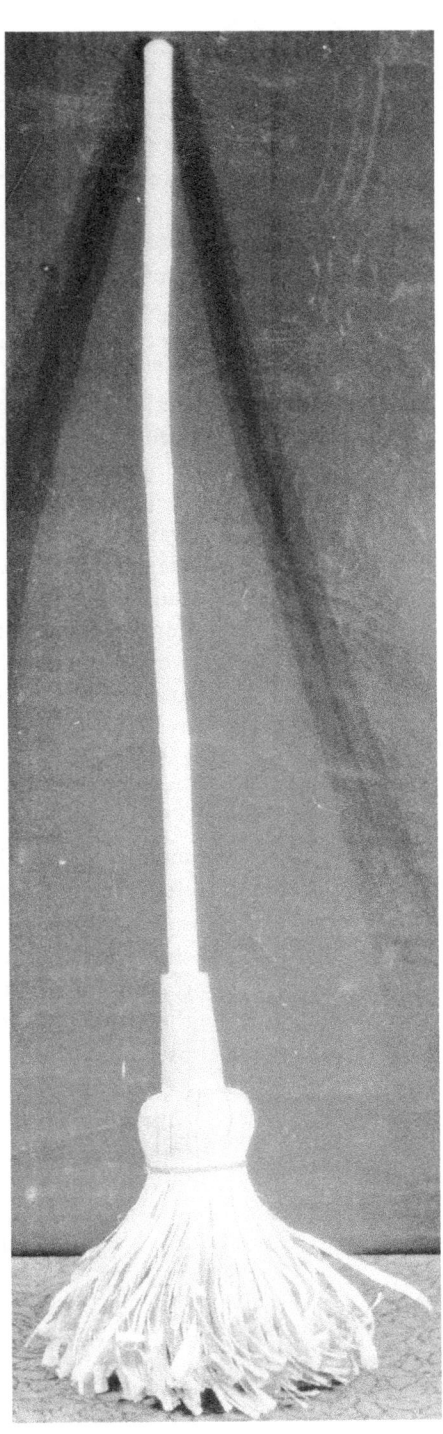

Eighteenth-century Indian or splinter broom. HHC.

ward from one end, after removing the bark. After the stick was peeled to the center of the branch, the heartwood stump was removed, the strips were gathered, reversing their direction upwards into a bundle, and binding them near their base with wood strips or rawhide. The original branch became a natural handle, long for brooms or short for scrubbers. Indian craftsmen sold or traded these brooms to colonists, who used them for a range of cleaning needs.

In 1797, Levi Dickenson, a farmer in Hadley, Massachusetts, handmade a broom for his wife using the tassels of a variety of sorghum, a grain he grew for seeds. Sorghum is a family of worldwide cereal crops with dozens of species that vary in sugar content. Some, such as sugar cane, are of high sugar content, and are used for sweet sorghum syrup, animal feed, biofuel or making alcoholic beverages, while others, such as millet and milo, are low in sugar. The plants have a single stem and ribbon-type leaves and tassels with seeds. Individual varieties range from two to ten feet in height, and tassels from six inches to two feet in length. The variety *Sorghum vulgare* var. *technicum* has the longest tassels. They are stiff and strong, even when wet, yet flexible and resilient when dry. Levi had rediscovered the most effective material for making brooms.

Levi's wife liked it so well she told her neighbors about it and word of Levi's excellent broom spread. He was soon growing more sorghum to keep up with demand for his brooms. By 1810, when the foot-treadle broom machine was invented along with other machinery of the industrial revolution, the sorghum Levi and other broom makers used was called "broomcorn." Levi's business became the C.D. Dickenson & Son Company, the first in the country to make tools for the manufacture of brooms, in about 1832. Holes were drilled in the handle for pegs, to which the broomcorn was lashed and secured with a woven pattern of flexible wood strips. Brooms were available with either long or short handles for a variety of cleaning applications.

The Shaker colony of Niskayuna (later called Watervliet), north of Albany, New York, also began making brooms around 1798. One of its members, Theodore Bates, is credited with inventing the flat broom by flattening the broomcorn with special vices and sewing it into the flat-sided brooms we are familiar with today, rather than the round brooms of the earlier besom type. Shakers developed foot-treadle machinery to wind the broomcorn around the handle while securing it firmly, and initiated the use of wire to sew the broom to its handle rather than the woven stems previously used. In the mid–1800s, it was found that sorghum made an excellent broom and farmers in the Mohawk Valley used much acreage for cultivation of broomcorn. The seed of the crop alone, it was said, paid the expense of cultivation, also being, when mixed with other good corn, good food for cattle.

As the industrial revolution evolved in the mid–1800s, more efficient and durable machinery replaced handcrafts. Greater quantities of brooms also enabled their cost to be lower. By 1876, C.E. Lipe of Syracuse, New York, invented the first power-driven broom sewing machine. In 1886, Lipe sold his patents to the Hand Stitch Broom Sewing Machine Company

Sorghum broom, ca. 1795.

of Pittsburgh, manufacturers of the McCombs machine, one of only two such machines made up to 1900.[3]

In 1884, Julius Wasserman founded the Amsterdam Broom Company in Amsterdam, New York, which soon became the center of the broom industry. The company was incorporated in 1909 with Julius as president, and it would become the largest broom factory in the world, although there were at least four major broom factories in Amsterdam, each employing more than 250 people. The company made an extensive line of household brooms, including the Gold Bond broom, 15 inches long, with four rows of stitching and a finely polished yellow handle. By the early twentieth century, most well-appointed households had a special broom rack to hold the many varieties and sizes required for different and specialized cleaning tasks. By then, the growing availability of vacuum cleaners and electricity reduced the need for brooms proportionally. By 1957, after years of declining sales, Amsterdam had only 30 employees, and the Wasserman heirs sold the factory to the Edy Brush Company, which was operated as the Amsterdam Brush Company into the 1970s.[4]

Brooms, of course, were good for sweeping dirt either out the door, or into a mound to be removed by dustpans or similar impromptu materials such as newspapers or cardboard. But brooms were relatively useless for removing dust. Their use largely just recirculated dust into the air, where it was breathed into human lungs or was redeposited in other locations in the home where the broom could not reach or be used. It settled on the floor, furnishings, shelving, dishes, or any of the numerous decorative objects that graced nineteenth-century homes.

Among these decorative furnishings were carpets, as they became available at lower cost to households. While enhancing décor with an appearance of comfort and luxury, they complicated enormously the cleaning process for the housewife, because dirt and dust settled or was tread into the woven fabric of the carpet, where it became deeply embedded. Loose surface debris could still be swept and gathered, as on hard floors, but much of the dust and fine dirt remained in the carpet itself,

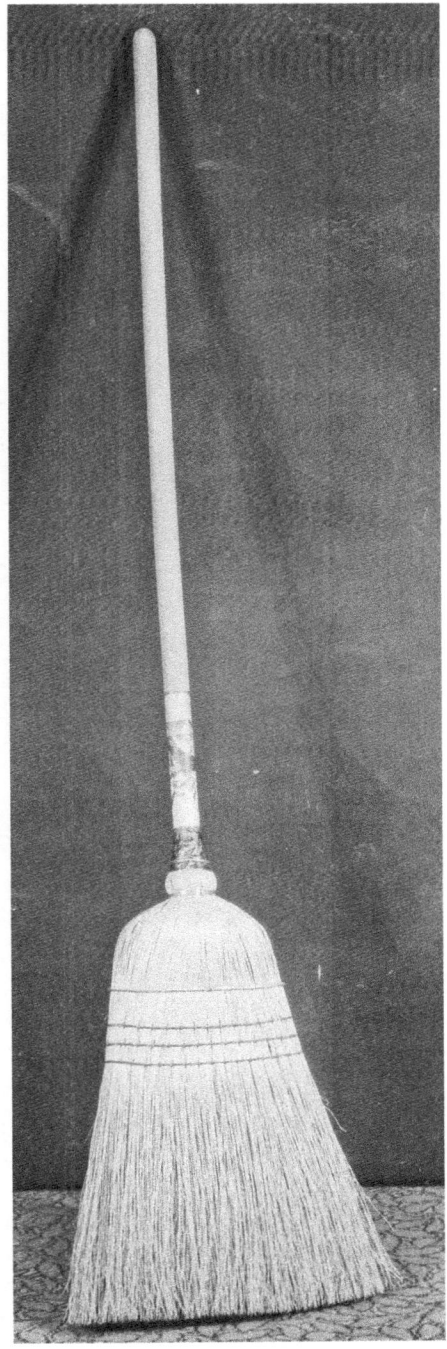

Modern corn broom, ca. 1876. HHC.

and could only be cleaned by laboriously removing holding tacks, taking up the carpet, hanging it outdoors, and physically beating it vigorously with stout sticks, paddles, or special rug beaters made for this purpose. This was a most unpleasant and exhausting task (despite

Carpet beater in use, ca. 1850. HHC.

the smile on the young lady's face), and in addition, led to weakened spots and ripped seams, shortening the life of the carpet. It is no wonder that carpets eventually would become central to the evolution of the vacuum cleaner.

Carpets, like brooms, have a long history. Wool was spun and woven by hand since about 6000 B.C. Marco Polo first noticed carpets in A.D. 1000 as he traveled in central Asia, where nomads expertly wove them from wool. The technique soon spread to Persia, India, China, and North Africa. The Moors introduced carpets into Spain. Pierre DuPont began weaving what were known as Oriental carpets in Paris in 1596, and was granted an 18-year royal patent by Louis XIII in 1627. A manufactory was established on the banks of the Bièvre River in 1664, in the factory originally established by the Gobelin family of dyers in the mid-fifteenth century. The Gobelins Manufactory was well known as the royal supplier of carpets, tapestries, upholstery, and furniture for Louis XIV, and its seventeenth-century buildings still exist today in Paris. Some French weavers fled to England in 1685 to escape religious persecution, settling in Wilton in Wiltshire, England, where they taught the new skills to local weavers. In 1750, a Mr. Moore was awarded a prize for the best imitation of "Turkey" (Turkish) carpets, and there was an establishment for such manufacture in Paddington. Subsequent carpets were made in Axminster in Devonshire.[5] There was a choice of many different weaves. Brussels carpet, originating in Tournai, Belgium, in the mid–1700s, was a superior texture made of worsted and linen, and had a rich, corded appearance. Wilton carpets were woven in a similar manner to Brussels, except the loops were cut open to form a velvet surface.

Carpet weaver, ca. 1770. From *The Complete Book of Trades*, by Nathaniel Whittock, 1837.

Tapestry carpets were manufactured by a process invented by Richard Whytock of Edinburgh in 1832, which colored the threads before the weaving process, rather than after. James Templeton of Glasgow invented Axminster carpets in 1839. His process produced an infinite number of colors that were dyed in the yarn before weaving and woven in a two-step weaving operation using chenille strips. By 1770, handwoven carpet making flourished in Britain, attracting top designers such as Owen Jones (1809–1874), but carpets remained a costly luxury, affordable only by the wealthy.[6] Nevertheless, hand-loom weaving was still being done in Northern Ireland as late as 1932.[7]

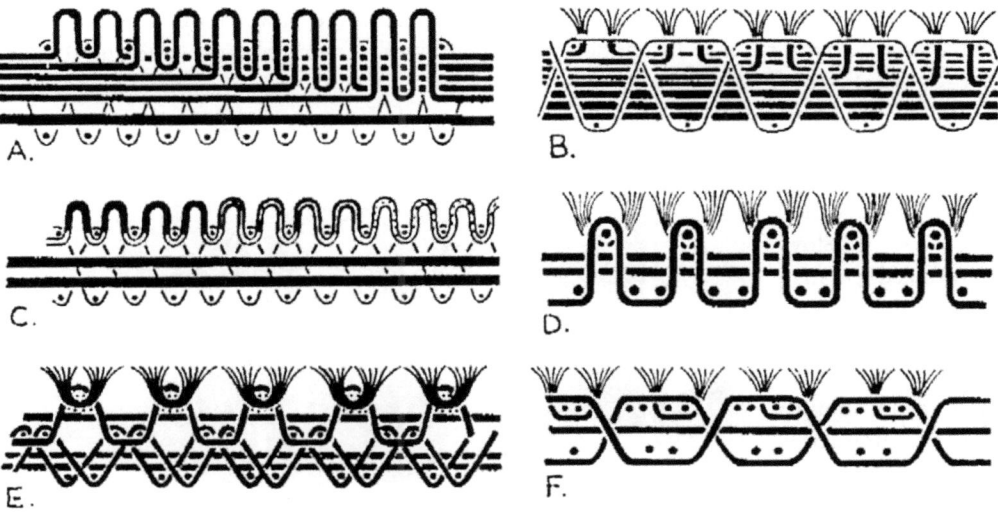

Carpet weaves, ca. 1850: (A) Brussels; (B) Wilton; (C) Tapestry-Brussels; (D) Tapestry-Velvet; (E) Chenille-Axminster; (F) Imperial-Axminster. From Derry and Williams, p. 581.

In 1791, William Peter Sprague opened the first U.S. woven carpet mill in Philadelphia. Sprague was born in about 1750 in the vicinity of Axminster, England, where he served as an apprentice of Thomas Whitty in his Axminster carpet factory. While serving in the British Navy against Americans in the Revolutionary War, Sprague deserted the Navy and relocated his family to Burlington, New Jersey, before the war was over. There he bought property in 1783. He began advertising carpets of any size woven in "the Axminster mode," which were priced "nearly as low as Wilton carpeting" but "of double its durability." By 1790 he had relocated across the Delaware River to Philadelphia, and the Senate Chamber carpet installed in 1791 on the floor of Congress (Independence) Hall, with a seal of the new United States, is often attributed to Sprague. Sprague relocated his carpet establishment a number of times in Philadelphia before his death in about 1808.[8]

By 1800, many other woven carpet mills were in operation, some of them making 27-inch runners that could be sewn together to make larger carpets. However, few people could afford handwoven carpets. A major breakthrough occurred in 1801, when Frenchman Joseph Marie Jacquard (1752–1834) invented a loom that programmed patterns with a chain of punched cards for the silk industry, but it was soon adapted for use with cotton, wool, linen, and eventually, carpet weaving. The Jacquard loom did not appear in the U.S. until 1825, but today, it is considered an important step in the history of computing hardware because of its ability to change the pattern of the weave by simply changing cards, and because it preceded Charles Babbage's (1791–1871) Difference Engine of the 1820s.

In 1839, Boston inventor Erastus B. Bigelow (1814–1879) invented the power loom, which was improved in England and first came into use by the Clinton Company of Massachusetts for both Brussels and Wilton carpets. The automatic insertion and removal of wires with looped ends raised the looped pile of the Brussels carpets, while thinner wires with a knife blade raised and then cut the loops for the Wilton pile. Bigelow applied for a patent for a Jacquard power loom mechanism in 1842 and had developed it by 1849. By 1850 the power loom had tripled production at the Clinton Company, which later became the Bigelow Company.[9]

In 1845 Alexander Smith started a carpet manufacturing plant in West Farms, New York. In 1876, the American Halcyon Skinner perfected the power loom to make Royal Axminster carpets. He and Alexander Smith combined their companies, and the Alexander Smith Carpet Company would by 1929 become the largest producer of carpets in the world. In 1877, Bigelow introduced the first broadloom carpets, which means that they were woven on very wide looms, thus enabling large carpets without the

Textile mill workers in Lowell, Massachusetts, ca. 1845. New York Public Library, Astor, Lenox, and Tilden Foundation, from Burke, *Connections*, p. 149.

seams of sewn-together runners. The next major step was the invention in 1895 of the automatic loom by J.H. Northrop of Massachusetts, called by many "the loom of the twentieth century," and which left the weaver with little to do but repair breakages of warp or weft and refill hoppers as needed.[10]

As with many products produced by mechanical means in the industrial revolution, prices of carpets fell as production quantity increased, and such goods became affordable for the growing middle class. Carpets were a luxury, but brought with them an enormous liability for households because, unfortunately for housewives, carpets, whether handwoven or machine produced, whether Axminster or Wilton, whether expensive or cheap, whether thick or thin, were the most perfect mechanisms for collecting and retaining dust and dirt ever devised by man. Carpets are the villains of this story or the heroes, depending on your point of view, because carpets were the fundamental reason that carpet sweepers and vacuum cleaners were invented. Carpets served as a nagging challenge for centuries, waiting for a solution to clean them effectively and easily. It would not be much longer.

As a result of the industrial revolution during the first half of the nineteenth century in America, not only brooms and carpets, but many manufactured household products became available at affordable cost, increasing both the quantity and quality of home furnishings, equipment, and goods. At the same time, immigrants were arriving in droves to populate the country as it expanded westward, and the number of new households increased dramatically with the proportional demand for manufactured goods. While the industrial revolution had begun in England in the early eighteenth century, well before the American Revolution, it did not impact this country until well after that war, when we were already at least 50 years behind the mother country in industrialization.

The industrial revolution in America began late and slowly. In his 1790 address to Congress, president George Washington (1732–1799) declared that the country had to develop independence in manufacturing for self-defense:

> A free people ought not only be armed but disciplined; to which end a uniform and well-digested plan is requisite; and their safety and interest require that they should promote such manufactories, as tend to render them independent of others for essential, particularly for military, supplies.[11]

He pleaded to Congress to give "effectual encouragement ... to the exertion of skill and genius at home." Congress responded within a month by establishing the U.S. Patent Office. Patents giving inventors exclusive rights for 14 years were immediately available for "any useful art, manufacture, engine, machine or device, or any improvement thereon not before known or used." Responsibility for the patent office was vested in a board that included then Secretary of State Thomas Jefferson.[12]

Early inventions patented included an automated gristmill by Oliver Evans (1790), a steamboat by John Fitch (1791), and the cotton gin by Eli Whitney (1794). By 1800, however, only 300 patents had been filed. It was not until after the War of 1812 that the federal government initiated efforts to unify the nation in agriculture, commerce, and industry. Legislation by Congress in 1816 rechartered the Bank of the U.S., established a single currency, and provided federal credit and land grants to individual states for the purpose of constructing thousands of miles of canals and roads over the next 30 years, thus linking states together in commerce and transportation and expanding agricultural markets. It was the first national public works program.

The 1816 legislation also was intended to strengthen the U.S. economy by reducing reliance on European imports that still continued unabated. England had not permitted its colonies to manufacture goods because to do so would have reduced the lucrative English colonial trade income. Congress now placed high tariffs on imported goods and provided credit to fund the establishment of indigenous American industrial manufacturing ventures. Within ten years, hundreds of new manufacturing companies were chartered in many states, making a variety of products for mass consumption.

Households could not only acquire inexpensive manufactured carpets and brooms, but also inexpensive furniture, cook stoves, utensils, farm tools and equipment, lamps, textiles, clocks, chandeliers, drapery, glassware, and many other artifacts that previously were available only via expensive English imports. By 1820, American patents had outpaced British patents, and by 1826, the Franklin Institute in Philadelphia held the first exhibition of American manufactured products. What would become known as the "American System" of manufacture featured the interchangeability of parts, an essential principle of mass production. Over the next 100 years, American industrialization continued to expand and intensify until it surpassed the rest of the world, profoundly changing our lifestyle and standard of living.

The impact of industrialization on individual households was gradual, but inexorable. The first communities affected were the larger urban centers, where most volume manufacturing occurred, and where over 30 percent of consumers resided. But over time, smaller communities in rural areas were impacted in the same way, as improved transportation of goods created a consumer market that both fostered business and encouraged the growth of a middle class. The purchase of manufactured goods was based no longer on bartering, but on cash. To obtain such cash, and to provide a better standard of living for his family, men needed either to sell their farm products on the open market, or become craftsmen, tradesmen, artisans, professionals, shopkeepers, or laborers. New manufacturing industries were providing jobs for a wide range of skills in factories. Schools in "mechanic arts" were founded, teaching new skills of engineering, invention, design, and mass production. Public works projects offered work for thousands on canals and roads. Demand for products and services increased along with population growth, as did the opportunities for employment and steady income. With such income, men could provide their families with not merely necessities, but with what were once considered luxuries.

Inexpensive manufactured products also seemed to make life easier for families. Mass-produced milled white flour reduced the need to hand-grind corn, rye, or wheat. Iron stoves, which were twice as efficient as fireplaces, required less wood to be cut, split, and carried. Coal, now economically transported by canal, required no preparation work at all. Manufactured leather goods, such as shoes and harnesses, eliminated the need for leatherwork at home. Ready-made cloth eliminated the need for home spinning and weaving. Chemical fertilizers, packaged seeds, and mechanical farm equipment increased productivity and reduced labor on farms. Sewing machines speeded up the making of family clothing. Kerosene eliminated the need for handmade candles. Each new mechanical marvel offered more of the same.

However, Ruth Schwartz Cowan, in her 1983 book, *More Work for Mother*, argues convincingly that it was men whose home tasks were mostly reduced by these early manufactured goods and were enabled to take part-time jobs to earn extra cash. Women's roles were increased, if anything — not reduced. Cowan concludes that fine-ground white flour enabled the baking of pies, bread and pastries, which was not possible with crude home-ground corn or wheat flour. So women baked more. Iron stoves allowed more varied and complex cooking procedures, creating more work than earlier single pot cooking, but also required regular cleaning with stove polish to prevent rust. Accumulation of more household goods, such as carpets, furnishings, and utensils, and larger houses to accommodate them, made cleaning much more extensive. With sewing machines, women made more family clothing to expand their wardrobes, which increased washing and ironing chores. Kerosene lamps, unlike candles, required frequent filling and cleaning.[13]

The net effect on households resulting from the transition to an industrialized economy was that many men now spent most of their time at places of work, and many women were left at home. Children, who once had shared in many household chores, now were attending school, or more likely in urban areas, working in factories, because many employers hired very young boys and girls for monotonous and dirty jobs.

What was once a shared division of household labor between husbands and wives now evolved into two separate domains. Men became committed to and defined by work places and women by home places. The home and all its attendant responsibilities, including the raising of children, moral family guidance, meal preparation, social entertainment, decoration, comforts of living, cooking, and, of course, housework, now became the primary job of women. A man might consider the home as his castle, but it was his queen who was fully in charge of all its creature comforts, inner workings, and maintenance.

No one understood the magnitude and importance of this challenge to women more than Catharine Beecher (1800–1878), an American educator and founder in 1823 of the Hartford (Connecticut) Female Seminary, a private girls' school. She was very religious, and all seven of her brothers became ministers, including Henry Ward Beecher, Charles Beecher, and Edward Beecher. By 1831, she had joined her father, Lyman Beecher, a Presbyterian minister in Cincinnati, Ohio, where he had been named president of the Lane Theological Seminary. There, she opened the Western Female Institute, which she hoped would become a model for a nationwide system of teacher colleges, and she began a fundraising effort to support this idea. Unfortunately, she was not well received in the area, where Cincinnatians, with slave state Kentucky across the Ohio River, were divided on the issue of slavery. Many disapprovingly regarded Catharine as an abolitionist, or almost worse, as a cultural elitist. Failing to obtain financial support, her school declined in enrollment until it closed in 1837.

That same year, she returned to Connecticut and began to write on subjects of slavery and the duty of women, and developed her theory that women could have a powerful influence on the character of the nation. Well aware of the many new technical schools being initiated to

Catharine Beecher (1800–1878), ca. 1870s. Courtesy Harriet Beecher Stowe Center, Hartford, Connecticut.

train men for careers in industry, Catharine was appalled at the suffering of young wives and mothers from "poor health ... and a defective domestic education." Catharine wanted to provide women with a scientific domestic education to elevate their status in society. She considered management of the home not just as a job, or as drudgery, but as a noble profession that was the basis of a healthy and moral society. Accordingly, in 1841, she wrote *A Treatise on Domestic Economy For the Use of Young Ladies at Home and at School* as a textbook for female schools and dedicated "to American Mothers." It was a handy, single source of household information that had not existed before, and there exists no better comprehensive description of the nature of housework in the early nineteenth century.[14]

Catharine's book delineated the almost infinite expectations of women's homemaking responsibilities in 371 pages and 37 chapters; it is probably the first evidence of what would 70 years later be well known as the study of home economics. There were individual chapters on healthcare; healthful food; helpful drinks; clothing; cleanliness; exercise; manners; habits of system and order; charity; economy of time and expenses; health of mind; care of infants; management of young children; care of the sick; accidents and antidotes; domestic amusements and social duties; fires and lights; washing, ironing, and cleaning; whitening, cleansing and dyeing; care of breakfast and dining rooms; care of chambers and bedrooms; care of kitchen, cellar and storeroom; sewing, cutting and mending; construction of houses; the care of yards and gardens; the cultivation of fruit; and the propagation of plants.

The book reads like an owners manual of the home and gives us a first hand description of the incredible intensity of manual labor required by nineteenth century housewives. Since our primary interest is in the historical techniques of cleaning floors, furniture, and carpets, I quote excerpts from Chapter 29, "On the Care of Parlors."

> To stretch the carpet, use a carpet-fork, which is a long stick, ending with notched tin, like saw teeth. This is put in the edge of the carpet, and pushed by one person, while the nail is driven by another. Sweep carpets as seldom as possible, as it wears them out. To shake them often, is good economy. In cleaning carpets, use damp tea leaves, or wet Indian meal, throwing it about, and rubbing it over with a broom. The latter, is very good for cleansing carpets made dingy by coal dust. In brushing carpets in ordinary use, it will be found very convenient to use a large flat dust pan, with a perpendicular handle a yard high, put on so that the pan will stand alone. This can be carried about, and used without stooping, brushing dust into it with a common broom.... When carpets are taken up, they should be hung on a line, or laid on long grass, and whipped, first on one side, and then on the other, with pliant whips.... Carpets can be best washed on the floor, thus: First shake them; and then, after cleaning the floor, stretch and nail them upon it. Then scrub them in cold soapsuds, having half a teacup of ox-gall to a bucket of water. Then wash off the suds, with a cloth, in fair water. Set open the doors and windows, for two days or more. Imperial Brussels, Venetian, Ingrain, and Three-ply, carpets can be washed thus; but Wilton, and other plush carpets, cannot. Before washing them, take out grease, with a paste, made of potters clay, ox-gall, and water.... Curtains, ottomans, and sofas covered with worsted, can be cleansed, by wheat bran, rubbed with flannel. Dust Venetian blinds with feather brushes.... Unless a parlor is in constant use, it is best to sweep it only once a week, and at other times use a whisk broom and dust pan. When a parlor with handsome furniture is to be swept, cover the sofas, centre table, piano, books, and mantelpiece, with old cottons, kept for the purpose. Remove the rugs, and shake them, and clean the jambs, hearth, and fire-furniture. Then sweep the room, moving every article. Dust the furniture, with a dust-brush and a piece of old silk. A painter's brush should be kept, to remove dust from ledges and crevices. The dust cloths should be often shaken and washed, or else they will soil the walls and furniture when they are used. Dust ornaments, and fine books, with feather brushes, kept for the purpose.[15]

Carpets and brooms and buckets, oh my! This text confirms that by 1841, brooms, carpets, dust pans, feather dusters, curtains, and upholstered furniture were all in common use. It is interesting to note the extraordinary efforts taken to clean carpets and to protect furniture and other articles in a room from the clouds of dust created by sweeping. Some dipped the broom in water, which trapped the dust somewhat. This was a huge problem that would only be solved later by the invention of vacuum cleaners. Catharine's book was highly successful. It went through 15 printings in as many years and brought her national fame. Young women across the nation were inspired by it, using it as a moral and pragmatic guide for the rest of their lives as mothers and housekeepers over the remainder of the century.

During this time, America was growing at a rapid pace. There was an unprecedented flood of immigration in the 1840s and 1850s, encouraged by the new network of interstate canals, and by 1845 the national population was about 20 million, about the same as the current population of the state of New York. An additional influx of immigrants would by 1850 increase the population to 23 million, and the annexation of the vast Western Territories would be disputed in regard to slavery, setting the stage for the Civil War.

Catharine's younger sister was Harriet Beecher (1811–1896), who was educated in Catharine's Hartford Seminary and in 1836 married Calvin Stowe, a widower professor in the seminary who was an ardent critic of slavery. They moved to Cincinnati to join Catherine and Harriet's parents. In 1842, a farmer near Cincinnati was arrested for transporting runaway slaves and was sued by the owner in violation of the 1793 Fugitive Slave Act. The farmer was found guilty. Harriet also became an avid abolitionist and in 1852, as Harriet Beecher Stowe, wrote *Uncle Tom's Cabin*, depicting the life of African Americans under slavery, including the story of the poor farmer in Cincinnati. Her book energized antislavery forces in the north, and she became famous as the country moved inexorably toward Secession by southern states in 1861. After the Civil War, in 1869, Catherine and her sister Harriet, both now quite famous, updated, revised and re-edited Catherine's 1841 book and as coauthors had it republished as *The American Woman's Home: or Principles of Domestic Science*.[16] The revised book was billed as

> a guide to the formation and maintenance of economical, healthful, beautiful, and Christian homes ... [dedicated to] the women of America, in whose hands rest the real destinies of the Republic, as molded by the early training and preserved amid the maturer influences of the home.

In the introduction, Catherine laments

> that women are not trained for these duties as men are trained for their trades and professions, and that, as the consequence, family labor is poorly done, poorly paid, and regarded as menial and disgraceful.[17]

There is an element of women's liberation in her tone. Religion was much more prominent in this new version, many housekeeping details were edited out, and there were numerous illustrations. Chapter II, "A Christian Home," described a well-organized and efficient kitchen. Chapter XXX incorporated several older chapters on individual rooms into a new one titled "The Care of Rooms," which contained a greatly abbreviated version of the "Care of Parlors" quoted above, but with no essential changes. Some of the significant technical advances since 1841, such as central heating and in medical knowledge, were recognized and included in the second book.

Central heating, apparently the latest in home heating technology, was addressed in a new Chapter V, "Construction of Stoves, Furnaces, and Chimneys," and in Chapter XXXVI,

Kitchen plan, Beecher, *American Woman's Home,* 1869. From Pulos, p. 175.

"Warming and Ventilation," which described the newly discovered scientific principles identifying different "microscopic plants" (later determined to be germs or bacteria) that were supposedly formed "by fermentation" in the blood of diseased persons. In 1859, just 10 years before, French chemist Louis Pasteur (1822–1895), had demonstrated that the fermentation process was caused by the growth of microorganisms, and that this growth was not due to spontaneous generation, as previously believed.

In Beecher's time of 1869, disease was believed to originate in impure air from swamps, called "miasma," in transmittal from sick people, or from "carbonic acid" (now known as carbon dioxide), believed to be a natural poison emitted from human or animal lungs and skin into the air. Beecher's book logically concluded that

> sickness and death are therefore regulated by the degree in which air is kept pure, especially in case of diseases in which medical treatment is most uncertain, as in cholera and malignant fevers.[18]

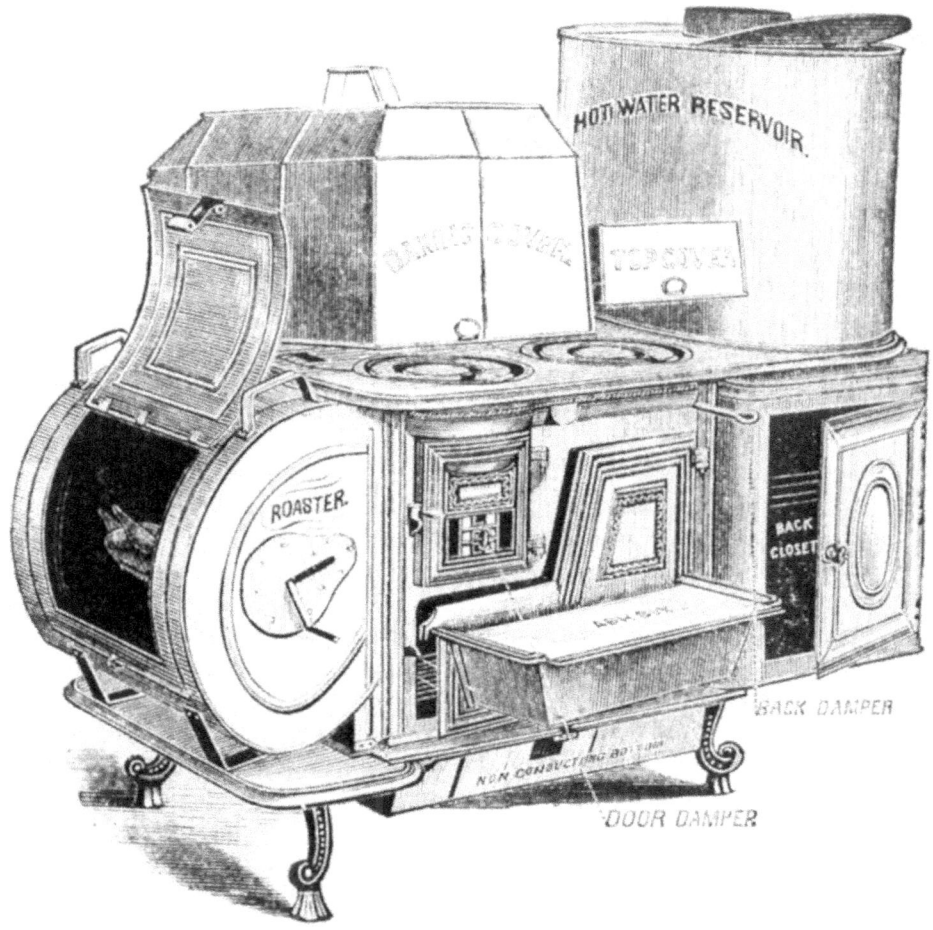

Cook stove, Beecher, *American Woman's Home,* 1869. From Pulos, p. 175.

Thus, according to the Beecher sisters, breathing of impure air was to be avoided through proper ventilation, with various registers connecting rooms (which were described in detail) and with slightly open windows at all times, to circulate air. In fact, the sisters believed that fireplaces were more healthful than central heating, because "carbonic acid" was more efficiently removed from the house by the natural ventilation of heated air out through the chimney. However, there was nothing in the Beecher books indicating any awareness of house dust as a specific hazard to health.

Advances in technology noted in the Beechers' revised book included cast iron stoves for cooking and/or heating, which became common about 1830. The cook stoves allowed a number of tasks to be done simultaneously, such as the heating of water, baking, roasting, frying, etc., and also controlled the heat more efficiently than fireplaces by dampers. The iron stoves for heating were, as mentioned earlier, twice as efficient as fireplaces. Furnaces for heating, introduced around 1860, provided a central heating system for the house; lamps using kerosene (which came into common uses about 1850) rather than sperm oil, tallow, or lard were more efficient; and the use of coal in stoves, rather than wood, saved a lot of cutting, chopping, and handling.[19]

One of the most surprising new technologies, described in colorful detail by the sisters' book, was the latest in toilets, called "earth closets." The device and its name were similar to another competing invention called a "water closet" (although the latter was not mentioned in her book). This earth closet was a typical toilet seat with a hole, behind which was a rather large wooden box holding the "earth." A release lever deposited a quantity of dry earth on top of human waste in either an outdoor privy, or in an interior commode, or closet, with a removable pan for emptying. The device was essentially a flush toilet, using earth instead of water. It was patented in 1860, 1869, and 1873 by the Rev. Henry Moule (1801–1880), a British priest in the Church of England, as a great improvement over the cesspit that also provided manure for farming. Apparently, it was effective in removing odors and for some time, was in stiff competition with the water closet. Both devices, however, required the manual filling of the "closet" with either earth or water, an additional laborious task for the household.[20]

What seems incredible about the Beechers' book is that in 1869, four years after the Civil War, every possible aspect of housework, including major tasks of washing, sewing, cleaning, cooking, and heating, was still done by what must have been exhausting manual labor. One must assume that Catherine and Harriet were describing housework requirements for a typical middle class home, even though they were both in the upper class. However, in this second book, it is difficult to discern many signs of advanced industrial technology in the 28 years since Catherine's first book.

In fact, the age of industrialization had progressed amazingly since then. Samuel F.B. Morse (1791–1873) perfected the telegraph in 1837, and by 1844 a working line was installed between Baltimore and Washington. In 1839, Cyrus McCormick (1809–1884) began production of his horse-drawn grain reaper. The federal government encouraged not only mechanical invention, but the invention of unique and marketable visual appearances. In 1842, Congress passed an "Act to Promote the Progress of the Useful Arts," patterned after an earlier, similar English act, which permitted the commissioner of patents to issue "design patents" for "any new shape or configuration of any article of manufacture not known or used by others before." Within a year, 14 design patents were issued, and within 10 years, the annual number had increased to 100. Design patents recognized that visual invention was just as important to protect as mechanical invention, and they encouraged originality by artists and designers.[21]

But mechanical invention was proceeding at a much more rapid rate. Samuel Colt (1814–1862) perfected his famous revolving gun by 1847. At the 1851 Great Exhibition of the Works of Industry in All Nations in the Crystal Palace, London, Cyrus McCormick's horse-drawn reaper won the gold medal. The Great Exhibition established America's international reputation as a serious competitor to England in the industrial revolution.

Although many of the technical examples above were obviously for the general benefit of society, rather than specifically for women, the exhibition also introduced an important invention that would in time become a standard household convenience. Featured was the first public water closet, which had attendants dressed in white, and customers were charged a penny for its use. Sir John Harington (1560–1612) had invented elements of the water closet in 1596 for his godmother, Queen Elizabeth, but water closets, or flush toilets, would not become popular in average American homes until indoor plumbing made them operational without the endless labor of manually refilling the closets.

Although the household implements described by the Beechers' books are a valuable first-hand record of nineteenth-century household responsibilities, it is puzzling to note important labor-saving household technologies that are not mentioned, even though they were already in common use. One of the most obvious was the domestic sewing machine.

Bostonian Elias Howe, Jr. (1819–1867), developed and produced successful versions of the sewing machine in 1846, and Isaac Singer (1811–1875) added foot power in 1851 to become the industry leader. By 1860 Singer was making 100,000 machines per year, and by 1866, three years before *The American Woman's Home*, several dozen companies were producing about a thousand machines per day. Surely the Beecher sisters knew of this. Why wasn't it mentioned, except as a footnote? The answer lies in Catharine's editorial at the end of *The American Woman's Home*:

> The sewing-machine, hailed as a blessing, has proved a curse to the poor; for it takes away profits from needlewomen, while employers testify that women who use this machine for steady work, in two years or less become hopelessly diseased and can rear no children. Thus it is that the controlling political majority of New-England is passing from the educated to the children of ignorant foreigners.[22]

This makes it clear that Catherine was not only an abolitionist, but was anti-industrialization. Indeed, as was the case with many early mechanical innovations in the industrial age, many despised the sewing machine because it took away jobs. French inventor, Barthélémy Thimonnier (1793–1857) patented a sewing machine in France in 1830, and by 1841, had 80 machines operating in a Paris shop. One night, enraged tailors and seamstresses broke into the shop and ripped apart all the machines. He died in poverty. New York inventor Walter Hunt (1796–1859) developed a successful sewing machine in 1834, but public pressure from needlewomen and religious leaders forced him to sell the rights to his machine. The buyer, George Arrowsmith, intending to patent the invention, became discouraged when his own daughter refused to use the machine because it "...would be injurious to the interests of hand-sewers."

Horace Greeley (1811–1872), famous editor of the New York Tribune, took up the needlewomen's cause in 1845, comparing their incredibly low pay and horrible working conditions to slavery. His articles convinced the public that such conditions were caused not by machines, but by unscrupulous manufacturers. Singer's success in 1851 was mostly due to his genius

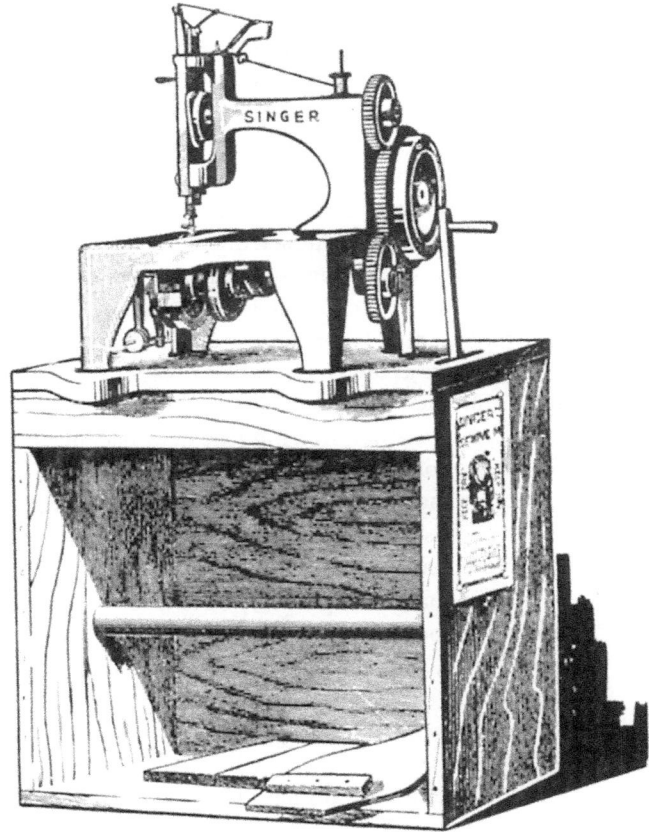

Singer sewing machine, foot-powered, 1851, U.S. Patent 8,294. From Smithsonian Institution poster for exhibition, 1985.

in marketing his machines directly to seamstresses and housewives, rather than to dictatorial factory owners, as all his predecessors had done. Domestic machines turned the tide of public opinion in favor of mechanization. Except, of course, for Catherine Beecher.[23]

A standard household product for cleaning, not nearly as technological as the sewing machine, but inexplicably not mentioned in the Beecher book, was the lowly mop. Since the vacuum cleaner industry later expanded into wet floor and rug cleaning, it is relevant to mention the origin of these cleaning implements, logical descendants of the broom. Mops had been in use for centuries by sailors, probably simple rags on sticks. The first American patent for one was Patent 241 in 1837 by Jacob Howe, called "Construction of Mop Heads and the Mode of Securing them upon Handles." One of the first patents for a mop with a built-in wringer was Patent 24,049 in 1859 by H. & J. Morton, and in 1869, Greenleaf Stackpole filed Patent 89,803 for one with a lever-operated clamp to hold the rag mop.[24] Mops, of course, were most useful on bare wood floors or floor cloths. Floor cloths were made of canvas hand painted with decorative patterns and used to protect rugs or floors in high traffic areas, such as kitchens, dining rooms or entrances. Called "crumb catchers," "grease catchers," or "oylcloths," they were used from colonial times until the late nineteenth century, when they were replaced by linoleum. Linoleum, a floor covering made of linseed oil varnish on canvas or cotton, and afterward printed, was invented by Englishman Frederick Walton (1834–1928), patented in 1860, and produced in quantity in 1869 in Staines, England, for export to Europe and the U.S.[25]

The Beechers probably were not aware of the first true vacuum cleaner being introduced that same year in Boston and Chicago, but they surely knew of another relatively new mechanical product not mentioned in the Beechers' book: the carpet sweeper, which by 1869, was evident in a number of patents and had been marketed over the previous ten years. The Beecher sisters were probably not interested in such a mechanical device any more than they were in the sewing machine. But one must wonder which workers Catherine Beecher might think were being denied employment because of the manufacture of carpet sweepers or vacuum cleaners, other than the very housewives she was advising to engage in exhausting manual labor.

Ineffective as it was, the carpet sweeper evolved from early rotating brush street sweepers and became the first such mechanical invention small enough and light enough to be used in the home. Carpet-sweeping devices had been evolving since 1811 through numerous inventions, most of which did not work very well, and some actually stirred up more dust than they removed. The carpet sweeper is of particular interest in our story of vacuum cleaners because it embodied many similar mechanical principles useful in cleaning carpets, the key one being the use of a brush.

Brushes have a long history back into antiquity, as do brooms and carpets. Early paintbrushes have been found in Egyptian tombs and were later adapted for use as hairbrushes. Greek writers, such as Homer, Sophocles, and Euripides, mentioned brushes. Brushes were generally less coarse and shorter than brooms, and were constructed of animal bristles or hairs from pigs, goats, badgers, squirrels, horses, bears, or even cows and camels. In some cases, other materials, such as whalebone fibers, were used. The principle functional advantage of brushes is that animal bristles have essentially a round cross section, allowing more flexibility and "spring-back" than the non-round cross section of the vegetable fibers of brooms, and so they retain their original shape and position almost indefinitely. In addition, they are much more durable than vegetable fibers, lasting much longer in use. However, animal bristles were much more difficult to obtain, sort, size, and fabricate into useful tools.

Making brushes was a laborious and tedious task. In the most basic process, tufts of bristles were inserted into individually drilled holes and fixed in place with molten glue. A more complicated process was used in "drawn" brushes, where a loop of wire was passed through the top of a hole through a board, and a number of hairs passed through the loop. The wire was then pulled through the hole from the top, pulling and doubling the hairs below into the hole. With the hole filled, the ends of the bristles were trimmed with shears, and the back of the hole covered with veneer to conceal the wire.[26]

Several English craftsmen pioneered in brush making. William Kent began manufacturing handmade hairbrushes in 1777, William Addis began making toothbrushes with bristles from pigs in 1789, and John Elin patented a brush arrangement that same year for sweeping chimneys. In 1830, a patent by Timothy Mason had accelerated the manufacturing process of brushes by cutting wide grooves in the base to receive bundles of bristles simultaneously. By 1854, U.S. inventor Hugh Rock had patented a hairbrush with a metal handle, and in 1857, H.N. Wadsworth was granted the first U.S. Patent (18,653) for a toothbrush. *Scientific American* in 1872 described the latest advance in brush making, the Woodbury machine, an American invention for bunching, wiring, and inserting bristles into the stock by means of a complex mechanical process.[27]

Since brushes were essential to the effectiveness of carpet sweepers, and later, in vacuum cleaners, we will now take a closer look at the development of a particular type of brush, the cylindrical rotating brush designed for the specific purpose of continuous brushing or cleaning of horizontal surfaces. In other words, a carpet sweeper.

The carpet sweeper is a mechanical device that combines the sweeping action of a broom with the collection function of the dustpan, both familiar implements in the 1800s. Instead of a broom, however, the primary operational element of the carpet sweeper is a brush. The brush, however, is not at all like the typical brushes described in the Beecher closet, but a special type of cylindrical, rotating brush that first made its appearance in 1699. That year, Edmund Heming took out a British patent for "a new engine for sweeping the streets of London or any city or towne."[28] We would call such a machine a street sweeper. It was a circular brush attached to the rotating wheels of a horse-drawn cart, but without a bin to collect the debris, it merely repositioned dirt elsewhere on the street. But the concept of a rotating brush to sweep debris was born. A rotating brush was nothing more than a cylindrical broom; in fact, such devices were called "mechanical brooms" in England.

Such brushes were intended to address a problem that Benjamin Franklin (1705–1790) observed in the 1750s and described in his autobiography regarding the streets of London, which

> when dry were never swept and the light dust carried away, but it was suffered to accumulate till wet weather reduced it to mud, and then after lying some days so deep on the pavement that there was no crossing but in paths kept clean by poor people with brooms, it was with great labor raked together and thrown into carts open above, the sides of which suffered some of the slush at every jolt on the pavement to shake out and fall, sometimes to the annoyance of foot-passengers. The reason given for not sweeping the dusty streets was, that the dust would fly into the windows of shops and houses.[29]

Over the years, a number of English patents, such as one to John Elin in 1789, were awarded for revolving brushes or mechanical brooms for sweeping chimneys, as well as for sweeping streets. One street sweeper patent placed a brush roller obliquely to the frame, to push the dirt to one side of the street, where, presumably, it could be more easily collected, shoveled into carts, and be removed. On November 1, 1825, a patent was granted to a

Joseph Whitworth's 1843 street sweeper, patent 3,124, U.S. Pat. Off.

W. Ranyard for a device with a number of brushes mounted upon two rims or placed upon an axis, which was raised on a vehicle or barrow. In 1828 a Bosse and Smith patent was granted for a "scraping, sweeping, and watering" device. From 1825 onward a succession of patents included a number by an English gentleman, Joseph Whitworth.[30] Whitworth's earliest patent, 5,275, of November 1, 1825, was extremely primitive, with brooms mounted between two carrying wheels similar to paddles in a water wheel.[31]

Whitworth's improved device was described in British Patent 8,475 dated April 15,

1840, as "machinery for cleaning and repairing roads and ways." He used an endless chain of brooms (Whitworth called it a circular broom) driven by the axle of the cart. A crossed sprocket chain was formed of open links and closed ones, which carried the brooms (or brushes). The brushes raised the dirt up an inclined carrier plate and dumped it into a container. It was probably the first to perform street sweeping mechanically on a broad scale. Whitworth was awarded a second British Patent, 9,433, dated August 2, 1842, for an "apparatus for cleaning roads." The first U.S. patent for street sweeping machinery was 3,124, dated June 1, 1843,[32] also granted to Joseph Whitworth. It shows a two-wheeled cart with shafts for hitching to a horse in front, and with a conveyor-belt-like assembly of brushes being driven by the wheels with chains, and the brush assembly angled down in the rear to contact the street, so that brushes could gather and drag the dirt back up an inclined ramp into the cart.[33]

In the meantime, the large and clumsy street sweepers had inspired inventors to develop smaller versions suitable for use in the home. The first documentation of a domestic floor sweeper was an English patent awarded to James Hume in 1811. He called it a "sweeping machine" adapted to sweep floors, and it consisted of essentially a box with wheels containing a rotating brush turned manually by means of a pulley connected to a handle or crank.[34]

America, lagging far behind England in patented inventions, finally made its mark internationally in 1851 at the "Great Exhibition" in Hyde Park, London, mentioned earlier. The U.S. was demonstrating that it was gaining ground on England in inventions and industrialization. American inventions such as Samuel Colt's revolver, Samuel Morse's telegraph, and Cyrus McCormick's horse-drawn reaper wowed the British, but England remained the global leader in technology and patents, including those in the sweeper category.

In 1853, James Hadden Young filed another English patent for a "hand sweeping apparatus," for "sweeping carpets, floors, and pavements," which had a brush roller, dustpans, and a handle in a frame covered in calico. The brush roller was rotated by a connection to the wheels, which eliminated hand cranking. It was the first of the English patents to mention the sweeping of carpets. The primary function of a brush roller was to loosen and dig dirt and dust out of the carpet. The bristles bent slightly as they dug into the carpet, and as they rotated, they sprang dust and dirt upwards into the pan above. Young also patented an improvement in 1854, which hinged or pivoted the dustpans, as modern carpet sweepers do, and replaced the calico-covered frame with a wooden box.[35]

One of the four English patents issued in 1858 was awarded on June 11 to Lucius Bigelow (no relation to the American carpet manufacturer) for "an improved machine for sweeping carpeted or other floors." Possessing all the elements of the modern carpet sweeper, its case was supported by two large wheels on one side, which were the driving wheels for the brush roll, and a swiveling caster wheel on the other side.[36] The Bigelow appears to be an early design that is substantially consistent with the classic definition of a carpet sweeper.

> A carpet sweeper typically consists of a small box. The base of the box has rollers and brushes, connected by a belt or gears. There is also a container for dirt. The arrangement is such that when pushed along a floor the rollers turn and force the brushes to rotate. The brushes sweep dirt and dust from the floor into the container. The sweeper usually has a long handle so that it can be pushed without bending over.[37]

There were a number of carpet sweepers manufactured that year in England, and some probably found their way to America. It seems more than coincidence that one of the first patents on carpet sweepers in America was essentially a copy of the Bigelow sweeper. English patents were legally valid in England, but nowhere else, a loophole that was exploited by

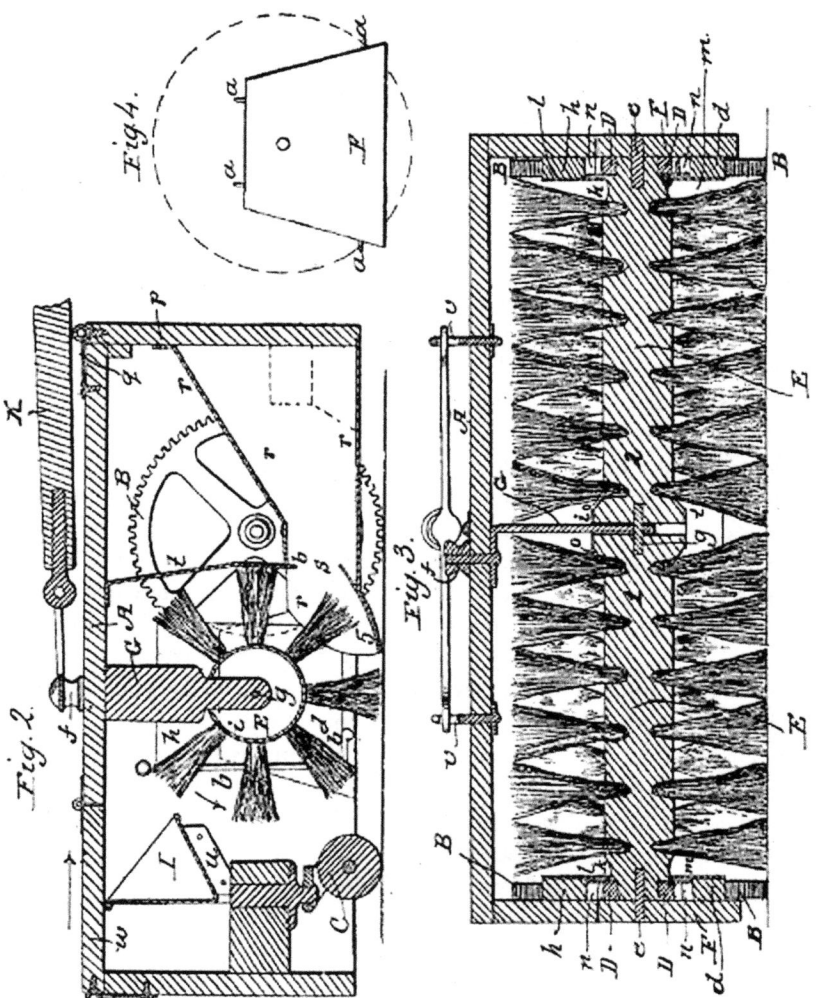

Hiram Herrick's 1858 carpet sweeper, patent 21,233, U.S. Pat. Off.

inventors elsewhere, and America was no exception. Moreover, by midcentury, most American homes already had a closet full of various brushes for dozens of cleaning tasks. A brush would have been the most obvious choice for a carpet-sweeping device. The number of brushes in the Beecher closet suggests that they were very popular, available in profusion, and used extensively. Note the arsenal of cleaning equipment, including brushes, which did not change from 1842 to 1869 in the two Beecher books:

In a closet should be kept, arranged in order, the following articles: the dust-pan, dust brush, and dusting cloths, old flannel and cotton for scouring and rubbing, large sponges for washing windows and looking-glasses, a long brush for cobwebs, and another for washing the outside of windows, whisk-brooms, common brooms, a coat-broom or brush, a whitewash-brush, a stove brush, shoe brushes with blacking, articles for cleaning tin and silver, leather for cleaning metals, bottles containing stain-mixtures and other articles used in cleansing.[38]

The American copy of the British Bigelow carpet sweeper was by Hiram H. Herrick of Boston, Massachusetts, one of the first men who applied for "an improved carpet sweeper" patent, which was granted on August 17, 1858, as U.S. Patent 21,233. The exterior was a simple rectangular wooden box but the interior consisted of "a revolving brush connected to driving wheels." His claims included: "dividing the brush in the center ... connecting to the driving wheels"; a "peculiar construction of the dust pan ... with a spring lip"; and "protecting the bearings from dust ... by means of plates."[39] Herrick's was one of a number of similar patents that year, and was actually manufactured and put on the market in 1859.[40]

Another patent for an improved carpet cleaner with a specific arrangement of components was granted the same day as Herrick's (August 17, 1858) to Augustine W. Noney, of Bridgeport, Connecticut, as U.S. Patent 21,211. In fact, the lower patent number suggests that the patent was granted even earlier than Herrick's. The inventor states in his patent: "I do not claim the revolving cylinder brush, nor the dust pan, nor the enclosing box, nor the combination of the three, all these having been long known and used."[41] This is interesting, because it suggests that the basic concept of a carpet sweeper was around long before 1858, probably as far back as James Hume in 1811, 47 years earlier. The Noney patent, as it appears from the Patent Office report, was a cumbersome affair and probably never went into use.[42]

The third patent was the Shaler improved carpet sweeper, patented as U.S. Patent 21,451 by Reuben Shaler, of Madison, Connecticut, on September 7, 1858, for a "combination of a brush with a traction roller."[43] The exterior form was much more refined than Herrick's, being more compact and curved. In an 1858 issue of the *Ohio Cultivator*, a semimonthly journal in Columbus, Ohio, devoted to "the improvement of agriculture and horticulture, and the promotion of domestic industry," is this descriptive testimonial for a patented Shaler Carpet Sweeper by the journal's editors:

> We have received from the manufacturers, a sample machine of the above, and having turned it over to the Household Department, are satisfied it will do what the manufacturers claim for it, viz:

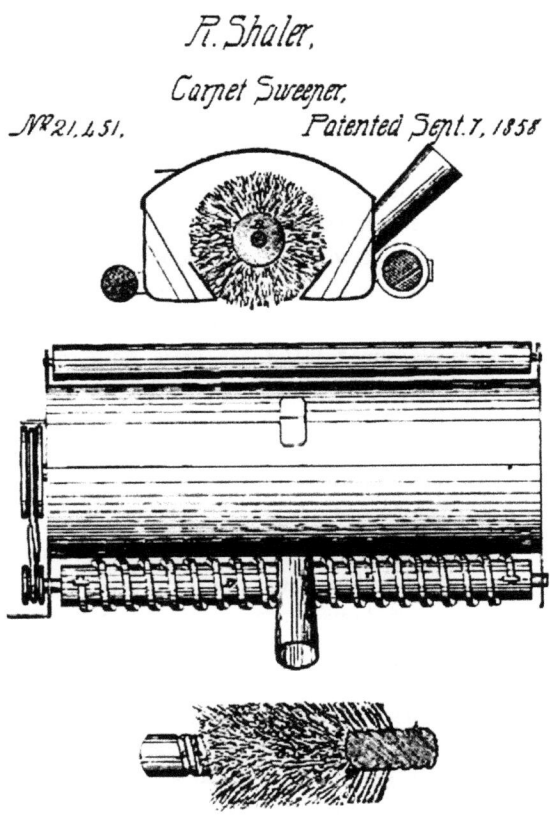

Rueben Shaler's 1858 carpet sweeper, patent 21,451, U.S. Pat. Off.

It sweeps cleaner than a Broom without injury to any Carpet. It makes no dust to soil the furniture or stifle the lungs. It is simple in its construction and not liable to get out of repair. It is easily worked, is durable and will last for years. It cleans a carpet thoroughly, taking the dirt, dust, lint, hairs, pins, needles &c., directly up into the box; whereas the common method of sweeping, drives the dirt along the entire surface of the carpet and thereby injuring it very much.

The Sweeper consists of a cylindrical hairbrush, pivoted in a tin case, the bottom of which is open, to let the brush upon the carpet. The case is pushed by a handle like a broom handle, runs upon rollers geared to the brush, so that when it is pushed over the carpet, the brush revolves rapidly and picks up any litter of dust or shreds, and deposits it in a chamber of the tin case.

Address Shaler Carpet Sweeper Company, No. 69, Fulton St. N.Y. Wholesale price, $3.50 each. We notice these Sweepers for sale at the excellent House Furnishing Store of DODDRIDGE & WHITE of Columbus. We advise housekeepers to try this nice little invention.[44]

This sweeper was obviously produced in quantity and marketed to consumers. A productive inventor, Shaler also invented a special bullet used in the Civil War (it split into three parts after firing), and a Wheel Skate, or as he called it, the "Parlor Skate" (1860), a type of inline roller skate.[45]

On October 5, 1858, Jacob Edson of Boston, Massachusetts, was granted patent 21,660 for an "improved carpet sweeper,"[46] and on the same day, Daniel Harris of Boston was granted patent 21,673, also for an "improved carpet sweeper."[47] On October 19, 1858, Augustus C. Carey of Ipswitch, Massachusetts, was granted patent 21,815, for yet another "improved carpet sweeper."[48] Another 1858 patent was granted to the Union Sweeper, which used geared wheels on one side of its body to turn the brush roller. There were no wheels on the other side.

Between 1858 and 1900, there were some 254 patents granted for improvements on carpet sweepers, and more than 50 manufacturers of them started, flourished, and then most passed out of business.[49] The Herrick machine, however, is generally recognized as the first to have been patented in the U.S., and it was on the market in 1859, or possibly as early as 1858. It also was publicly recognized with a gold medal and diploma at one exhibition, and it was made in the city that would become famous for its carpet sweepers: Boston. There was high demand for sweepers in Boston, and for the next 12 years, almost all sweepers manufactured in the U.S. were made there, except for a few in Connecticut. In 1860 a New York merchant contracted for 30,000 Herrick sweepers to be delivered in 1861, but the outbreak of the Civil War caused the contract to be canceled,[50] along with many other civilian products, as industry turned to manufacture of military materials for the next four years.

The sweeper business spread gradually to New York, New Jersey and Rhode Island, but Boston remained the primary manufacturing center until about 1882, when Midwestern states became the center. Other Boston sweepers were the Weed, the Boston, and the Welcome. The latter was manufactured extensively and had a tin top and cog gearing which tended to damage carpets, but it was produced several years after the other two had folded. In 1900, an early Herrick sweeper, the one that started it all, was still in use in a New England town as a possession of the Herrick family.[51]

There were nine patents for carpet sweepers in 1859,[52] including one by Henry Davis with a rotating brush roll driven by a wheel (U.S. Patent 24,103, by letters, granted May 24) and another by J.B. Baker with the brush roll driven by a crank on the handle (U.S. Patent 92,929 granted July 27). Other similar appearing devices of the time were called sweeping boxes, which rather than wheels, had sled runners on the extended sides of the box to glide along the floor.

The intense demand for carpet sweepers was driven by the onerous task of cleaning carpets manually: moving furniture from on top of them, removing the tacks that held the carpet in place, rolling the carpet up, lugging it outside, hanging it over a line, beating the dust out with a carpet beater not unlike a tennis racquet, and then repeating the foregoing labor in reverse to return the carpets to their proper position in the house.

Of course, one could avoid all of this by cleaning the carpet in place, following these specific instructions of the mid–1800s, which were still being used up to 1900:

> Use a whisk to rid the corners and the edges of the carpet of dust, then gently, but with a steady stroke, sweep all the dirt into the middle of the room, and take it up with a dust pan. Repeat this operation to secure any dust you may have blown back. Should the carpet be very dusty, moist tea leaves scattered over the floor before beginning to sweep will gather up most of the fine dust and prevent its rising and settling on the walls, etc. It freshens and cleans a carpet wonderfully to wipe it thoroughly with a woolen cloth rung out of water mixed with household ammonia.[53]

Another method was to use a shampoo made by cutting Fels Naptha laundry soap bars into chunks and boiling into a thick jelly, then immersing overnight in cold water until the gelatin jelled. This substance was then applied to the carpet with a hand brush. When the nap was raised, the rug was hung to dry on an outdoor line or on poles in a cleaner's shop, where the heat from a stove encouraged the drying process.[54]

There was another mechanical alternative to carpet sweepers, but it still required the removal and transport of carpets. Three patents were taken out in 1860 for carpet beating machines: U.S. Patents 27,730; 28,389; and 30,590. The bulky machines imitated the human motions of arms swinging carpet beaters. They were installed in special cleaning centers for carpets, often connected to laundries. As late as 1902, an English handbook on laundry management would survey the various systems of carpet beating systems without any mention of vacuum cleaners.[55]

Speaking of vacuum cleaners, it is important to mention at this point that some of these early inventors recognized that carpet sweepers were inadequate to the task of removing the desired amount of dirt from the thick carpet pile and began to experiment with design versions that added air movement, creating some suction, and thus, technically, primitive vacuum cleaners. We tend to regard vacuum cleaners as a later development, when electric motors made them more practical. But as early as 1860, patents for sweepers began to include hand- or manual-powered mechanisms for generating limited suction. Since these early developments led technologically more directly to the familiar electric vacuum cleaner, we will describe them in more detail later.

In 1876, 100 years after America had declared its independence, President Ulysses S. Grant, by proclamation, invited other nations to participate in an exhibition

> designed to commemorate the Declaration of Independence of the United States, [with a] display of the results of Art and Industry of all nations as will serve to illustrate the great advances attained, and the success achieved, in the interests of Progress and Civilization, during the century which will have then closed.[56]

This, of course, was the Centennial Exhibition held in Philadelphia. Indeed, there had been spectacular technological progress in America by this date, as the industrial revolution here caught up with England. The exhibition, running from May to November, featured a Corliss steam engine of 1,400 horsepower, which drove miles of leather belts to power machinery throughout the exhibition. It was at this exhibition that Alexander Graham Bell

(1847–1922) introduced a prototype of his telephone, and Remington displayed and demonstrated its first typewriter. Dry plate photography had been perfected. An array of utilitarian, domestic, manually operated mechanical devices were exhibited that included apple corers, cherry stoners, coffee mills, corn poppers, egg boilers, flour sifters, knife sharpeners, lemon squeezers, dishwashers, and stoves. Singer had its own pavilion and was already selling 600,000 foot-operated sewing machines per year. American manufacturing was approaching a volume and variety equal to England's.

Among the items exhibited at the Centennial Exhibition was a new carpet sweeper called The Lady's Friend, invented that year by Gore & Edgecomb, of Goshen, Indiana. It featured an innovative rubber friction drive but with only two wheels and a runner. Well-made and popular, it challenged Boston manufacturers in sales and was successful for many years.[57]

That same year, the most famous name in carpet sweeper history appeared. On September 19, 1876, Melville R. Bissell (1843–1889), of Grand Rapids, Michigan, obtained his first patent for a carpet sweeper (patent 182,346). He had purchased a Welcome sweeper from Boston, was not satisfied with its performance, and decided to improve it to better pick up the sawdust from the carpets in the crockery shop he operated with his wife, Anna. They realized its marketing potential, and began to make more.

The Bissells began to assemble sweepers in a room over the crockery shop. Women working in their homes prepared components. They secured hog bristles with string, dipped the tufts in hot pitch, inserted the tufts into brush rollers, and finally, trimmed them with scissors. Anna gathered the parts in clothes baskets and returned them to the shop for assembling. Melville and Anna sold the sweepers on the road with their buggy, each taking opposite sides of the street, and demonstrating them by throwing a handful of street dirt on carpets.[58]

The sweeper Bissell patented had a brush roller driven by reduction gears or cogs, and because these were in the center of the brush roll, instead of at its end, could clean close to baseboards. Later, he modified it by replacing the cog gears with a belt friction drive, and it became known as the Bissell Centre Bearing Sweeper. It was said by many to be the first sweeper that picked up more dust than it managed to stir up.[59]

Through aggressive sales efforts by the Bissells, demand soared, and soon they were making 30 sweepers each day, shipping them to retailers in Michigan, the Midwest, and throughout the U.S. Early manufacturing improvements resulted in renaming the sweeper the Bissell Grand Rapids. Bissell began making sweepers in exotic woods, such as walnut, maple, and mahogany,

Bissell's 1876 carpet sweeper. Courtesy IHA.

and began promoting the sweeper as a Christmas gift. Within a few years, annual sales reached 65,000 units, even though there were over a dozen factories making sweepers in the country by then.[60]

All sweepers mentioned thus far were two-wheeled sweepers. The first four-wheeled sweeper was the Aurora, patented in 1877 by E.T. Pringle, of Aurora, Illinois. It was well made with a round leather belt that transmitted power from the wheels to the brush. Sold at a high price, the Aurora survived only until around 1881, when it succumbed to the strong competitors in Grand Rapids.[61]

Six competitors, inspired by Bissell's success, had sprung up in Grand Rapids, including the Michigan Carpet Sweeper Company, the Grand Rapids Carpet Sweeper Company, the Grand Rapids Brush Company, and the Plumb & Lewis Manufacturing Company. The latter firm made a sweeper invented in April 1880 by A.D. Plumb of Grand Rapids. Plumb had a year earlier begun manufacturing a sweeper invented by G.W. Gates and Benjamin F. Potter for The Grand Rapids Carpet Sweeper Company. Plumb's new design, called the Mystic, was immediately implemented, and doubled sales within a year. The most innovative feature on Plumb's patent was a spring between the sweeper case and the drive wheel, allowing the user to lower the brush by pressing down on the sweeper handle. This was called "automatic adjusting," and within a year, every manufacturer in

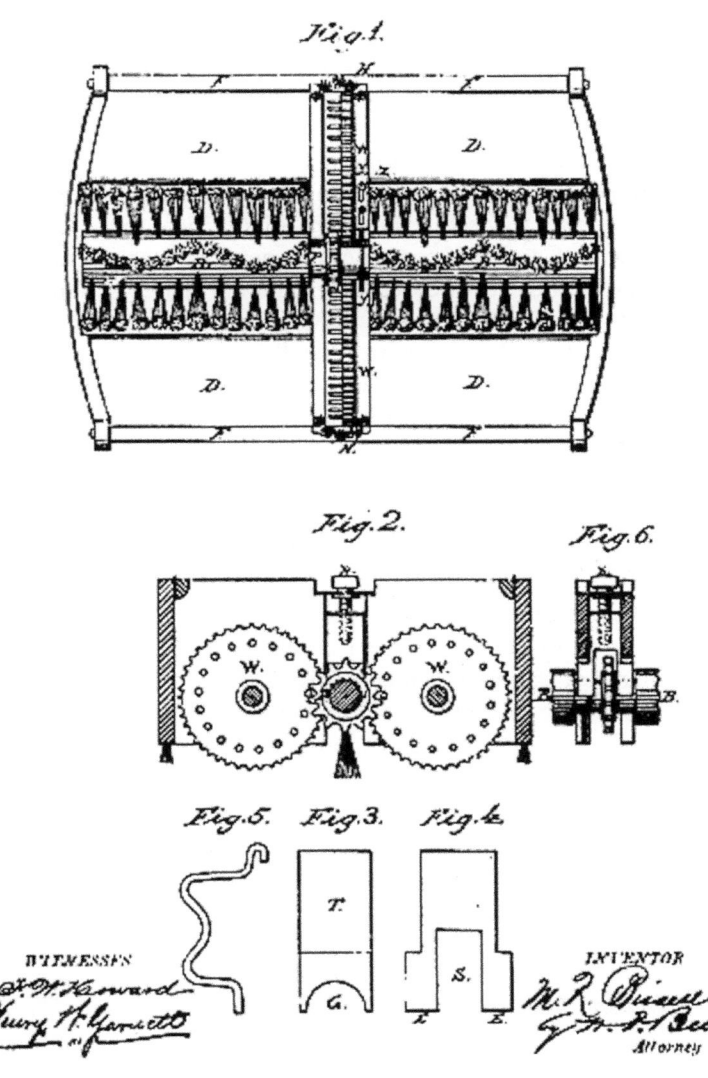

Melville Bissell's 1876 carpet sweeper, patent 182,346, U.S. Pat. Off.

Grand Rapids was using the feature in some form or another. To avoid patent infringement, some used an ingenious "cam-action" arrangement, patented in 1881 by G.W. Gates, which replaced the spring by connecting the bail (the part of the handle that is widened into a bracket that connects to the sweeper body at both sides) directly to the brush roller, so it could be raised and lowered by tilting the handle. But this arrangement was never as popular as the Plumb.[62]

Bissell sold more sweepers than all his competitors combined and stayed ahead with new features, but they were not all successful. In 1878, he introduced the Bissell Cog Wheel Sweeper, which sold well but proved impractical and was replaced sequentially that same year by an iron, then a rubber, friction drive. Bissell then debuted a sweeper called the New Idea, with a rubber friction drive and two wheels, but it also proved unsuccessful and was soon outclassed by four-wheeled rubber friction drives.[63]

A two-wheeled Paragon sweeper, an attractive well-made sweeper by the Grand Rapids Brush Company, appeared about this time, as well as an odd-shaped, two-wheeled sweeper called the Nickel Plate, made in Ashtabula, Ohio, and introduced in 1881. In 1882, the L.B. Lay & Company of Milwaukee, Wisconsin, debuted the Minneapolis, featuring a leather belt driving cog wheels, but it was too complex and soon disappeared. Only a few sweepers competing with Bissell appeared after this: the Castle, in 1884, which had a rope furniture protector; and the Diamond, made in Chicago in 1894, which had only two wheels.[64]

In 1883, Melville and Anna Bissell incorporated their company as the Bissell Carpet Sweeper Company, constructed a new five-story factory, and absorbed all Grand Rapids manufacturers of sweepers except Plumb & Lewis, with which it would merge in 1886. By 1889, the Bissell name had become synonymous with carpet sweepers, and there were few competitors. But that year, Melville died of pneumonia at the age of 45, and Anna Bissell (1846–1934) took control of the company, becoming one of the first female corporate executives in the United States, and one of the first employers to provide employee health benefits. She expanded Bissell's international markets to 20 countries, and British homeowners responded enthusiastically by buying Bissell sweepers after Queen Victoria allowed a Bissell to be used in her palace. By then, any carpet-sweeping activity was known as "bisselling." From about 1898 to 1914, the primary technical features advertised by Bissell were "Cyco" ball bearings and "Dust Proof Axle Tubes: easy running, no noise, no oiling." In 1900, Bissell sweeper models included the Boudoir, the Triumph, and the American Queen. Prices ranged from $2.75 to $5.75. Since the demand for sweepers was limited, and they lasted for many years, the problem was that too many companies made too many sweepers for the available

Anna Bissell (1846–1934). Courtesy of Bissell, Inc.

market. Most dropped out, and the major survivor was Bissell, leader in both quality and cost; Bissell alone made enough sweepers to supply the world.[65]

By this time, the tables of international competition had been turned. England was now looking to America for mechanical innovation, instead of the other way around. Richard Walton Kenyon (b. 1851), representing the Ewbank Company of Accrington, Devon, manufacturer of water meters, mangles (ironing machines), wringers, and wash machines since 1874, was visiting the U.S. in 1882 to buy wooden blocks for Ewbank mangles. Kenyon met the owner of a small carpet sweeper manufacturer in Chicago and visited his factory. He thought that if the carpet sweeper could be so successful in America, it could also be so in the U.K.

So in 1889, the first Ewbank brand carpet sweeper, designed by Kenyon and made by the firm Entwisle & Kenyon, Ltd., was sold to Mr. Hiram Waddington, a joiner and builder of Accrington. The Ewbank soon became the most popular sweeper in England and was used in many homes, including the palaces of royalty. Just like with the Bissell, the act of using a carpet sweeper was also referred to as "ewbanking." Models came in all sizes, with Miniatures for children (a common practice of the day), followed by the larger Standard and Parlour Queen, boasting a "very powerful pattern for the thickest piles." Ewbank was run by the Kenyon family for many generations and evolved into a major floor care appliance manufacturer after World War II. Earlex Ltd. purchased Ewbank in 2003, but a full line of floor care products, including carpet sweepers, is still being produced under the Ewbank brand today.[66]

While carpet sweepers were being refined and came into common usage throughout the world, other inventors were experimenting with ways to improve their efficiency beyond the mere sweeping action of brushes. The most successful method was to introduce the rapid movement of air through mechanical means to remove more dust from carpets. This would add another new functional dimension to sweepers, one that would be essential to all vacuum cleaners. We said earlier that we would return to 1860, when inventors began adding air movement to carpet sweepers, and in the next chapter, we will follow the development of this essential feature.

CHAPTER 2

Suction Cleaners After 1860

Where did inventors get the idea to add air movement to carpet sweepers? By 1850, several companies in London offered commercial cleaning services from the back of carriages. One source says "they used carriage-mounted steam engines or hand operated fans, bellows, or piston pumps to create suction. They would back the carriage up to a home, and pass a hose through the window to vacuum a room."[1] Such an early use of suction has not been substantiated by other sources, but as we know, mechanical brush sweepers were developed in England early in the century. It is more likely that a combination of moving air and rotating brushes were the mechanics driven by steam, pumps, or bellows. In any event, this appears to be the birth of commercial carpet cleaning, which later also became popular in America. After the addition of lighter gasoline engines increased their efficiency, these carriage-mounted cleaning services flourished until about 1910, when home vacuum cleaners began to be available, and even today, similar commercial carpet cleaning services are readily available.

By itself, the concept of creating airflow to move dust is not a great leap of science. Anyone familiar with the obvious effect of a high wind blowing dust across a dirt road would have recognized such a principle as an effective mechanism. To generate airflow, it was also well known that a rotating fan would do so, as in a ceiling fan. Around 1860, ceiling fans were becoming common, but they were not powered by motors, as they are today, but by a stream of running water combined with turbines (devices that spin in the presence of a moving fluid such as water, steam, or air). The turbines drove belts, which turned fans that originally had only two blades. The turbines could power many fans, and such mechanisms soon became popular in many public places such as office buildings, cafes, and stores for their cooling effect. Some of these belt-driven fan mechanisms can still be found, particularly in the South, but have been converted to electricity, thanks to Philip H. Diehl of New Jersey. In 1882, he used a Singer sewing machine motor to power such a ceiling fan and sold it as the Diehl Electric Fan. Ceiling fans have been electrically driven ever since and are still quite popular in the South. In fact, fans would become the first common household electric appliances. In 1887 Diehl founded Diehl & Company, which would in 1928 become the manufacturer of vacuum cleaners and motors for Singer.

The fundamental principle of creating suction is quite simple. You create suction when you drink liquid with a straw. When you draw air out of a straw by sucking, it lowers the air pressure within the straw to zero. Natural atmospheric air pressure is approximately 15 pounds per square inch, and this presses down on the liquid in your glass, pushing it up through the straw into your mouth. Now imagine a closed container with a hole on opposite sides. Inside the container, behind one relatively large hole, is a fan, which is blowing air *out of* the container. As air is removed, the air pressure within the container is lowered, and natural atmospheric air pressure pushes air *into* the container through the hole on the opposite side to refill the container. As the speed of the fan increases, the air moves faster, both out of one hole and into the other. If you make the air inlet hole smaller, the air must move faster to keep up with the larger amount that is being blown out, and the suction is increased.

This basic principle would have generated significant suction in large commercial devices and was presumably known to inventors in America. They might easily have concluded that adding suction to a carpet sweeper would be just as effective, but the problem of adapting this commercial, large-scale technology to small, portable devices for domestic application was enormous. Still, many tried.

Bellows were another alternative principle for moving air. Bellows were devices constructed with a closed container of pleated canvas, similar to an accordion, which, when deflated by wooden members on each side with handles, emitted a sharp outward blast of air through a narrow nozzle at the forward end. They were used from early times, including ancient Egypt, to increase the heat of fire for smelting metal or blacksmithing, or simply to intensify a fire in fireplaces. In the case of early suction cleaners, it was the inward sucking of air into a bellows while being inflated that created the desired suction. This would later become the dominant means for producing suction in manual suction cleaners.

The first known concept of using suction in a carpet sweeper was the July 10, 1860, U.S. Patent 29,077 by Daniel Hess, of West Union, Iowa, who proposed a carpet sweeper as follows:

> The nature of my invention consists in drawing fine dust and dirt through the machine by means of a draft of air, forcing the same into water or its equivalent for the purpose of destroying it substantially as will be hereinafter specific.[2]

Hess proposed a rotating brush and an elaborate bellows mechanism to draw air in around the brush and to deposit fine dirt and dust sucked up into two water chambers. Hess also stated that the air would be cleansed as it passed through the device. It was the first concept to propose a bellows to produce suction in a carpet sweeper and the first to mention water as a filtration means, although there is no evidence that his machine was ever produced for sale.[3]

The other means of creating suction, as described earlier, is with a fan in a box. On June 8, 1869, U.S. Patent 91,145[4] was issued to inventor Ives W. McGaffey of Chicago, Illinois, for a sweeping machine that bears an amazing visual resemblance to a modern upright vacuum cleaner (see group photographs of vacuums at the end of this chapter). It had a wide nozzle that was pushed along the floor and stirred up the dust. The dust was pulled in by suction through a four-bladed fan, contained in a wooden box, and into a cloth dust bag, which held all the dust that did not escape through a small hole that was 1 inch wide by 12 inches long. Suction was produced by a belt and transmission connecting a pulley on the fan rotor, with a second pulley mounted on the handle providing a crank handle for the operator. The faster the operator turned the crank, the faster the fan went, and, presumably, the greater the suction. It was much work with little result, but it seemed better than a carpet sweeper, some of which used similar cranks and pulleys simply to rotate the

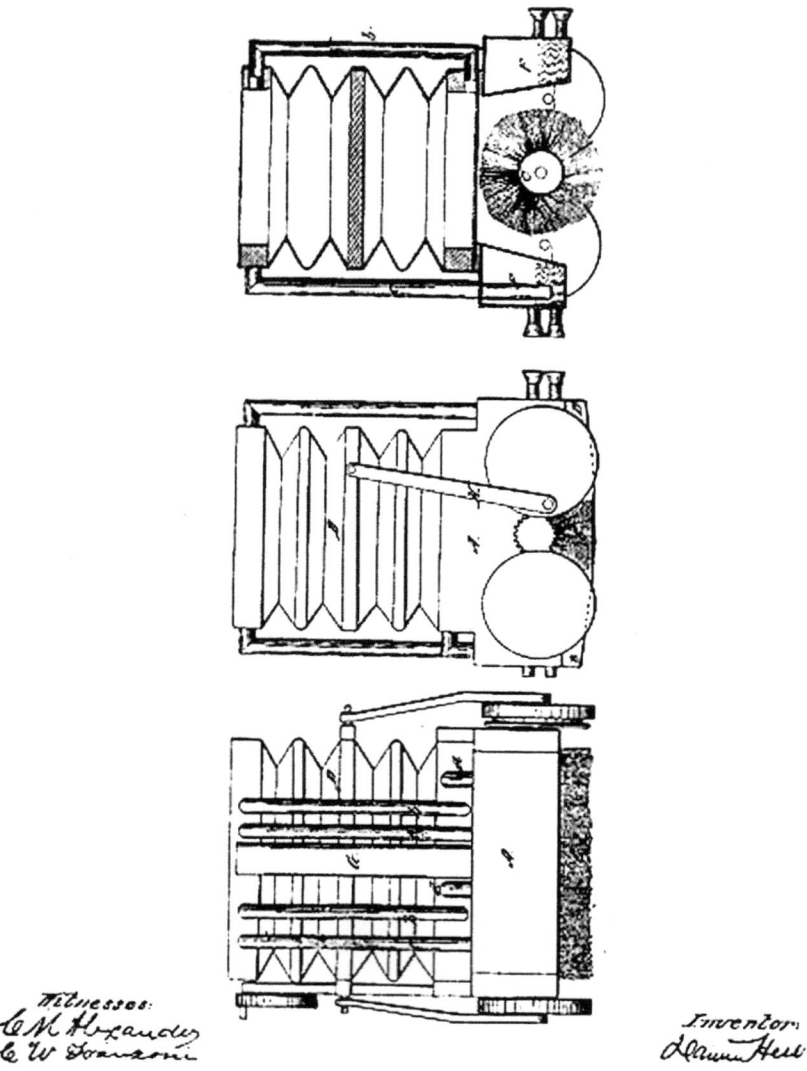

Daniel Hess's 1860 carpet sweeper, patent 29,077, U.S. Pat. Off.

brush. The American Carpet Cleaning Company of Boston, Massachusetts, produced and sold McGaffey's device, optimistically called the Whirlwind, at a retail price of $25. At the time, Boston had been the center of carpet sweeper manufacture since the Herrick sweeper of 1858. Unfortunately, the Whirlwind sold for only a short time and was a business failure; the company reported losing $60,000.[5] Most Whirlwinds were lost in the Chicago fire of 1871, or the Boston fire of 1872. Only two of the original machines have survived: one found in a barn in New Hampshire, now in the private collection of Robert Tabor,[6] and the other at the Hoover Historical Center in North Canton, Ohio. McGaffey, later in 1871, constructed an industrial suction cleaner powered by a steam engine.[7] McGaffey would try to electrify the machine in 1900, but the device didn't catch on.[8]

Several manually operated carpet sweepers with hand cranks on the handles with pulleys soon followed the Whirlwind; however, in these, the hand cranks did not produce suction, but only drove a rotating brush, and thus fall into the manually powered carpet sweeper category, rather than suction cleaners. One, U.S. Patent 92,929, was granted on July 27, 1869, to J.B. Baker. Another was "Willey's cylindrical carpet and floor duster" (see group photographs of vacuums at the end of this chapter). Freeman O. Willey and Daniel B. McEnery, of La Fayette, Indiana, were granted U.S. Patent 125,369 on April 2, 1872, for "an improvement in Carpet Sweepers." It had a hand crank on the handle with a pulley to rotate the rotary brush. Another, the Hatlinger Champion (by Joseph J. Hatlinger, U.S. Patent 165,730, July 20, 1875) also had a rotary hand crank high on the handle, connected by a belt to a pulley at the bottom of the machine that was connected to the rotating brush.

Eventually, the concept of a hand crank on the handle with

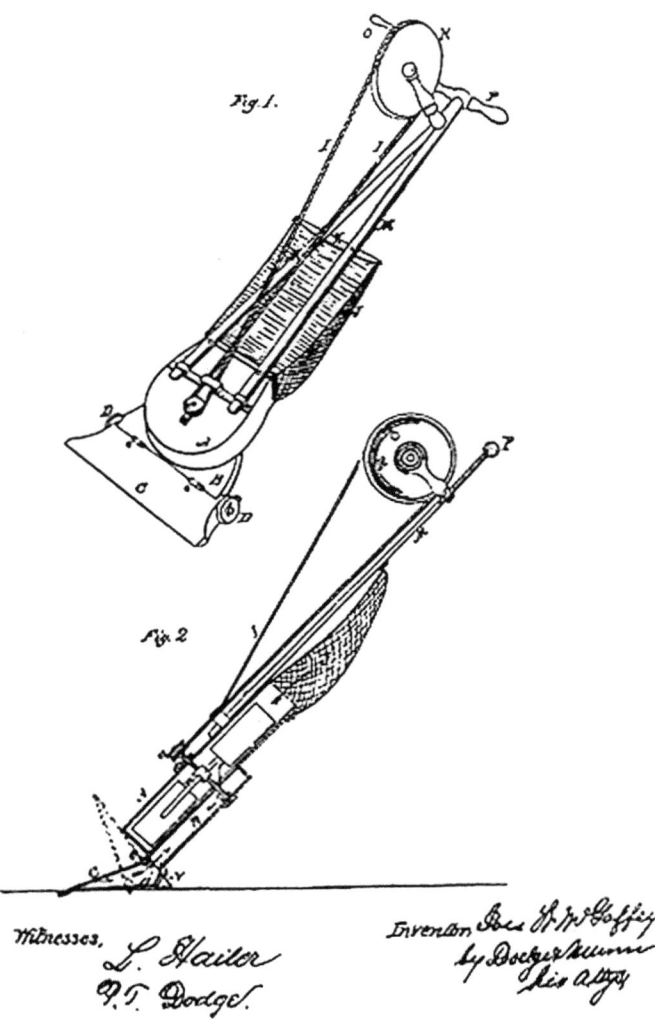

Ives McGaffey's 1869 suction sweeping machine, patent 91,145, U.S. Pat. Off.

pulleys to the nozzle evolved to drive both a fan to produce suction and to rotate the brush roll. In 1875, the Agan sweeper (see group photographs of vacuums at the end of this chapter), a wooden machine that was the first to combine both, was introduced. Similar to the Whirlwind, a series of pulleys and leather belts turned not only a fan for suction, but a brush roll to sweep the carpet as well. The Agan had all the basic components and functions of a typical vacuum cleaner, except instead of a bag to contain the dust, it had a box.[9]

Hiram C. Agan (1843–1918), a Civil War veteran and an active inventor of Fayetteville, New York, at the time, probably invented the device, although the patent cannot be found.

So by the time of the 1876 Centennial International Exhibition in Philadelphia, the first official World's Fair of the U.S., which opened on May 10, there existed commercial household carpet sweepers with manually generated suction and brush rolls. Although the term "vacuum" was not yet in use, these were technically vacuum cleaners. It is not known if any were on display at the Centennial Exhibition, but even if they were, they would have been upstaged by demonstrations of the first telephone by Alexander Graham Bell and by the Remington typewriter, both of which would soon greatly impact communications. Ironically, at the same time these technical innovations were being exhibited, the country was reminded that out on the frontier, in eastern Montana, it was still the Wild West. News arrived on June 25 and 26 that at the Battle of the Little Bighorn, Indians massacred George Armstrong Custer and 268 men of the U.S. 7th Cavalry in the most famous action of the Great Sioux War of 1876.

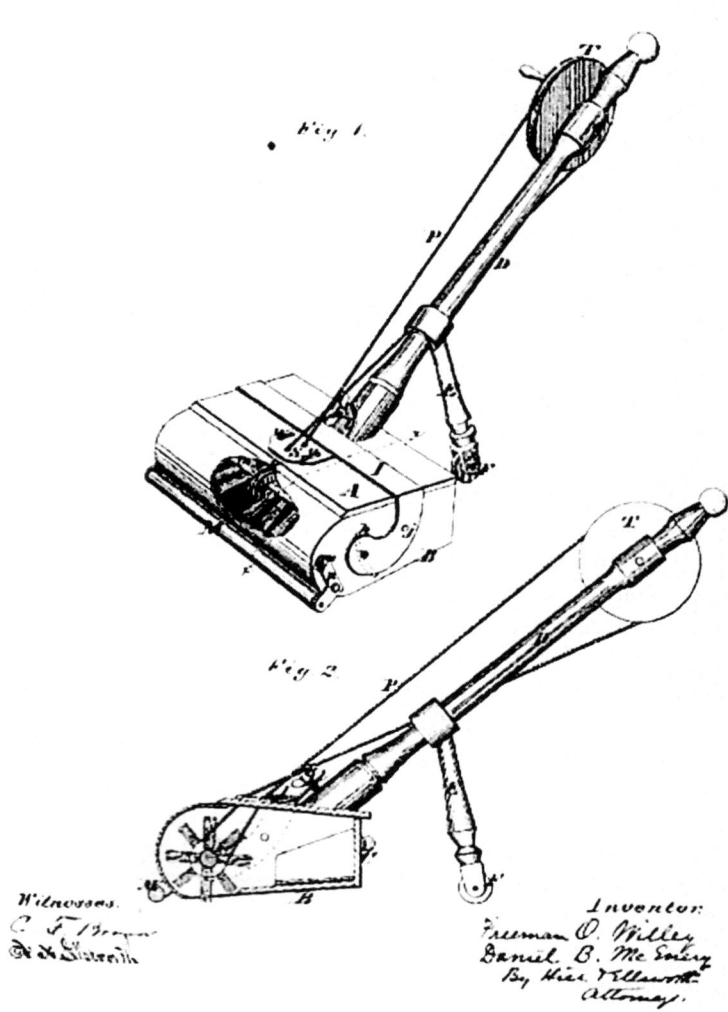

Willey/McEnery 1872 carpet sweeper, patent 125,369, U.S. Pat. Off.

After the exhibition in Philadelphia, manual suction cleaners continued to appear along with carpet sweepers. On March 1, 1881, the Favorite was patented. Patent 279,572 was issued to E.S. Leaycraft in 1883 for a "pneumatic sweeping apparatus" that moved the dust by air through a water filter. In 1884, English Patent 14,050 was issued to John Newton for "an improved method for removing dust, together with any infectious matter, from furniture, carpets, rooms, and thoroughfares." In 1890 the wooden Baby Daisy was made in England, which was the first working utilization of a bellows to create suction. The Baby Daisy was pumped using a long upright handle and included a floor nozzle, as well as an upholstery nozzle with a six-foot "lengthening tube." G.L. Cummings of England patented another pneumatic carpet cleaning concept in 1891.[10]

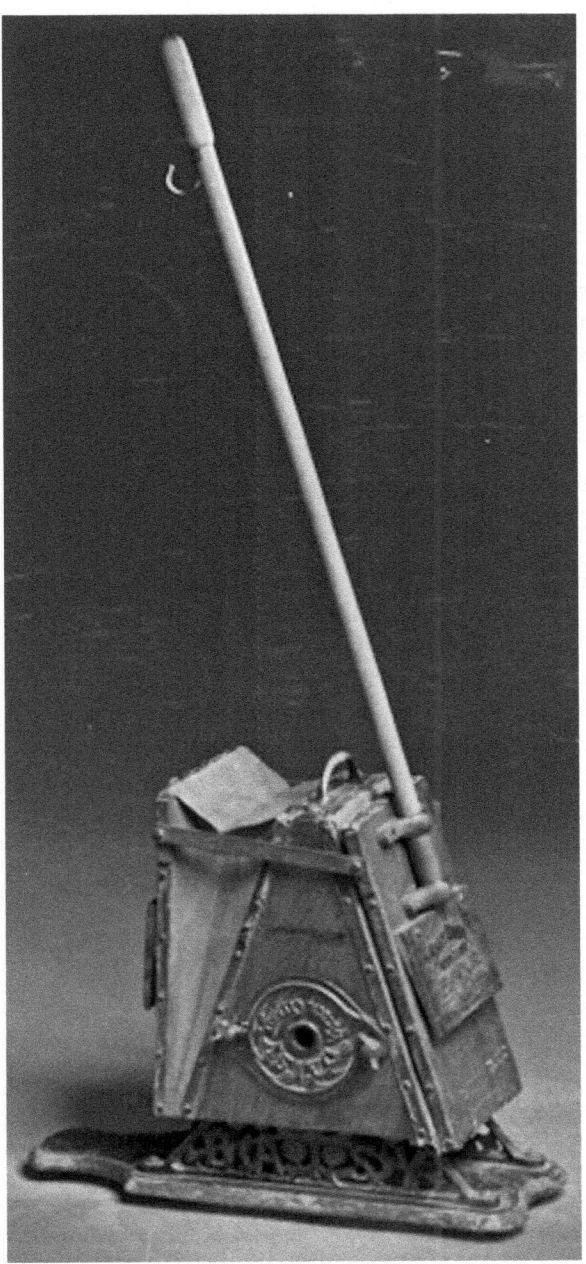

The Baby Daisy, an 1890 suction cleaner. HHC.

Prior to about 1903, the term "pneumatic" was commonly used to describe suction cleaners, but the term simply meant that moving air was involved in the process. There was no distinction between suction or pressure, both methods of moving air. The principle of compressing air by water pressure was developed in the 1860s by French engineer Germain Sommellier (1815–1871) to power pneumatic drills used in the building of the Mont Cenis train tunnel through the Alps to connect Italy and France. Steam engines later compressed air for similar drilling tasks, which led to the invention of the air brake for trains in 1869 by George Westinghouse (1846–1914). In the early 1890s, compressed air was often used in factories and foundries to clean and produce certain finishes on metals or to remove dust from carvings or relief work. An article in the July to December, 1894, issue of *House Furnishings Review* described a new invention with a hose that was charged with 50 pounds of air pressure to the square inch, which was used to clean a room and its contents "similar to a garden hose." This was probably some sort of compressed air equipment, and leaves to the imagination where all the dust went. Early air compressors were heavy, cumbersome, steam-driven machines.[11]

Another story of the same year reported that a St. Louis railroad porter often attached a hose to the compressed air from the steam engine in order to blow dirt out of railroad coaches. In one instance, it was said, a bag accidentally caught on the end of the hose, and as he drew the hose along the aisle, he noticed that the bag inflated and caught the dirt. He soon patented a device where a strong current of air was forced through a small opening, blowing the dirt into a surrounding bag that collected the dirt and let the air through.[12] The powerful force of compressed air was obviously so much more effective in moving dirt than the meager suction generated by hand or foot operated suction cleaners that it attracted inventors to use it in commercial cleaning devices.

In the last half of the nineteenth century, America was awash with new inventions at every turn, and it was leading the world in industrial capacity and technical innovation. Samuel Morse had invented telegraphy in 1844. Alexander Graham Bell invented the telephone in 1876. Edison, among his many other devices, invented the phonograph (1877), the microphone (1878), the incandescent light bulb (1879), the electric distribution system (1880), and the kinescope (1891). Frank J. Sprague invented electric streetcars (1887) and electric elevators (1892). George Eastman invented the Kodak camera (1888). The Duryea brothers manufactured the first American automobile (1893) using a gasoline engine (invented by Karl Benz in 1876). Wireless telegraphy was developed by Guglielmo Marconi in 1897; it would be perfected to carry voice as radio in 1906 by Reginald Fessenden.

It was a time of enormous growth in the U.S. The population, which stood at 35 million after the Civil War, had increased to 53 million by 1880, an increase of 50 percent. Immigration had risen to 800,000 annually. The U.S. birthrate in 1882 was 39.8 per thousand, almost twice that of Great Britain and triple that of France. The Gross Domestic Product (GDP) had doubled since the Civil War and was in 1882 the largest in the world: one third larger than Britain, twice that of France, and three times larger than Germany. Coal production had tripled, and steel production had risen from 20,000 tons in 1867 to 2 million tons in 1882.[13]

The electrical age had arrived, although up to the 1880s, electromagnetic current had been used only for telegraph, telephone, and lighting purposes, not for powering mechanical devices. Edison had established a power station in New York in 1882, with steam engines powering generators of direct electric current (DC), a continuous flow of electrons in one direction, for lighting about 225 homes. The electric motor, an essential component in making vacuum cleaners practical, was only now becoming useful for practical applications. Electric motors were not really considered for industrial power since so many large industrial machines were effectively powered by steam, and the invention of the gasoline engine in 1876 provided a new power source that was lighter and smaller than steam engines. In fact, gasoline engines powered early vacuum cleaners. But electric motor technology had been slowly evolving over the century.

In 1821, English physicist Michael Faraday (1791–1867) had discovered that electromagnetism could produce motion by electricity running through one circuit inducing a current in another. He built the first experimental electric motor — essentially a rotating needle.[14] In 1828, Hungarian Ányos Jedlik built a small-scale model car powered by an electric motor. In 1835, Brandon, Vermont, blacksmith Thomas Davenport (1802–1851) developed a small, battery-powered electric motor for a small scale car and is credited with the invention of the first American DC electric motor. In 1845, Charles Wheatstone developed the first electric generator. Electric storage batteries were developed in France from 1865 to 1881 and used to power electric vehicles.

A former Edison employee, Frank J. Sprague (1857–1934), who founded the Sprague Electric Railway and Motor Company in 1884, in 1886 invented a DC, constant speed, non-sparking motor with fixed brushes for electric streetcars, proving that electric current could be used for motors with stationary and rotating magnets. But these required a moving commutator to reverse the direction of the current and brushes to transmit the current to moving parts, which wore out quickly. Sprague streetcars obtained their electric power from overhead wires, but the great disadvantage of DC current was that it could not be transmitted any great distance, because it lost electrons through friction as it traveled through a wire. This became a huge problem for Edison as well, since his lighting systems used direct current.

A solution was soon found. Serbian inventor Nikola Tesla (1856–1943) had worked for Edison upon his arrival in the U.S. from Europe in 1884, but he established his own Tesla Electric Company in New York in 1887 to perfect a completely different alternating current (AC) system he had originally conceived in 1882 in Budapest. In the AC system, the movement of the electric current itself periodically reversed direction, and the voltage could be increased or decreased with a transformer. The same amount of power could be transmitted with a lower current by increasing the voltage (often hundreds of kilowatts), which led to significantly more efficient transmission of power through wires, thus enabling much greater distances for the distribution of electricity.[15]

Just as importantly, in 1889 Tesla invented an AC motor that consisted of only two basic parts: an outside stationary stator to produce a rotating magnetic field, and an inside rotor attached to the output shaft that was given a torque by the rotating field. The rotating field was created by two or more alternating currents out of step with each other, and was called the polyphase system. This was referred to as an induction motor, because an *induced* current created the magnetic field on the rotor. The advantage over the DC motor was that it eliminated the need for the commutator and brushes required by DC motors, providing a much longer motor life, and making AC motors much more feasible for industrial use.[16]

Between 1889 and 1891, Tesla was granted 40 patents on his polyphase system. He leased his patents for royalties to George Westinghouse, who wanted to use the motors on trains and streetcars. This brought the AC system in direct conflict with Edison's DC system and instigated the so-called War of the Currents. Edison claimed that AC was much more dangerous because of its high voltage, and publicly referred to the electrocution of a condemned murderer in 1890 by Westinghouse AC current as being "Westinghoused." Edison essentially asked Americans, "Is this the invention you want your little wife to cook dinner with?"[17] Actually, the first electric stove was demonstrated at the Chicago World's Fair in 1893 (see below).

Nevertheless, Westinghouse successfully refuted Edison's accusation of danger with facts and figures from respected scientists, and AC soon became a superior option for electric transmission as well as for motors. But both systems would remain for some twenty years, and the net effect on the average person was that depending where you lived, your local electric system could be AC or DC, and you would have to buy only electrically operated equipment that matched your local system.[18]

Westinghouse had underbid ($399,000 vs. $554,000) General Electric (created in 1892 from Edison's and other electric companies using DC current) and received the contract for installing all the AC current power and lighting equipment for the 1893 Chicago World's Fair (also known as the Columbian Exposition, the White City, or the first Electrical Fair), which dazzled the world with electric lighting, streetlamps, an electric car, the first Ferris wheel, and all the latest technologies from around the world. The fair, which was attended by 25 million, then a third of the total U.S. population, used Tesla's alternating current system, and Tesla himself presided in the electrical building, demonstrating one electrical miracle after another.[19]

In 1896, the hydraulic power station at Niagara Falls, based on Tesla's inventions, was completed by Westinghouse, and encouraged many new industries to use electric motors to drive their machines. Westinghouse became so successful that by 1897 the royalties accrued to Tesla mounted to a rumored $12 million and would have grown to billions, ruining Westinghouse, if continued. George Westinghouse offered to buy Tesla's patents outright for about $216,000 to avoid payment of royalties, and out of gratitude to Westinghouse for developing his inventions, Tesla generously agreed. Unfortunately, Tesla would die in relative poverty.[20]

Despite the advantages and availability of electric motors since the mid–1880s, and because of the two different systems, they initially could not compete in manufacturing and industry on a cost basis with well-established steam-power. In 1900, most factories still used huge steam engines that drove noisy and cumbersome systems of overhead shafts, with pulleys and belts to drive each individual machine. Only then did they begin to build electric motors into machine tools.[21] At the same time, electric automobiles outsold those powered by steam and gasoline, and a Belgian-built electric racing car, *La Jamais Contente* (The Never Satisfied), driven by Camille Jénatzy, set a land speed record of 100 kilometers per hour (62 mph) in 1899.[22]

Engineering as a profession had become famous over the previous generation with the creation of dramatic, large-scale public wonders such as bridges, skyscrapers, giant steam engines, electrical turbines, steamboats, locomotives, and automobiles. The many creative inventors and their world-famous inventions in the latter part of the nineteenth century inspired many youths to enter careers in engineering as the wave of the future. To the new generation coming of age, engineering was the stepping-stone to mechanical invention, technology, and public fame. In 1870 only 17 schools taught engineering in the U.S., but by 1890 there were 110. Between 1870 and 1914, the annual number of engineering graduates leaped from 100 to 4,300. By 1900, there would already be 45,000 trained engineers ready to advance the mechanical and electrical age into the twentieth century, changing the world.[23]

During this same era, the general public also became aware of microbial bacteria as the cause of disease, rather than the "bad air" of the traditional "miasma theory" of Catherine Beecher's day. In 1877, Prussian physician Heinrich Koch (1843–1910) proved that endospores embedded in soil were the cause of anthrax. In 1882, he discovered the bacteria that caused tuberculosis, and in 1883, identified the bacteria that caused cholera. People began to realize that there were invisible threats hiding anywhere dirt and dust could accumulate. Cleanliness acquired a new meaning other than Godliness. For the first time, people began to focus on hygienic and healthful practices as preventative insurance against disease.

These concerns also inspired social change in ways that would affect housekeeping. Ellen Swallow

Ellen Swallow Richards (1842–1911), photograph ca. 1881.

Richards (1842–1911) had virtually created the new field of home economics, picking up where the Beecher sisters had left off, but in a more scientific and businesslike way. She was the first woman admitted to the Massachusetts Institute of Technology, and the first American woman to receive a degree in chemistry. She believed that women's work in the home was a vital aspect of the economy. In 1881, Richards published *Cooking and Cleaning: A Manual for Housekeepers*, applying scientific principles to nutrition, clothing, physical fitness, sanitation, and efficient home management. In 1893, she had conceived an exhibit building at the Chicago World's Fair Department of Hygiene and Sanitation called "The Rumford Kitchen," demonstrating the usefulness of domestic science in the home. The name honored Massachusetts born Benjamin Thompson, also known as Count Rumford of Bavaria, who in about 1793 was the first to apply the term "science of nutrition" to the study of food.

Throughout the first decade of the twentieth century, Richards would write and publish many books to further the cause, including *The Chemistry of Cooking and Cleaning* and *The Cost of Cleanness, Sanitation in Daily Life* (1907). Young housewives of the twentieth century were anxious to use the new technologies to make their homes cleaner, safer, more healthful, and less labor intensive. In 1909, Richards would become the first president of the American Home Economics Association.

These advances of public health awareness spurred inventors in both England and the U.S. to increase cleaning effectiveness. In 1893, English Patent 546 was issued to James J. Harvey for "improvements in apparatus for removing dust from the surface of books, ornaments, pictures, and the like." Harvey's Pneumatic Dusting Machine was a suction machine with bellows operated by two persons, one of them working a bellows and the other pushing the dust collector from place to place.[24] A number of English patents included 2,577 to J.E. Howard and J.C. Tate in 1896, for improved apparatus "to remove dust from cushions, carpets, and other articles." An ejector operated by compressed air generated air suction. Patent 5,050 was issued in 1898 to A.F. Gue and P.J. Bonner for improvement in a carpet sweeper and beater, using a fan to produce suction. Patent 9,350 was granted George Wilton in 1899 for a surface dust remover, operated by a hooded brush with a water filter. It utilized air suction produced by an ejector. Patent 13,298 was issued to J. Eaton-Shore for a device "for withdrawing and sterilizing dust and other minute or light particles of matter." Essentially, it was a hooded broom with suction. American patents included 614,832, issued to Franz Burger in 1898 for a "machine for cleaning fabrics" (filed December 10, 1897). A "vacuum chamber" powered by steam generated a vacuum, and the dirt was deposited in the chamber via a hose connecting to a rectangular box placed upon the fabric. This is the earliest reference to the term "vacuum" found for a cleaning device. In 1899 Patent 628,505 was issued to George L. Westman for a "pneumatic carpet sweeper," comprised of a hand-cranked exhaust fan connected by a hose to a moveable carpet sweeper with a handle. It is not known if any of these machines were built or sold.[25]

Now, at the turn of the century, thousands of newly-trained engineers were applying their ingenuity to small household products or appliances that reached the everyday lives of individuals, including cameras, gramophones, electric irons, electric toasters, and electric fans. Entrepreneurs in small towns formed startup companies in basements and sheds across the country, hoping to become a part of the mass-market mechanical manufacturing boom. Hundreds of small appliance industries, including that of suction sweepers, were advanced by a new generation of mechanically minded inventors and college-educated engineers who invented and improved mechanical devices to accomplish their tasks faster, cheaper, and more efficiently.

Compressed air still seemed to be the best hope for effective commercial cleaning. Inventors had seen it in use in shops to blow away debris. If you have ever used a pressure washer and experienced the sense of power by the blast of water, you can imagine how impressed they were with the blast of compressed air. Now, they thought, if only they could figure out a way to collect and retain the dirt, rather than just blowing it away, the problem of carpet cleaning would be solved.

John Thurman's 1899 pneumatic carpet renovator, patent 634,042, U.S. Pat. Off.

This was the vision of John S. Thurman of St. Louis, Missouri. He invented a carpet cleaner for the General Compressed Air Company, which appears to have been the first powered by a gasoline engine. Such an engine had been invented by German inventor Nikolaus Otto (1832–1891) in 1876 and was rivaling the steam engine because of its lighter weight and portability. Otto's engine inspired the invention of the first automobile by Karl Benz (1844–1929) in 1886, and by 1893, Charles and Frank Duryea made the first American automobile. On October 3, 1899, when there were already 2,500 motorcars in the U.S., Thurman was issued U.S. Patent 634,042 for a "pneumatic carpet renovator."

Thurman's machine consisted of a heavy metal frame having mounted on its longer axis a wedgelike nozzle extending the entire length of its lower edge. The nozzle was pivoted so that it always would be inclined in the direction the renovator was being moved. The top of the renovator was a large canvas bag. Air introduced into the nozzle and issuing from the nozzle struck the carpet at an angle, and being carried up into the bag, carried with it the dust dislodged from the carpet. It was supplied with air from a portable unit, consisting of an air compressor driven by a gasoline engine mounted with the necessary gasoline and air storage tanks on a small horse-drawn wagon. The wagon was drawn up in front of the building to be cleaned, and a large hose, about 1¼ inches in diameter, was carried into the house and attached to an auxiliary tank, from which ½-inch diameter hose lines were carried to two or more renovators. The renovators were very heavy to carry about and their operation was quite complicated, requiring skilled operators.[26] In an advertisement in the *St. Louis Dispatch*, Thurman offered his horse-drawn cleaning system with door-to-door services in 1903, charging $4 per visit. By 1906, he was offering built-in central cleaning systems that used compressed air but featured no dust collection.[27]

Many sources erroneously credit Thurman with the invention of the first powered vacuum cleaner. Technically, his invention was not a vacuum cleaner at all, because it, like many other commercial cleaning systems of the time in both England and America, used a blast of compressed air to blow dirt into a large canvas bag, rather than using suction. The U.S. Circuit Court of Appeals confirmed this fact in a later 1916 ruling for the Second Circuit, in infringement litigation by David T. Kenney's Vacuum Cleaner Company (described below) in New York against the Innovation Electric Company. The court concluded that Thurman's later patents of November 29, 1901, and December 17, 1901, were not applicable to the case being conducted; however, the judge, Augustus Hand (1869–1954) went further and questioned Thurman's technical competence in his 1916 ruling:

> The Thurman patent is clearly not applicable. The specification describes a blast nozzle, and the inventor does not appear to have attempted to design a vacuum cleaner, or to have understood the process of vacuum cleaning.[28]

Of course, by 1916, it had become obvious that "vacuum" was the most practical way to clean carpets, and in fact, was already the foundation of a new technology and industry. It was easy in retrospect to regard poor Thurman as an inept inventor who did not understand such an obvious and proven principle. But in 1899, when Thurman patented his invention, this was not very obvious to anyone. Besides, we should look for the silver lining in Thurman's cloud: the principle of blowing air, rather than sucking it, would eventually result in leaf blowers, the wake-up call of Saturday mornings in suburbia, and snow blowers, both polar opposites of vacuum cleaners.

The difference between moving air by pressure (blasting or blowing) and moving air by suction is a critical distinction in the history of cleaners. In the former, the major problem is the almost impossible task of collecting the dust or debris in a container, such as a box or bag, in front of the air stream, since much of it flies in uncontrollable directions. Suction is far superior, because it is much easier to collect dirt in a container that itself is the same closed container in which a vacuum must be created. Suction is created by a fan blowing air out. The dirt can be filtered out within the container, before the air is exhausted from the container by the fan. Thus, the container itself can become the dirt receptacle.

Still, inventors tried to blow dust and dirt away by moving air under pressure, rather than by suction. On October 3, 1899, patent 634,042, and on December 18, 1900, patent 664,135 were issued to Corrine Dufour of Savannah, Georgia, for "an Electric Carpet Sweeper and Dust Gatherer," which used sponges to absorb dust. *House Furnishings Review* described it well:

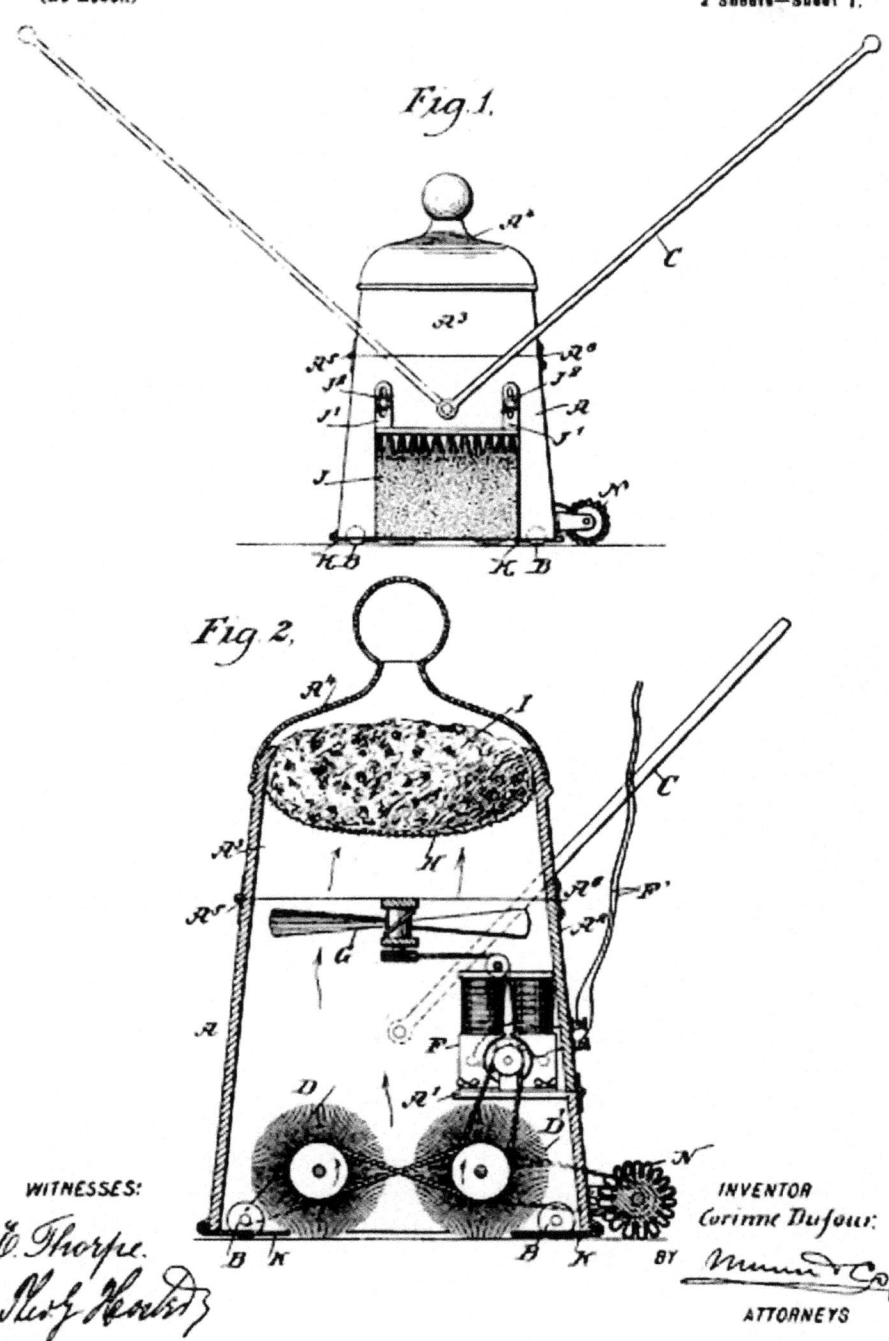

Corrine Dufour's 1900 electric sweeper and dust gatherer, patent 664,135, U.S. Pat. Off.

> The machine comprises a pair of brush cylinders rotated by connection with a fluted rubber roller in frictional contact with the floor, and there is also an electric motor inside the casing which is utilized to run a rapidly revolving fan, which takes up the dust and drives it against the sponge. The latter being saturated with water, retains the dust and aids greatly in cleaning the room ... when the sponges become coated with dust, the cover of the sweeper is lifted and the sponge taken out for cleansing.[29]

This appears to be the first reference to the use of an electric motor in a cleaning device, although the name and type of motor is unknown. As seen earlier, although electric motors were available at the time, they were relatively huge and heavy. B.G. Lamme, chief engineer at Westinghouse, had just developed early alternating current motors in 1897, which were highly successful and used for electric railway motors.[30] The only electric power available to households in 1899 was by privately owned utility companies in a few major cities, and then used primarily for lighting. Vast rural regions in the U.S., as we know, would not be electrified until the 1930s.

Commercial cleaners using compressed air continued to be refined, using numerous types of apparatus to produce suction, including air injectors and rotary pumps. Eventually, a piston-type vacuum pump with very light poppet valves, and mounted tandem with the air compressor, was adopted. These machines were mounted on wagons, similar to their predecessors. This system was fairly effective, since the dust and germ-laden air was removed entirely from the room and purified by means of separators before being discharged into the outside atmosphere. The foulness of the water in the separators clearly showed the amount of impurities removed from the air. The development of this successful system paved the way for the stationary plants, which would soon be installed in many large buildings and private homes, and would be used until the straight vacuum system was adopted.[31]

It is almost impossible for us today to imagine the incredible sense of change experienced by people living in the early twentieth century. It must have seemed to them as if an entirely new world of technology had suddenly been discovered and that the familiar nineteenth century of their childhood had disappeared overnight.

America had finally freed itself from the economic influence of the British Empire, and as a result of the Spanish-American War, was an international power with the global reach of a navy that exceeded England's. The United States was building the Panama Canal, the largest construction project in history, and skyscrapers with elevators were being built in Chicago and New York. America led the world in technical innovation and industrial development. Technological change was at top speed. Mass production generated high volumes of consumer goods at lower prices and employed millions in factories to do so. Wages were high, so that ordinary working-class consumers could afford products that previously were only available to the wealthy. In addition, they could easily obtain them with mail order catalogs from Sears Roebuck & Company or Montgomery Ward.

The automobile was replacing the horse and wagon, allowing people to travel when and where they wanted. By 1908 there would be 24 U.S. manufacturers of automobiles alone. Electric streetcars were operating in cities, steam trains were spanning the country, fast steamships were crossing oceans, and the gasoline-powered airplane had just been invented to open an exciting new frontier of transportation. Huge water turbines were creating infinite electric power, and that electricity was not only lighting homes, but was beginning to power a myriad of laborsaving devices to improve middle class life styles. The wealthy were building enormous Victorian homes, which were being imitated in size and style by working people. More carpets to clean!

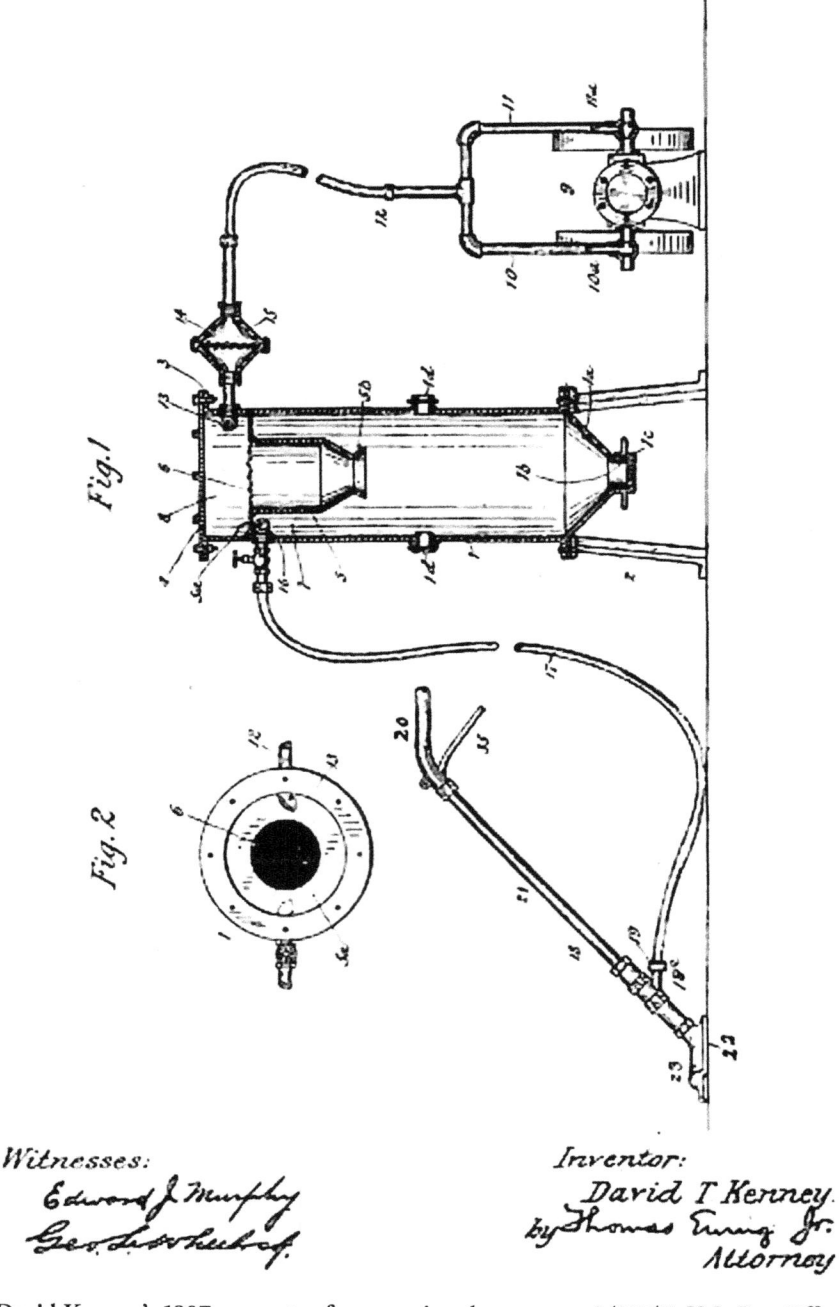

David Kenney's 1907 apparatus for removing dust, patent 847,947, U.S. Pat. Off.

It was within this new scientific and technological world that the new vacuum cleaner industry developed, producing hundreds of manual suction sweepers for homes in rural areas without electricity, as will be described, and perfecting new, more powerful electric vacuum cleaners for those in large cities with electricity. Inventors, entrepreneurs, and trained engineers, the latter now numbering 45,000, were using the new energy source of electricity to develop all kinds of laborsaving devices. Unfortunately, it would take some time for these devices to benefit many ordinary households, due to the slow progress in developing and expanding electrification around the country.

On November 19, 1901, U.S. inventor David T. Kenney (1866–1922), who had operated a plumbing business in Plainfield, New Jersey, since 1891, filed a patent application, granted on March 19, 1907, as U.S. Patent 847,947, for an "apparatus for removing dust" that was to be installed in a building. It was the first mechanical system in which vacuum alone was used. A stationary steam-driven power vacuum pump in the basement connected via a pipe outlet to every floor, where a flexible hose and nozzle were used for carpet cleaning.[32] Kenney called this nozzle a "renovator," which had a slot about 12 inches long and 3/16 of an inch wide, attached to a metal tube which served as a handle.[33] This is the same principle used today in central vacuum systems. Kenney's subsequent patents included patent 739,263, applied for February 8, 1902, as a separate improvement to his initial 1901 patent, and granted September 15, 1903, for a "separator for apparatus for removing dust"; patent 781,532, applied for May 28, 1904, and granted January 31, 1905, also for an "apparatus for removing dust" (the critical nozzle detail patent)[34]; and patent 1,057,347, for still another "apparatus for removing dust," which detailed the nozzle assembly. He applied for this patent in June, 1906, and it would be granted on March 25, 1913, after lengthy litigation.[35] None of these mentioned the term "vacuum cleaner," but these were the Kenney patents that would be litigated and upheld by the courts, and would dominate the domestic and commercial vacuum cleaner markets in the U.S. until 1923.

Accomplished English engineer Hubert Cecil Booth, S.C.G.I., M. Inst., C.E. (1871–1955) is often credited with inventing the first powered mobile vacuum cleaner. In fact, he only claimed to be the first to coin the term "vacuum cleaner" for devices of this nature, which may explain why he is so credited. As we well know, the term "vacuum" is a misnomer, because there exists no vacuum (technically the absence of any air whatsoever) in a vacuum cleaner. Rather, it is the air moving through a small hole into a closed container, as a result of air being blown out of the container by a fan on the inside. But I suppose a "rapid air movement in a closed container to create suction" cleaner, would not sound as scientific or be as handy a name. Anyway, we are stuck with it historically, and it is hard to find any references to "vacuum" prior to Booth, other than Franz Burger's 1898 patent, mentioned earlier, although there are numerous references to "suction-creating" or "pneumatic" (moving air) machines in the nineteenth century. Interestingly, Booth himself did not use the term "vacuum" when he filed a provisional specification describing in general terms his intended invention, the modest object of which was "improvements relating to the extraction of dust from carpets and other materials."

Booth, as an engineer, also designed and built suspension bridges, factories, and Ferris wheels at Black Pool in England, Earl's Court in Paris, and in Vienna. What inspired this prominent engineer of huge structures to become interested in vacuum cleaners? As Booth recalled in 1936, "My attention was first directed to the mechanical removal of dust from carpets in 1901 through a demonstration of an American machine by its inventor."[36] This event occurred at the Empire Music Hall in London. The American inventor is not named, but Booth's description of the machine appears to describe American inventor John Thurman's patents as modified in 1901. Booth much later described his experience:

> The machine consisted of a box about a foot square, having a bag on the top, to which compressed air at 90 pounds pressure was supplied; the air was blown down in the carpet from two opposite directions, as the inventor trusted to the reflection from the surface underneath the carpet to drive the dust and air into the box. I remarked that I could not see how one could get the dust out effectively in this way, as much of it must be blown out sideways; further, a cushion or seat where there was no box to reflect the air could not be cleaned. I asked the inventor why he did not suck out the dust, for he seemed to be going around three sides of the house to get across the front. The inventor became heated, remarking that sucking out dust was impossible, as it had been tried over and over without success; he then walked away.[37]

Thurman was correct in a sense. As we have seen, suction had been tried in over 30 American and British patents over the previous 40 years in dozens of manual power sweepers that "sucked out dust," with little success. But the problem was not the principle, as Thurman assumed, but the degree of power needed to increase suction to an effective level. Booth went on to say,

> I thought over the matter for a few days, and tried to experiment by sucking with my mouth against the back of a plush seat in a restaurant in Victoria Street, with the result that I was almost choked. I came to the conclusion that I could construct a machine to work by suction. A friend offered to finance the experiments, and I went ahead. I designed and produced a machine incorporating and combining suitably the various elements mentioned above, to which I gave the name of "The Vacuum Cleaner."[38]

Booth's quotes are from his own reminiscences, titled *The Origin of the Vacuum Cleaner*, published in the *Newcomen Society Transactions*, V. 15, p. 93, London, 1936. By that time, the term "vacuum cleaner" had been in common use for several decades, so his claim to have given his 1901 cleaner that name is questionable. In his 1936 treatise, he also enumerated and illustrated nineteenth-century patents preceding his own, though he did not know of them in 1901, and therefore probably invented his own machine independently.[39]

But did he invent the term "vacuum cleaner"? Booth's theatrical description of his discovery of vacuum 35 years after the fact seems a bit contrived, if not self-serving, since even in 1901, as an accomplished engineer, he would have been quite familiar with the technical principle of creating suction, such as already existed in London for some time with commercial cleaning equipment that used steam power or bellows to create suction; with dozens of manual household sweepers with suction-creating devices; and with the vacuum pump, the mechanical device that used suction to raise a liquid in a tube since antiquity. A dual-action suction pump had been found in Pompeii, proving that Romans were already aware of such technology. Suction pumps had been perfected in Europe around 1635 to raise water for irrigation projects, mine drainage, and water fountains, but because air pressure could support only a certain weight of water, nine meters was the limit such pumps could raise water. At the time, however, no one knew why.

Booth may have known of Italian physicist Evangelista Torricelli (1608–1647), who had used the suction pump principle to develop the first mercury barometer in 1643. But up to this time, it was not known what caused these devices to function, other than what was called "suction." French mathematician and physicist Blaise Pascal (1623–1662) solved the mystery in 1647 when he proved that it was indeed a vacuum that enabled these mechanical instruments to work.

This was a scientific revelation, because for centuries before, it was believed that a vacuum (a space empty of matter) either did not exist, or was "against nature." Greek philosopher Aristotle wondered how nothing could be something, and he therefore concluded that "nature abhors a vacuum." In medieval Europe, the Catholic Church also regarded the idea of a vacuum as being against nature, and therefore heretical, because the "absence of anything" implied the absence of God.

Booth probably knew of German scientist and inventor Otto von Guericke (1602–1686) of Magdeburg, who in 1650 invented a vacuum pump, which consisted of a piston and two-way flaps to pull air out of whatever vessel to which it was connected. In a dramatic experiment to demonstrate the enormous power of air pressure, von Guericke joined two copper hemispheres (called Magdeburg hemispheres) and pumped the air out of the resulting enclosure. Then he harnessed a team of eight horses to each hemisphere and showed that they were unable to separate the hemispheres. When air was again let into the enclosure, the hemispheres were easily separated. Von Guericke proved that substances were not pulled by a vacuum, but were pushed by normal atmospheric air pressure.

Booth must have known of English ironmonger Thomas Newcomen (1664–1729), who in 1710 developed a low-power steam engine to pump water from flooded coalmines, the invention that most historians agree started the industrial age in England. The Newcomen engine used vacuum instead of pressure to drive the piston. Booth would have known of German physicist Joseph Toepler (1836–1912), who in 1850 invented a mercury piston pump, which operated by atmospheric pressure and vacuum, and he would have known of German Chemist Hermann P. Sprengel (1834–1906), who in 1865, while working in London, achieved the highest vacuum achievable at that time with a mercury pump. The Sprengel pump was the key tool that made it possible in 1879 for Thomas Edison (1847–1931) to sufficiently exhaust the air from a light bulb so a carbon filament incandescent electric light bulb would last long enough to be practical.

Surely, because it happened just nine years earlier, Booth must have known of the "vacuum flask" invented by Scottish physicist and chemist Sir James Dewar (1842–1923) in 1892. It was a container made of two thin-walled bottles, nested one within the other, and sealed together at their necks, with the space between evacuated of air to create a partial vacuum. This absence of air avoided heat transfer between the walls of the bottles, and thus slowed the warming or cooling of the contents of the inner bottle. We now call such a device a Thermos bottle, because a German company, Thermos GmbH, introduced the first commercial application of the invention in 1904. Public knowledge of this recent, almost magical invention could have made "vacuum" a high-tech subject of conversation of the day, and thus may have inspired Booth to adopt the term.

Booth proceeded to build his first machine. It was a massive, gasoline-powered, horse-drawn vacuum cleaner; Booth called his initial model Puffing Billy after an early steam locomotive (another high tech invention in its day). Vacuum in Booth's device was initially created with a piston suction pump, but later versions used a "multi-stage turbine fan." Booth rented his machines, rather than sold them, through his company, the British Vacuum Cleaner Company. When put to use in a bright red van, a Puffing Billy was pulled to the exterior of a home or business, and workers extended hoses inside through doors and windows to vacuum the interior. Its most prestigious engagement was cleaning the carpets of Westminster Abbey before and after the coronation of King Edward VII and Queen Alexandra in 1902. This led to the purchase of two of his cleaners, one for Buckingham Palace, and one for Windsor Castle, which greatly increased his prestige. Later, during World War I, a number of Booth cleaners removed 26 tons of dust from London's famed 1851 Crystal

Palace exhibition hall, mostly from the girders, a feat that was credited with helping to stop an outbreak of spotted fever among naval reservists billeted there. This is evidence of the early reputation vacuum cleaners acquired for reducing the dangers of disease and health problems.[40]

Booth filed his patent on February 18, and it was published August 30, 1901, as British Patent 17,433.[41] In 1902, Hubert Booth filed an application for his American patent, submitting complete specifications with drawings of his machine that used suction for cleaning instead of compressed air. Hubert's brother, Stone Booth, who had been in Boston for some time on personal matters, returned to England and saw Hubert's vacuum system.

Hubert Cecil Booth's 1905 Puffing Billy.

Stone was impressed and took two of Booth's larger, four-horsepower machines back to Boston. Soon, both David Kenney and Stone Booth were conducting public demonstrations.

Kenney's first system was installed in the Henry Clay Frick building in Pittsburgh in 1902, consisting of a stationary 4,000-pound steam engine in the basement, which powered pipes and hoses to all parts of the building. It cost $2,100. It also included a "wet extractor" that sifted out dirt drawn in through the nozzle. That year, Kenney's invention was taken over by the Vacuum Cleaner Company of New Jersey, which used positive piston pumps made by the International Team Pump Company to produce vacuum. Later, the Vacuum Cleaner Company of New Jersey was liquidated and was succeeded by the Vacuum Cleaner Company of New York. A version of the Booth vacuum machine, called the B.B. Vacuum Cleaner was manufactured by the McCreery Manufacturing Company in Toledo, Ohio, and sold to people who wanted to get into the housecleaning business.

Kenney saw the Booth machine demonstrated in Boston in late 1903 and purchased it and its American application in early 1904. By 1906, Kenney's company claimed installations at the White House in Washington, D.C. and the Times Building in New York. Kenney's most significant patent, which he had applied for in 1901, was essentially the concept of sucking air through a narrow nozzle to remove dirt from a carpet, a very basic essential to all vacuum cleaners, since the smaller the opening, the greater the airflow. The Vacuum Cleaner Company of New York, holder of the Kenney patent, had brought suit against the Innovation Electric Company for infringement of this patent. The result was that the Kenney patent was held valid and infringed, and it was granted on March 19, 1907, as U.S. Patent 847,947. By 1909, The Vacuum Cleaner Company of New York had turned its manufacturing interests over to the McCrum-Howell Company of New York and continued only as a patent holding company owning the Kenney and other patents, under which licenses were granted to the most important companies making and selling vacuum cleaners.[42]

The first decade of the twentieth century became a hotbed of technical innovation for vacuum cleaners. The earliest, already described, were primarily intended for the commercial cleaning market, and they cost much too much and were way too large and unwieldy to be considered for domestic use by homeowners.

For example, around 1905, Ira Hobart Spencer (1873–1928), founder of the Organ Power Company in 1892, which had made a multi-stage turbine blower for organs called the Organblo, founded the Spencer Turbine Cleaner Company in Hartford, Connecticut, to make and market a stationary installed vacuum cleaning system using a modified multi-stage turbine blower to produce suction. It operated with 5 inches of water suction (today's vacs create 80 inches) that was much higher than other systems.[43] The machines were constructed with sheet metal casings riveted on and mounted on horizontal shafts. The separators were sheet metal receptacles for catching litter. Lightweight hose was used to connect to the carpet renovators, which had cleaning slots varying from 10 × ¾ inches to 20 × ¼ inches, and were connected to pipe lines.[44]

The United Vacuum Appliance Company, in Connersville, Indiana, marketed its No. 350 home installed system for $225, and the Connersville Blower Company offered a portable Auto Vacuum Housecleaning system on a truck for home servicing. In France by 1903, vacuum machines driven by electric motors and mounted on wheel carts were used to clean theater seats.[45] Although these commercial systems were very popular despite their cost, smaller, lighter vacs were needed for domestic use.

These would soon appear. Dr. William Noe of San Francisco constructed the first such self-contained, electric portable vacuum cleaner in 1905. It had a mechanically driven rotary brush similar to carpet sweepers for loosening dirt from the carpet. The dust was sucked up by large, two-stage turbine fans and discharged into a rigid metal dust container mounted on the handle, similar to dust bags on compressed air cleaners. It was mounted on wheels and had a large DC electric motor.[46]

Noe became associated with business partners Benjamin J. Skinner and Alonzo E. Chapman of the Skinner Manufacturing Company, located in the Starr-King Building at 121 Geary Street, San Francisco. They made and sold Noe's cleaner to commercial services in and around Oakland and San Francisco in 1905, which took the machines from house to house. The services priced their work by the square yard and also sold a few machines to users.[47]

That same year, on October 31, Chapman filed an application for a patent covering a "pneumatic sweeper and renovator." The device consisted of a two-stage turbine and motor-driven fan, the inlet of which was connected to a thick floor nozzle containing a rotary brush driven by the motor and discharging into a dust separation device. It weighed an incredibly heavy 92 pounds (small, lightweight motors had not yet been developed) and used a fan 18 inches in diameter. It had no wheels, but skidded along on a board, requiring significant physical effort. In March of 1906, Skinner filed a patent application for a "pneumatic carpet sweeper."[48]

The Skinner Manufacturing Company, its building, and its machinery, were all destroyed in the San Francisco earthquake and fire of April 18, 1906, but one of its vacuum cleaners was in Southern California at the time, and is now on display in the Hoover Historical Center, in North Canton, Ohio (see group photographs of vacuums at the end of this chapter). Deciding to relocate and reorganize, the Skinner organization contracted with Robbins & Meyers in Springfield, Ohio, who built some sample machines, but Skinner was not successful in establishing its organization there. On January 1, 1907, Skinner and Chapman organized a new company called The Electric Renovator Manufacturing Company in Pittsburgh, Pennsylvania. There, the machine based on Chapman's and Skinner's patents was produced and called the Invincible Renovator. It still weighed 92 pounds. Due to years of litigation after application, the patents were pending in the patent office until May 23,

1916, when Chapman's patent 1,183,952 was granted, and May 30, 1916, when Skinner's patent 1,185,354 was granted.[49]

There were numerous other vacuum cleaner developments during the early part of this first decade of the twentieth century. In 1906, The Burke Electric Company in Erie, Pennsylvania, produced a small, high-speed universal motor attached to a centrifugal fan blower that had all the features of the modern suction-creating unit, and in fact, could serve as a suction cleaner if the filtering bag was placed on the discharge nozzle and a floor nozzle was fitted on the intake opening of the fan. An Ash Can vacuum, used to clean offices and weighing 60 to 70 pounds, was introduced between 1905 and 1907.[50]

A number of new manufacturers entered the new field of vacuum cleaners including: the Vacuum Engineering Company of New York (1906), the Palm Vacuum Cleaner Company of Detroit (1906), the Dunn-Locke Company of New York, and the Blaisdell Machinery Company of Bradford, Pennsylvania (1906). The son of the founder of Blaisdell was George G. Blaisdell (1895–1978), who in 1932 invented the famous Zippo cigarette lighter,[51] still being produced today. A patent was filed in 1906 by German immigrant engineer Hermann Bogenschild of Milwaukee, Wisconsin, for a mechanical "dust removing apparatus" that was mounted on wheels for portability, and its hose connected to a filter system.[52]

The major technical innovations of vacuum cleaners in these early years of the twentieth century were successful primarily because of the tremendous suction power created by the latest new sources of energy, gasoline engines and electric motors. Only the latter were small enough, light enough, and without noxious fumes to be reasonably practical for normal domestic household operation. However, they were still quite unwieldy, and despite their power, their use was feasible only in the few large cities that had an electric power system.

In 1900, 60 percent of the population in the U.S. lived in rural areas without electricity. But because the widespread news of the successful development of the new, powerful vacuum cleaners had excited the public to the very idea of "vacuum" as a new and effective cleaning principle, there was an increasing demand across the country for manually operated vacuum cleaners, the latest hot technology (just as a mere decade or so ago, we all rushed to buy cell phones or the rediscovered technology of cyclonic vacuum cleaners). Many manufacturers rushed to fill this consumer demand for vacuum cleaners. There were literally hundreds of new manufacturers that made manually operated vacuum cleaners, starting about 1904. From 1869 to 1920, mostly after 1904, there were more than 250 manual suction cleaner manufacturers, and at least double that number of models.

Some were cast iron, others sheet metal or wood, and all were available in a range of shapes, sizes, and colors. Although functionally impaired by lack of power, they were lightweight, inexpensive, and probably performed a bit better than carpet sweepers because of the added air movement. Many required significant physical exertion, and thus served additionally as in-house, private, physical fitness centers. Still, they made carpet cleaning easier than the grueling ordeal of beating the carpets outdoors. These innovative devices all sought to create suction using a vacuum, but through vastly different mechanical means.[53]

The use of bellows had been pioneered in 1890 by the British Baby Daisy machine and this principle was utilized in America by many early manual machines. Devices with bellows created suction as air rushed into the bellows when it was pulled open. Some were mounted on a board or sled runners, operated with a tillerlike handle, and were dragged around the room. With a single bellows, there was no suction when the bellows closed. But with bellows in pairs, allowing one to open while the other closed, each back and forth stroke of the tiller would operate bellows sequentially and create more or less continuous suction. The foot-operated Griffith (1903) had two bellows operated by a person treading on them alternately, as if climbing stairs, while another person operated the separate hose and nozzle. It was probably the first home Stairmaster.[54]

Another two-bellows model, Herman Kotten's 1910 Kotten Vacuum Cleaner (see group photographs of vacuums at the end of this chapter), used two bellows powered by a manual rocking action, and was operated by the user standing on a skateboardlike platform with a "see-saw" action. Another similar product made by George Webster, of Christiana, Pennsylvania, was actually called the "See-Saw Suction Cleaner." Some models incorporated bellows in a cylindrical plunger-type body. Others used hand-cranked wheels to operate the bellows, but most looked much like carpet sweepers without a rotating brush. One inventor attached bellows to the bottom of shoes, so the user could power the bellows by walking about the room. Another submitted a patent powered by a bellows connected to a rocking chair, so that the man of the house could enjoy the evening paper rocking in the chair, while the wife performed the vacuuming. Ladies loved this one, I'm sure.[55]

Other examples of the bellows type include the Duntley Pneumatic Sweeper, made by the Duntley Pneumatic Sweeper Company of Chicago (1909, $9.75) (see group photographs of vacuums at the end of this chapter); the Williams Combination Sweeper, made by the Frank W. Williams Company of Chicago; the Victor made by C.L. Cornell Manufacturing Company in Randolph, New York; the Little Witch (1901), made by United Manufacturing and Distributing Company in Chicago; the Banjo, made by the Twentieth Century Vacuum Cleaner Company in New York (1902); the Keystone, available from the Lanning-Stone Sales Company in Chicago; the Keller Hand Power Cleaner, made by the Keller Manufacturing Company in Philadelphia (1908); and The Queen Louise, made by the Sterling Vacuum Cleaner Company in Sebring, Ohio.[56] A patent was granted to Quist and Blanch in 1910, covering a device that consisted of a suction-creating bellows that was operated by a pitman (connecting rod) connected to a crank pinned on the supporting floor wheels. The operator supplied the energy required for creating suction and its degree depended on the speed the machine was operated.[57]

Hand pumpers were another type of manual vacuum, generally mounted on a wooden board or sled runners for stability and portability, and were operated with a long handle. They had two cylindrical diaphragm chambers: one to create suction, and the other a compartment to hold dirt. The easiest method of operation was with one person operating the pumping handle while the other used the hose or wand to vacuum up the dirt. Some channeled the exhaust air into a blower tube, so that the user could blow dirt out of narrow crevices. Examples include the Lehman, made by Lehman and Son, Inc. (1900, See group photographs of vacuums at the end of this chapter); the Challenge and the Hercules, made by the Flower City Purchasing Company of Rochester, New York; the Busy Bee; and the Regina A, made in Rahway, New Jersey, New York City, and Chicago (1908). This latter was the first pneumatic cleaner made by the Regina Music Box Company, licensed under the Kenney patent (847,947). Gustav Brachhausen founded the company in 1892, but by 1909, the music box industry was being replaced by Thomas Edison's Gramophone and by the Victor Talking Machine Company. Other cleaners included the Dustkiller, sold by Sears (1910); the Junior, made by the Blaisdell Machinery Company in Bradford, Pennsylvania; and the Rochester, made by the Rochester Vacuum Cleaner Company.[58]

Another type of manual vacuum, called "plungers," consisted of long, slender cylinders tapered at the bottom with a nozzle that contacted the floor. A plunger handle was drawn upward with one hand, while the other hand stabilized the machine with a handle on the main body. This action created suction at the nozzle, like a hypodermic needle, and the dust was sucked into the body and was trapped in a funnel-like device. A cloth air filter was incorporated into the body. Some models used bellows, dust collection bags, or twin handles, side by side, which when pulled apart, pulled the plunger up and created suction. There were dozens of manufacturers, but some examples include the National (1905); the Success (1910, See group photographs of vacuums at the end of this chapter); the Jem, made by the Guarantee Sales Company of Chicago; the New Home, made by the R. Armstrong Company in Cincinnati, Ohio; the Home (1910, See group photographs of vacuums at the end of this chapter); and the Allen (1910), made by the Allen Company in Chicago.[59]

"Wheel operated" drives were the most powerful of manual vacuums and created a continuous suction. Many used belts and pulleys for mechanical advantage. The power wheel was set in motion with a cranking handle. The large wheel drove a belt that went around a smaller pulley, which spun a shaft to turn a fan, or to pump a bellows. The easiest method was to have one person provide the manual power, and another to operate the hose and nozzle for cleaning. Examples include the Agan Vacuum, made in Ludlow, Vermont (by Frank W. Agan, patent 862,369, 1907); and the 1908 Santo Vacuum Cleaner, a hand crank wheel operating an air compressor in reverse, made in Philadelphia. The 1910 Doty, made by Doty Manufacturing in Dayton, Ohio (see group photographs of vacuums at the end of this chapter), had wheels powering a crankshaft that moved push rods connected to two bellows, which were synchronized to produce continuous suction when pushed back and forth.

"Friction" drives used the rear wheels of a sweeper for power when pushed by the operator. The energy was transferred from the wheels touching the floor to a fan via a worm drive during forward motion, and the front wheels were connected to a brush roller to create a sweeping action. They looked much like early electric upright sweepers, but had no motor, cord, plug or switch. They were lightweight, easy to use, and were used into the 1940s. Examples

1910 Allen manual vacuum. HHC.

Early vacuum cleaners in the Hoover factory fourth floor storage room. Photograph ca. 1950. (A) Success Hand Vacuum Cleaner (1910); (B) Home Vacuum Cleaner (1910); (C) Richmond Vacuum (1909); (D) Skinner (1905); (E) Willey's (1872, full size & miniature); (F) Skinner Invincible vacuum cleaner (1909); (G) Dayton vacuum cleaner (1914); (H) Sweeper #5 (probably 1913); (I) Hatch & Goesser, experimental model (1905, never sold); (J) Leasure (1910); (K) Kotten vacuum cleaner (1910); (L) Brilliant Vacuum Cleaner (probably 1911); (M) Duntley (1909); (N) Doty (1911); (O) Hickory Broom (1650); (P) Hoover Laundry Cleaner (1909); (Q) Hoover Model O (1908); (R) Old Patent Office Model; (S) Automatic (1908); (T) Hoover Side Outlet Cleaner (1909); (U) Twig broom (replica of type known to be used as early as ca. 2300 B.C.); (V) P&W vacuum cleaner (probably 1913). HHC.

include the 1910 Scott & Fetzer Grasshopper; the 1919 Scott & Fretzer Vacuette Model B; the Viking, made by the Vital Manufacturing Company of Cleveland, Ohio; and the Kwick-Kleen, made for Sears Roebuck Company by the Standard Vacuum Manufacturing Company of Cleveland.[60]

There seemed no end to the ingenious mechanical options to meet the escalating consumer demand for low-cost household cleaning devices with suction that seemed to work better than carpet sweepers. Among the many manual vacuum sweepers were the Holland; the Jobier; the Feeney; the Star (1907); the Daisy 64 (1907); the Vacuo (1907); the Wizard (1907); the Cyclone (1910); the Domestic (1910); the Marvel Vacuum Cleaner (1910); The Leasure (1910, See group photographs of vacuums at the end of this chapter); the Aerio (1910); the U.S. (1911); and the Everybody (1913).[61] One has to admire this extensive litany of creative names for vacuum cleaners, which is so much more interesting than the ubiquitous multidigit model numbers we see today that it deserves to be documented for posterity.

Early vacuum cleaners, left to right: 1875 Agan; 1910 Kotton; 1905 Skinner; 1869 Whirlwind; 1872 Willey's; 1900 Lehman; 1910 Leasure. HHC.

The first decade of the twentieth century marked the gradual transition from manually operated vacuum sweepers to sweepers powered by the new, but slowly distributed, source of energy: electricity. By 1913, manual devices would become much less popular, because more efficient electric sweepers, requiring much less physical effort, had become available, and because more major cities and regions were being provided with electric power to use them. However, since rural electrification was not even begun until the 1930s, many manual vacuum sweepers would be still be around until then. This transition was only one small example of cultural change in a country that was at the same time undergoing enormous changes in international leadership, transportation, economics, and modernity. The most dramatic change was the influential expansion of electricity distribution systems for manufacturing and for new consumer products for the home.

CHAPTER 3

Electrics 1900–1920

In addition to its relatively high initial cost, there were restrictions on the use of electricity that slowed the purchase of many of these new appliances by the general public. It is difficult for us today, accustomed as we are to 24/7 availability of electricity, to understand the gradual electrification of the country that took place starting about 1903, and which was only half completed by the mid–1930s. The generation of electricity, and its complex infrastructure of transformers, poles and wires, required enormous capital investment, considerable time, and Herculean effort, including that of the federal government. In the early part of the century, only privately owned utility companies provided electric power and then only in major cities, populated by only about 40 percent of residents. Even there, many companies provided power only at night, for the important task of lighting. This was of absolutely no use, however, for any electrical home appliances normally used during the daytime.

Earl H. Richardson, a meter reader at the Ontario (California) Power Company, who independently developed an electric laundry iron in 1903, found a solution to this problem by convincing his employer to generate power all day on Tuesdays (ironing day for most households) so his irons could be used. Customers loved the unique "hot point" at the tip of his iron, to iron pleats and buttonholes. It became the first successful electric laundry iron, and in 1907, his successful business became the Hotpoint Electric Heating Company.[1]

From the beginning of electrical service, similar regional power companies would be instrumental in promoting the use of electricity, and this often would include not only the print promotion of electric appliances such as vacuum cleaners, but outright massive sales campaigns directly to consumers. From 1909 to the 1960s, the Commonwealth Edison Company of Chicago would sell portable electric appliances in company shops and from home delivery vans.[2] Westinghouse and General Electric, the largest national electric companies, would maintain immense local retail stores, which would sell all kinds of electric appliances. Examples of this practice, which would aid measurably in developing markets and sales for vacuum cleaners, as well as other appliances, will be found throughout this book. Even today, many regional, public electric cooperatives formed during the 1930s give away free appliances as incentives for those who attend their annual meetings.

Most rural communities were not electrified until the government established public utility companies in the mid–1930s, and even then, only 10 percent of farms had power, compared to 90 percent of urban households. This explains why manual vacuum cleaners remained in use well through the 1920s, despite the availability of electric vacuum cleaners, and why the latter were so expensive and were considered a luxury until the 1950s. So it is important, as we describe early electric vacuum cleaner development, to understand that the national consumer market for electric vacuums, and many other electrical appliances, was relatively very small, for a very long time.

To sell these new electric appliances was therefore not an easy matter. The business world had to change to more aggressive advertising and sales techniques. The man who set the pace in 1906 was Alfred C. Fuller (1885–1973), who essentially invented the techniques of door-to-door selling and market research by the way he sold the brushes he made. He personally demonstrated how effective his brushes were, instead of just telling customers about them, and he sought their suggestions of how his product could be improved. Customers loved his new, practical designs, and by 1909 he had hired 270 dealers throughout the U.S. to work for him as "Fuller Brush Men."

More and more new electric vacuum cleaner manufacturing companies were being organized, adding to those manufacturers of manual vacuum cleaners and early electric vacuums already described. In 1907, the Sweeper-Vac Company was organized in Worcester, Massachusetts, and the Pneu-Vac Electric Vacuum Company introduced a 60-pound mahogany cabinet cleaner that looked like a piece of expensive furniture.

Another newcomer in 1907 was the Santo-Keller vacuum cleaner, made by the Keller Manufacturing Company in Philadelphia. Although Keller had made earlier manual vacuum cleaners, and still did, this was its first electric. The new Santo-Keller had a vertical cylinder shape that stood upright, without wheels, used a bellows driven by a ¼-horsepower induction motor, and was what might be called a "canister," "cylinder," or "tank" type of cleaner, although such upright cylinder styles were often called "fire plugs" for their similarity to fire hydrants. This particular Santo-Keller would later become an inspiration to the well-known European entrepreneur who founded Electrolux, as we will later see.

Invincible Vacuum Cleaner Manufacturing of Canal Dover, Ohio, in 1907 came out with a commercial cleaner with small bicycle wheels and a 110-volt motor. The Invincible was a canister-type vacuum with a ten-foot long hose, offered in choices of two, three, or four turbine fans; the four-turbine model was mounted on a truck for commercial cleaning of churches or factories. Patent dates for the Invincible range from March 1907 through May 20, 1911.

Then there were the individual inventors and entrepreneurs who were so essential to the nascent industry. James B. Kirby (1884–1971) was born in Richfield, Ohio, a suburb of Cleveland. When he was 13, in 1897, a Cleveland newspaper article hailed Kirby as "the Wizard of the West Side" who had a "flair for gadgetry" and was a "child born with a knowledge of electricity." This was an obvious favorable comparison to Thomas Edison, called "the Wizard of Menlo Park" by national news media.[3] By 1900 Kirby was the owner of a motor repair shop in Richfield. Kirby invented his first vacuum cleaner in 1906, inspired by a mobile truck device cleaning his neighbor's carpets with the latest technology of vacuum cleaning. After examining the equipment, Kirby said he "could make a machine do that without all that apparatus."[4]

His first effort in 1906 was essentially an upright drum at the bottom of which were two pails of water and on top of which was mounted a hand-powered pump. Some referred to it as a "garbage can" type. A flexible hose connected the cleaning nozzle to the water tank in which the dust-laden air was drawn up in bubbles and filtered before reaching the suction chamber and pump. It was essentially a two-person manual pumper similar to those described earlier, and it required frequent filling and emptying of the water buckets. Displeased with the task of dirty water disposal, in 1907 Kirby would add an electric motor to drive the pump and a system that used centrifugal action by shooting the dust-laden air into the lower portion of the drum through a tube tangential to the drum, while dust particles were caught by a cloth bag filter surrounding the drum and motor.[5] It was the first application of the "cyclonic action" principle. The Domestic Vacuum Cleaner Company offered Kirby's machines for sale at $85 with a motor and $25 without one.[6]

Among these many vacuum cleaner developments came James Murray Spangler (1848–1915), a part-time, 59-year-old inventor who lived near New Berlin (now North Canton), Ohio. He had previously invented a velocipede wagon, a combination hay tedder and rake (a

James Murray Spangler (1848–1915), photograph ca. 1910. HHC.

machine used to spread hay to dry after cutting), and a grain harvester, but each new invention used up income from the older, which left him in constant need of funds.

Spangler had recently moved to Canton, Ohio, to take a temporary job as a janitor in the Folwell Building, the main occupant being the William R. Zollinger Department Store, with miles of carpet to be cleaned. After a few years of operating a large Bissell carpet sweeper (3,000 per day were then being produced) he developed a cough, and soon found that the dust aggravated his asthmatic condition. In early 1907, Spangler was watching a street sweeper in operation and decided to build a similar device for his work at Zollinger's. He started by attaching an old electric fan motor to drive the Bissell brush roll, which made it easier to push, but the dust still made him cough. He wanted to remove, and somehow, contain the dust.[7]

That summer, Spangler built a working model to explore his idea, using a wooden soapbox for the body, with adhesive tape over cracks and crevices to make it airtight. The motor was from an old electric fan, to which he attached a new fan with blades cut from an old stovepipe. The rotary brush was a piece of broom handle with goat hair attached with staple tacks, but which only rotated by contact with a carpet. One of his wife's pillowcases served as a dust bag, and a broom handle was used to push the device.[8]

After successfully testing his rough model, Spangler constructed a more substantial prototype made with a metal drum, which differed in principle from the first only in that

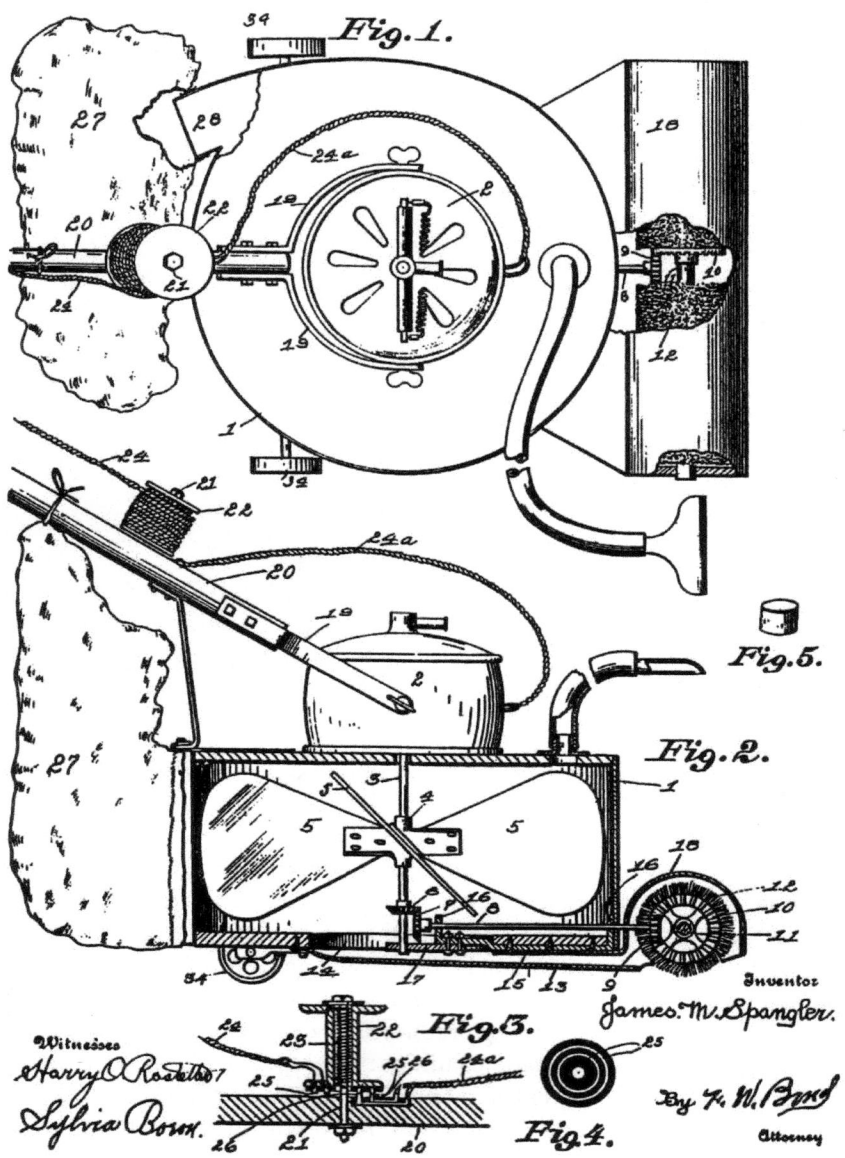

Spangler's 1908 carpet sweeper and cleaner, patent 889,823, U.S. Pat. Off.

the rotary brush roll was powered by a belt attached to the motor shaft. It worked pretty well, and others thought so, too. Spangler soon formed a partnership with C. Ray Harned, and on September 14, 1907, Spangler applied for a patent (889,823)[9] on his "carpet sweeper and cleaner." A month later, on October 15, 1907, an agreement was entered into between Harned, Spangler, F.G. Folwell and W.H. Folwell, whereby the Folwells agreed to furnish $5,000 to manufacture and sell the Spangler sweeper. Thus, the Electric Suction Sweeper Company of Canton, Ohio, was born.[10]

Spangler, who had quit his job at the Folwell Building, took one of his machines and visited a family friend in New Berlin, the former Susan Troxler, now Mrs. William H. Hoover, hoping for a sale. Spangler demonstrated his machine in her dining room and hallway, and the results were so impressive that she bought it. When her husband and sons came home that night, they noticed that the front hall carpet was not covered with the usual protective newspapers. When they commented on this, Susan wheeled out Spangler's clanking monster. They laughed as they inspected the whining machine in action. Her oldest son Herb kidded her.

Mother! You've been taken to the cleaners by the cleaners! Why, this thing will fly apart in your face someday! Whenever you want to do the upstairs carpets, just call over. We'll pick out a few good strong men, and we'll have it upstairs in no time.

1908 Electric Suction Sweeper Company Model O. HHC.

Susan replied, "That's all right boys. You and your father run the factory. I'll run the house."[11] Susan was reaffirming the role of women everywhere by asserting her supreme authority in matters of household responsibility. The factory she referred to was the W.H. Hoover Company, a leather goods manufacturer founded in 1875 by her husband, William H. Hoover (1849–1932), which he ran with his three sons, Herbert, Frank, and Dan.

Spangler's sweeper went into production in the fall of 1907 with the wooden and tin body and brush housing painted in black with decorative stripes and decals. On the brush housing, in gold and red, was printed "Trade Mark, Electric Suction Sweeper Company, Canton, Ohio." Two bags were provided — an inner one of coarse cheesecloth, and an outer one of red sateen, supported by the broomstick handle. On Thanksgiving Day, Spangler gave a public demonstration of his cleaner, performing as a salesman. Murray was more than a salesman. He was the entire working staff of the company. He designed the parts, ordered them from local suppliers, picked them up by streetcar, and assembled them in his basement at the rate of two or three per week. Spangler had used his own money to buy Emerson Electric (DC) motors, offering his own house as collateral. Ray Harned was supposed to have been selling the sweepers, but sales were soon lagging far behind production. So Spangler, 59 years old and in failing health, had to help with that, too.[12]

Not surprisingly, the new company soon encountered financial difficulties that threatened the business, which was going nowhere by mid-1908. The partners concluded that they needed more capital, as well as more technical experience and sales organization, to successfully manufacture and market Spangler's invention.[13]

Spangler's original patent application was granted on June 6, 1908.[14] By this time, at a rate of two to three per week since the fall of 1907, Spangler had built from 80 to 120 cleaners with many yet to sell and with creditors at the door. Things were so bad that Murray was in jeopardy of losing his home.

By coincidence, the W.H. Hoover Company in New Berlin, with William H. "Boss" Hoover as president, was also encountering some business trouble. The company had been started as a tannery in the 1840s by Daniel Hoover, William's father, that later made harnesses, Sensible Irish horse collars, and saddles, and by 1905, employed 200 workers. New Berlin, with a population under 700, had become an incorporated town, and "Boss" Hoover had become its mayor. There was no one in town with a better credit rating, as typified by this January 10, 1907, credit report.

> He is an old time merchant who has been in this line for many years. He is a shrewd, capable merchant doing a good business, meets his obligations in a satisfactory manner, each year showing a healthy gain, and he is in good credit for his business wants....[15]

The problem for the W.H. Hoover Company was that horse collars and saddles were of no use to an expanding national proliferation of automobiles, which reduced the use of horses. By 1908, annual car production was 63,000, with some 40 auto manufacturing companies competing for a growing public demand for more. Henry Ford had just introduced the Model T, selling for

William H. "Boss" or W.H. Hoover (1849–1932). HHC.

an incredibly low price of $845. It was advertised with a slogan that hit at the heart of the Hoover business: "stronger than a horse and easier to maintain." Moreover, the 1907 failure of the Knickerbocker Trust Company caused a national financial panic that lasted a year during which the stock market fell by 50 percent. With Hoover's sales of horse equipment falling steadily, he began seeking new business opportunities.

Somewhat begrudgingly, because it was the automobile that was killing his horse accessory business, W.H. began to make leather straps to secure automobile canvas tops, as well as door straps, hood straps, trunk straps, fan belts, and driving cushions. But steel and rubber were rapidly replacing leather, a material no longer in high demand. Hoover needed new products that were part of the new mechanical industrial world of the twentieth century.

When W.H. learned that creditors were closing in on a family friend, James Murray Spangler, and when his top salesman, T.F. Albee, told him Murray's vacuum cleaner was something he could enthusiastically sell, he began to see the business possibilities of the device. His wife Susan, an enthusiastic early owner of an original Spangler sweeper, also may have offered a word of encouragement. So on August 5, 1908, W.H. and his sons Herb and Frank met with Murray on W.H.'s front porch on Maple Street in New Berlin (now North Canton) and discussed the details. By August 8, papers were drawn up. The Electric Suction Sweeper Company was reorganized, taking over the debts and assets of Spangler's firm, including all patent rights and debts. Three hundred and sixty shares of stock at $100 par value were issued. W.H. "Boss" Hoover became president, his eldest son Herbert W. Hoover, Sr. (1877–1997), was appointed secretary and general manager, and Murray Spangler became superintendent on salary and royalties.[16]

The manufacture of electric cleaners was at first a sideline for the W.H. Hoover Company. A room in the leather goods factory was set aside for a small force of 20 people, including manufacturing, office, and sales force. In August 1908, the Hoover Model O went into production, colored black. It weighed 40 pounds, less than half the weight of other "portable" cleaners. Both the Westinghouse Electric (AC) and Emerson (DC) motors weighed 25 pounds; had a speed of 1,750 revolutions per minute (rpm), the highest speed recommended as safe by Westinghouse[17]; and produced $\frac{1}{8}$ horsepower. Electricity was provided to homes, depending on their local power source, in either alternating or direct current until it would be standardized as AC in 1910, so both versions had to be produced. Model O was priced at $60 ($65 in the West), with cleaning attachments for an additional $15. A large, 50-pound, commercial version with an 18-inch nozzle, Model 18, was also produced in the fall of 1908 and sold for $100, but would be quickly discontinued in 1909.[18]

In November, 1908, as William Howard Taft (1857–1930) was elected president, succeeding Theodore Roosevelt (1858–1919), the Model O production color was changed to gray, with fancy red striping decalcomania, and on December 5, a small ad was placed in the *Saturday Evening Post* (a popular national magazine), offering a free trial in the home and a free booklet: "Modern Sweeping by Electricity." The campaign not only identified prospective customers but established local dealers to complete the sale in each area that expressed interest. Hundreds of inquiries came in. The plant capacity was only four or five cleaners per day, and by the end of December 1908, only 372 had been made.[19]

When H.W. was assigned to be general manager of the Electric Suction Sweeper Company, he was asked by his father what it meant. He responded, "Yes, sir. Sink or swim."[20] So he began to swim. The start-up operation was not without complications, both in manufacture, sales, distribution, and in legal matters, all requiring expertise and implementation. Since the key Kenney patent had been upheld as valid in 1907, and since the Model O had

Hoover assembly line, ca. 1909. HHC.

to use the basic principle of sucking air through a narrow nozzle to remove dirt from the carpet, Hoover bought a Kenney license from the Vacuum Cleaner Company of New York to protect itself from litigation. However, Hoover questioned whether that company had the right to sub-license in this manner, since licenses were limited to those in the original agreement, and no more were available. Accordingly, in order to get a valid license from someone who already owned one, Hoover went to Detroit to buy out the Palm Vacuum Cleaner Company, founded 1906, which held an original Kenney patent license.[21]

W.H. "Boss" Hoover was not satisfied with the quality of production manufacturing. He had five objections to the Model O cleaners: "They cost too much, were too heavy, too noisy, too hard on floor coverings, and too difficult to operate."[22] Another problem was the return of sweepers for repairs. The brush roller "bearings" were simply waxed string wrapped around each end, which wore out quickly. Herb Hoover recognized that the sweepers were

> ... not good enough for a nation of housewives who aren't used to mechanical household tools. We've got to get someone in here who's accustomed to engineering around the housewife's inexperience with gadgets.[23]

And so he did. Herb Hoover established the company's first engineering department by hiring mechanical engineer Francis Mills Case from the firm of Buchard & Case in Cleveland, Ohio, who initiated a number of mechanical revisions in 1909. Case realized that carpet agitation was necessary for effective cleaning, and so designed the motor-driven rotary brush so that the carpet underneath the nozzle was free to rise and be pulled against the lip of the nozzle and vibrate at 1,000 times a minute, in a beating action not dissimilar to the traditional process of beating a rug with a carpet beater.

To accomplish this, the front wheels were placed behind the suction nozzle and enclosed for the first time on any cleaner. This provided an additional functional advantage, as the

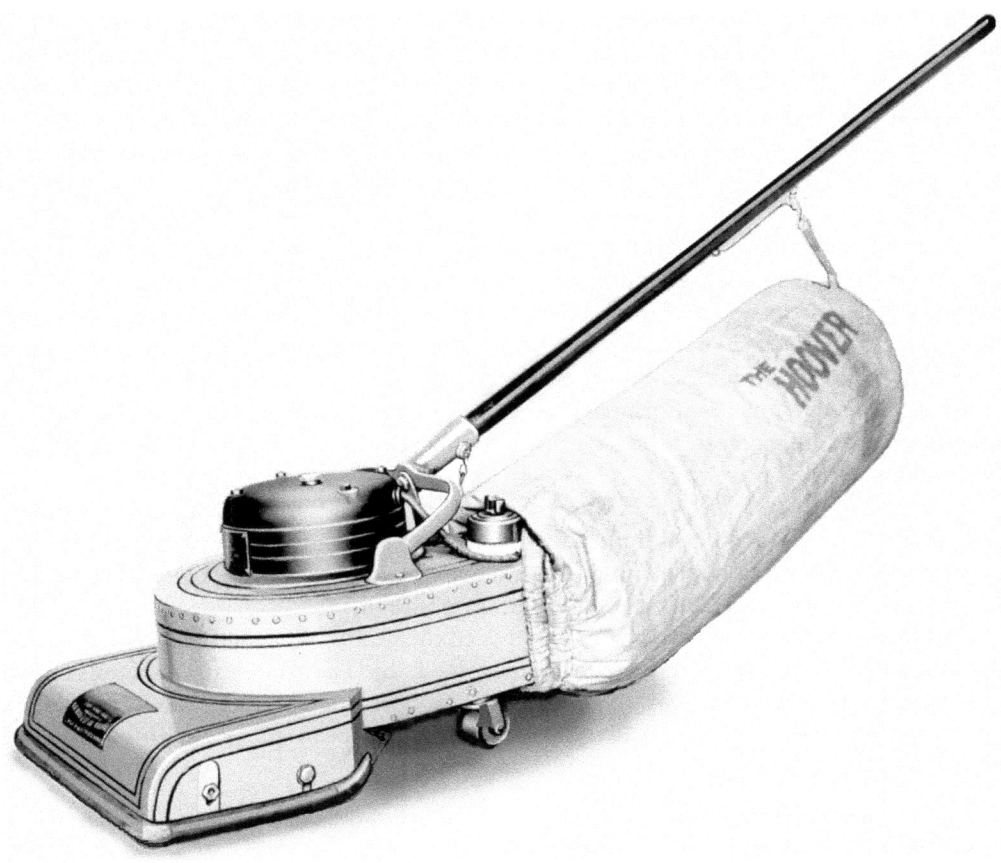

1909 Hoover Senior Model 2. HHC.

nozzle was now able to clean a swath 12 inches wide, instead of the 8 inches when restricted by wheels on either side. Case's patent for this, U.S. Patent 1,147,307, was filed in September 1910 and granted July 20, 1915. An application for a reissue was filed in 1916 and granted on October 23, 1917, as Reissue Patent 14,383.[24]

This unique, patented feature was incorporated into Case's design for the Hoover Model 2 Senior cleaner, manufactured from March 1, 1909, to May 31, 1909, and priced at $100. The Senior was replaced by an earlier, alternative, and more costly Side Outlet version (called Model 3 by some), made starting January 1, 1909, and priced at $110. It had an aluminum sand-cast body with an exhaust outlet at the left hand side, and also included an elastic bag attachment. The elastic blew loose when the bag pores became clogged, so a later version clamped the bag on with fasteners. The Side Outlet would be made until 1916. Case was also busy with a number of other sequential, overlapping models.[25]

On January 1, 1909, an Improved Model O was introduced with brush adjustment knobs to lower the brush as bristles became worn, and the cheesecloth inner bag was removed.[26] In the Spring of 1909, demand was so heavy that Hoover withdrew its advertising, but would renew ads in the October 9, 1909, *Saturday Evening Post*, with the price increased to $70. The lead of the ad was "Sweep with Electricity for 3¢ a Week." The ad claimed: "Dust is Full of Disease," "the motor will outlast the house you live in," and "we are turning out hundreds of machines a week. The demand is enormous."[27] A larger, heavier 12-inch Senior with swivel casters was introduced October 14, 1909, for $110, and was also made until January, 1916. In 1909, 2,382 Hoover sweepers had been built and marketed. With 1908 production added, Hoover had produced 2,754 to date.[28] The Model O had become the first commercially successful portable electric vacuum cleaner, and would be made until January 10, 1910, when it would be replaced by Model 1, priced at $60, and produced until May 31.[29]

Vacuum cleaners were already a growing and highly competitive market, although primarily for the upper class, and primarily for stationary, built-in systems. A typical advertisement of a built-in system showed the master of the house in the parlor having his butler give his coat a few strokes of the vacuum; on the floor above, the maid was doing the same to the lady's hat; other domestics were cleaning the furniture and carpets all the while. The vacuum cleaner and financial prosperity were soon identified as a single concept, a perception that would retard the portable industry until 1945, when, for the first time, the middle class finally became prosperous.[30]

In 1909, *Good Housekeeping*, a women's magazine with a circulation of nearly 300,000, established its Seal of Approval for products advertised in the magazine, certifying that the product had been tested by the Good Housekeeping Research Institute's Domestic Science Laboratory, and was backed by a two-year limited warranty.

Many vacuum cleaner manufacturers stressed the healthful attributes of their product. Typical was the Ideal canister vacuum cleaner, made by The American Vacuum Cleaner Company in New York City, which made machines operated by hand as well as those run by electric motors. A May 1909 advertisement for the Ideal in *House Beautiful* magazine enthusiastically described what its product accomplished.

> It eats up the dirt: literally sucks out all the dust, grit, germs, moths, and eggs of vermin that are ***on*** the object as well as ***in*** it — gobbles them down into its capacious maw, never to trouble you again.[31]

Labor saving was also stressed. That same year in June, the Ideal electric canister cleaner was advertised for "$55 or $60" in the *Ladies Home Journal*.

> Housecleaning this Spring is Different! You Don't Have to Pound the Dust Out! Keep your carpets on the Floor! Keep your wall decorations hanging! Keep your upholstered furniture in its place![32]

Despite the proliferation of different configurations and mechanics of electric vacuum cleaners, there were two distinct types: Upright cleaners, the most popular for carpet cleaning, were comprised of a floor cleaning head or nozzle, which usually contained a rotating brush roll, and a motor, onto which an upright handle and bag were attached. Uprights had a large impellor (fan) driven by the motor, through which the dirt passed before being blown into the bag. This would be called a "dirty" or "direct" air system. A separate fan usually cooled the motor.

Cylinder, tank, or canister cleaners, primarily for above-floor cleaning, had the motor and dust container in a unit, usually on wheels, that was separate from the vacuum head or floor nozzle. A flexible hose connected the two, allowing for a more flexible movement of the nozzle above the floor or behind furniture. Since the air was pulled directly into the dust container and filtered by it before passing through the fan, this would be called a "clean air" system.

It wasn't only electric motors, still large and heavy, that were used to power vacuum systems. A water motor, the Water Witch, made by the Vacuum Hydro Company of New York, was advertised as having light, "almost entirely aluminum built suction pumps by means of a water wheel" that could "be set temporarily in the kitchen sink or bath tub." The sucked-up dust was drawn through the tub and directly carried away by the water. Such advertisements assured buyers that they only needed to "push a very light tool over the floor" to accomplish "better results without any machinery; no dirty bags with germs to empty"—both pointed references to the heavy, portable, electric motor types.[33] Also, as we know, there were literally scores of lightweight, human-powered vacuum cleaners on the market at this time, due to the highly limited distribution of electricity. Portable electrics, as we now know them, were in a distinct minority, were still extremely heavy and cumbersome, and were limited to the wealthy.

In 1909, a dramatic innovation in motor technology took place at Arnold Electric Works (formerly the McCrum-Howell Company) in Racine, Wisconsin. George C. Schmitz had partnered with Frederick Osius in 1904 to found the company, and they used the name Arnold because Schmitz and Osius were not familiar-sounding American names.[34] Since 1904, Chester A. Beach and Louis H. Hamilton, employees working at Arnold, had been developing the first small, lightweight, high speed, fractional horsepower electric motor. It was capable of 7,200 rpm, and was powerful and safe. More importantly, it could operate using either alternating or direct current, both of which were then used in various parts of the country, and thus was known as a Universal motor. Having developed the motor, Arnold proceeded to introduce a suction cleaner in 1909 sold under the trade names Arnold, Magic, Richmond (see group photographs of vacuums at the end of Chapter 2), and several others,[35] under Patent 1,116,850, awarded to George C. Schmitz on November 10, 1914. The design was sticklike, with a vertical motor and nozzle, and a bag beside them.[36]

But it was the smaller, lightweight motor, not the vacuum cleaner, that was the new invention that would revolutionize not only vacuum cleaners, but would initiate vast new national markets in small home appliances. The motor would make Racine the small electric capital of the world, and would enable Beach and his colleagues at Arnold, Louis H. Hamilton and Frederick Osius, to form a new company, the Hamilton Beach Manufacturing Company, in Racine in 1910, after purchasing Arnold Electric for $300,000. Hamilton and Beach would leave the company in 1913 to found the Dunmore Electric Company.[37]

More vacuum manufacturers entered the market. The Birtman Electrical Company, started by L.F. Birtman, was incorporated in Chicago and Rock Island, Illinois, in March 1909,[38] manufacturing the Bee vacuum cleaner, a vertical tank type, for $65. In Toledo, Ohio, under the name of the Bissell Motor Company

1909 Birtman Bee cleaner. IHA/Lifshey.

(not the Bissell carpet sweeper company), the National Super Service Company was organized. The Clements Manufacturing Company of Chicago (incorporated in 1911) made a Cadillac cleaner of Clements' own design, marketed with the slogan "Save Your Back with a Cadillac."[39]

New vacuum patents in 1909 included patent 921,632, by John W. Smith, filed on May 29, 1908, "for a semi-portable type cleaner consisting of a small, motor-driven, suction-creating device together with a dirt separating means mounted as a unit that could be transported within the house to which was attached a flexible hose equipped with a nozzle." Another applied for that year was by Tracy B. Hatch and Edwin W. Goeser, proposing a machine in which the dust collector, the fan, and the motor were all enclosed in a horizontal metallic cylinder, supported by one forward small wheel and two large rear wheels for transportation. The cylinder had a handle on one end and a cleaning nozzle on the other. It differed from prior designs of the dumping or bucket type, and was a forerunner of tank-type cleaners. Hatch and Goeser patent 980,944 was granted January 10, 1911.[40]

In Detroit, Fred Wardell, a salesman for a device known as the Eureka Massage Vibrator, convinced friends at the Stecher Electric & Machine Company to make his massager, as well as a vacuum cleaner, and to use him as its national sales agent. Stecher is said to have produced the first universally wound AC-DC motor.[41] Wardell named the cleaner the Eureka, the Greek word for "I found it!" It was first manufactured in 1909 as the Eureka Model 1. Wardell had seen the Richmond or Arnold machines and realized that they could be changed from the typical upright cleaner configuration by rotating the motor into a horizontal position and arranging the nozzle to reach the floor from that position, creating another forerunner of the tank-type cleaner. It had no wheels at first, but later they were made available as attachments. This led to Wardell founding the Eureka Vacuum Cleaner Company in Detroit in August 1910, when its first upright debuted. Within a year, annual sales totaled about $42,000.[42]

In 1910, the three Frantz brothers, Edward L. (1882–1971) and Clarence G. (1888–1981), in the building materials business, and Walter A. (d. 1970), a mechanic, organized the Premier Vacuum Cleaner Company in Cleveland, Ohio. They had met inventor James Kirby in a real estate deal, and had all been impressed by his new lightweight vacuum cleaner design, which was, according to Kirby, a "broomstick with a gadget on the end." It had a single caster at the rear, a narrow nozzle with a rotating beater driven by the airflow to agitate rug pile, and a cloth bag suspended from the handle.[43] It also had a glass bowl "dust exhibitor" on the housing, giving the operator a continuous visible indication of the cleaning action. This was probably the earliest lightweight vacuum cleaner that we later would call a "stick" cleaner. It also had a cut-off mechanism that permitted attachments to be used with it, a unique innovation at the time.

The Frantz brothers bought the rights to manufacture Kirby's design, and Premier began selling it as Model B in 1911 for $25. Walter handled the manufacturing, and the goal was to make 100 machines per week. A Philadelphia electrical and hardware jobber named L.D. Berger agreed to take all the cleaners they could produce. Walter originally used a G.E. 3,000 rpm motor, but Premier soon began making its own motors, designed by engineers Milton H. Spielman and Walter H. Poesse.[44] The company name soon was changed to the Franz Premier Vacuum Cleaner Company to make the company sound more personalized to the public. Premier Model C introduced a wide nozzle and brush, and Model D had an air-driven brush. Premier would later manufacture Jim Kirby's Ezee cleaner he would design in 1914.[45]

3. Electrics 1900–1920

The year 1910 marked the beginning of what many call the "second phase" of the industrial revolution, where mass production extended into every corner of the manufacturing world. U.S. electric current was standardized at 60-cycle AC and 120 volts. It was transmitted at high voltage, but stepped down by transformers before entering the household. By then, only one in ten American homes, mostly urban, had been wired for electricity. The standardization of electricity was a boon to manufacturers, which now needed only to make one electrical version of their product, and to consumers as well, as they no longer had to worry that when they moved to a different community, they could not use the same appliances due to a different type of electric system.

Electric appliances were being promoted as the "modern servant" because of the diminishing number of real servants in the domestic scene, due to their cost and the difficulty of finding them. For example, in New York City, the number of servants for every 1,000 families declined from 141 in 1900 to 66 in 1920, a decrease of over 50 percent. This trend obviously helped the sale of vacuum cleaners as well as other electrical appliances. A million and a half Singer sewing machines were being sold each year. New electrical appliances on the market in 1910 included laundry irons, fans, toasters, ranges, wash machines, and Victrolas (a phonograph concealed in a furniture cabinet). Middle class families were growing in number, as well as economic power, and they were spending freely on telephones, motion pictures, and automobiles. Annual car production in 1910 was 181,000; there were nearly a half-million cars registered in the U.S., and 1,000 miles of concrete roads had been built.

New vacuum cleaners continued to enter the market. A cleaner with a nameplate stating "Colonial Fan and Motor Company, Warren, Ohio" was allegedly made in 1910, and called the Bell cleaner.[46] Also in 1910, the first electric vacuum cleaner in Europe was patented by P.A. Fisker, and was produced in Denmark under the name of Nilfisk, the telegraph address of the company, Fisker and Nielson, founded a few years before. Nilfisk would survive to the present day. The Brilliant electric vacuum cleaner in the U.S. came out in about

Clarence G. Frantz (1888–1981). IHA/Lifshey.

Edward L. Frantz (1882–1971). IHA/Lifshey.

Hoover Suction Sweeper factory, 1910. HHC.

1911 (see group photographs of vacuums at the end of Chapter 2) and the Dayton vacuum cleaner entered the market in 1914 (see group photographs of vacuums at the end of Chapter 2).

In 1910, the B.F. Sturtevant Company began making vacuum cleaners of the horizontal tank type, one of the earliest of this configuration, and produced an identical line under the Western-Electric label. The company was founded in 1860 in Boston by Benjamin Franklin Sturtevant (1833–1890), inventor and mechanical genius, known as the father of the centrifugal fan, or fan blower, a hand-operated device that he invented in 1863. Another entry into the vacuum cleaner market in 1910 was the Duntley pneumatic cleaner, made in Erie, Pennsylvania.

The P.A. Geier Company had been incorporated in Cleveland, Ohio, in 1905. The company sold metal stampings and punch presses, as well as mixers, hair dryers, washing machine units, and a very popular vibrator called the Royal vibrator. Phillip Geier, owner, founder, and designer, began building vacuum cleaners by hand in his back yard garage, and soon would make vacuum cleaners under the Royal name. The first Royal Model 1, with a snap-on brush, a six-inch nozzle, and wheels, would be produced in 1910, based on a patent Geier applied for in 1911, and by 1912 the company would be focused solely on vacuum cleaners.[47]

On January 3, 1910, the Electric Suction Sweeper Company was reorganized as the Hoover Suction Sweeper Company, and Murray Spangler was assigned royalties for his invention that started the company. Spangler continued to invent, evidenced by his November 29, 1912, application for a brush roll gear-driving mechanism, in which he described "the object of this invention is to provide a construction which will be composed of few simple parts, put together in a practical mechanical manner, the parts being so exactly and peculiarly adapted to each other that they combine to produce a very superior, substantial, and efficient carpet sweeper." His application was granted August 31, 1915, as U.S. Patent 1,151,731, seven months after his death on January 22, 1915. It was, of course, assigned to the Hoover Suction Sweeper Company, as all Hoover employees, then and in the future, have done.[48]

Cleaners did not sell themselves just by being produced and made available to the public. H.W. Hoover, now the firm's key salesman, made an enlightening discovery, relating later that

> I would stock up a hardware store with cleaners, go out two months later and find none of them moved. I would get busy and demonstrate them to housewives and move the stock. Quite unwittingly, I stumbled on the fact that specialty demonstrations were the correct way to sell vacuum cleaners.[49]

By 1910, 55 percent of the country was rural and without electric service, so sales were limited and difficult. Hoover organized sales efforts by training salesmen who were placed in retail stores. The store was to guarantee 15 demonstrations weekly, paying the salesman $1 per demonstration. The salesman sold a sweeper to a dealer for $90 (retail price was $125). Cleaning services bought Hoovers and cleaned homes door-to-door.

From January 10 to May 31, 1910, Hoover produced Model 1, replacing the Improved Model O of 1909, but still priced at $60. Model 1 Improved, with the exhaust moved from the side to the center, followed Model 1, and was produced from January 30 to December 1, 1912. Also in 1912, Hoover introduced its first lightweight model, the Hooverette, priced at $45 with an eight-inch nozzle and a ⅛-horsepower universal motor, the latest in current technology. It would be made

Detroit cleaning serviceman M.J. Summers cleaned 120 to 150 homes per month using a Hoover. Photograph ca. 1911. HHC.

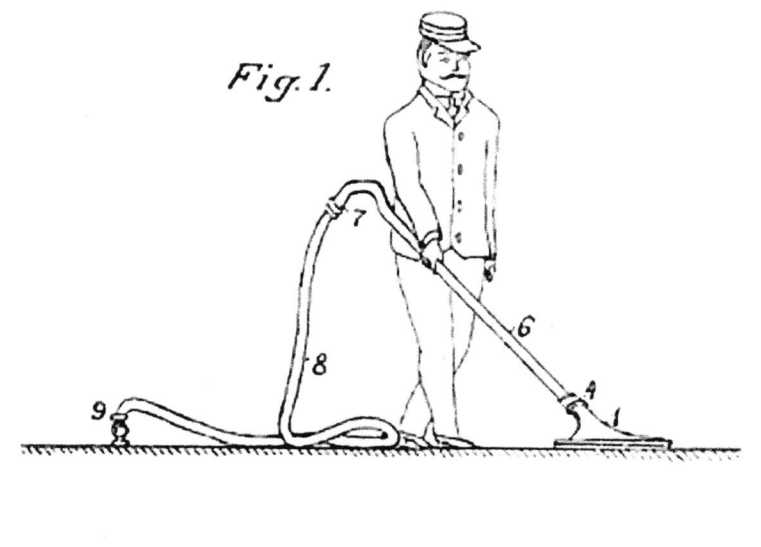
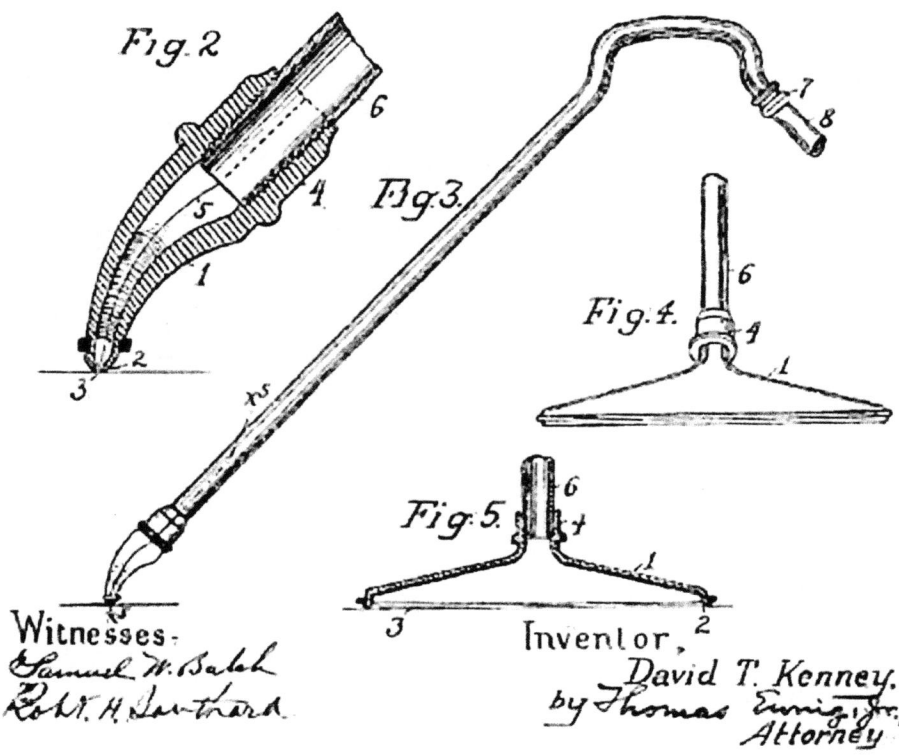

David Kenney's 1913 patent 1,057,347, U.S. Pat. Off.

until January 1914. From May 1912 to May 1914, Hoover produced its Junior cleaner, which sold for $75 with new lightweight aluminum brush rolls, and with a ten-inch nozzle. $75 sounds quite reasonable until you realize that in 2011 dollars, it would be $1,692, another obvious reason why selling them was so difficult. Hoover's key competitors at this time were Santo, Duntley, Premier, and Eureka.[50]

David Kenney's patent 847,947, filed in 1901 and granted in 1907, dominated the market and required all manufacturers to purchase licenses and pay a percentage of their sales to use his essential "narrow nozzle" principle. Manufacturers were paying dearly and were determined to reduce their costs. In 1911, four vacuum manufacturers met in Melville Church's New York office to form a defense association to fight the Kenney patent. The four were: Mr. Church, representing the Blaisdell Machinery Company in Bradford, Pennsylvania and the Spencer Turbine Cleaner Company of Hartford, Connecticut; Mr. George Ray, representing the Invincible Manufacturing Company of Pittsburgh; and Mr. Harry Frease, representing the United Electric Company of Canton, Ohio, which made the Tuec cleaner. In later meetings, the group of four grew to seven, with the addition of the B.F. Sturtevant Company, Hyde Park, Massachusetts; The American Radiator Company, Chicago, Illinois; and the United Vacuum Appliance Company, Connersville, Indiana. This group was a forerunner to the Vacuum Cleaner Manufacturers Association (VCMA). They entered into an agreement to accept Kenney licenses at 1 percent, but on June 28, 1912, they were offered licenses at 2 percent and reluctantly accepted.[51]

After lengthy litigation, Kenney's 1906 patent application for his "apparatus for removing dust" (a nozzle) was finally granted on March 25, 1913, as U.S. Patent 1,057,347, finally confirming Kenney's industry rights. Six months later, on October 21, 1913, the first meeting of the newly formed Vacuum Cleaner Manufacturers Association, met in Toledo, Ohio, at the Hotel Secor. It was the first trade association in the houseware industry. At the meeting, 11 firms were represented, among them Bissell Motor Company (no relation to the Bissell Carpet Sweeper Company), Clements Manufacturing Company, Eureka Vacuum Cleaner Company, Hoover Suction Sweeper Company, and Premier Vacuum Cleaner Company. Fred Wardell of Eureka was elected chairman, and Fred Bissell of the Bissell Motor Company was elected secretary.[52] Membership was confined exclusively to licensees under the Kenney patents, and this was the primary subject of the business agenda. Section VII of the Kenney license agreement stated that the owner of the basic Kenney patent "…shall not grant any … further license during the continuation of this license, without the consent of three-quarters of the then existing licensees in good standing." The second VCMA meeting would be in December, 1916, and at subsequent meetings in 1917, 1918 and 1919, there would be questions raised regarding the legitimacy of the organization as an association, since only licensees were permitted to be members.[53]

In 1912, Clarence and Walter Frantz left Premier and went on their own to form a new company, which made washing machines as well as vacuums. On April 4, 1913, they organized Apex Electric Manufacturing Company, and William Orr designed the first vacuum cleaner, an upright, introduced in 1914 under the Apex name A3, which had a divided nozzle to spread suction to its ends.[54] When Apex could only get 100 motors per day, the new firm began making its own motors.

Eureka in 1913 was making six different vacuum models with special attachments for bare floors, walls, upholstery, and crevices. They even could be used for drying hair. Hoover made 4,659 cleaners that year,[55] and debuted its lightweight (16 pound) and low-cost Friction Drive Baby, with a wooden brush roll driven by friction, a ten-inch nozzle, exposed carrier wheels, and a Robbins & Meyers motor capable of 6,000 rpm, but without a furniture guard. The Baby sold for $35, half the price of the original Model O. Also in 1913, the Pneuvac company of Boston introduced its "3 machines in one Sweeper-Vac," which could be used as a carpet sweeper, a vacuum cleaner, or both in combination. Hoover also introduced its larger, 12-inch Junior cleaner in June 1914 for $75, which would be made until June, 1922.[56] Regina made its first ambulant (portable, with wheels) cleaner that year.

In 1915, the original Frantz Premier Vacuum Cleaner Company was purchased by General Electric and became the Electric Vacuum Cleaner Company,[57] with Julius Teuter as its first president. This was the basis by which GE entered the vacuum cleaner business, but it continued to make cleaners under the Frantz Premier brand until World War II. The 1916 Frantz Premier was a very successful model K-14-B[58] with a rotating metal brush roll with rows of rubber fingers. Air rushing through the slotted nozzle caused the brush to revolve at a high rate of speed.[59] In 1915, former Premier engineers Milton H. Spielman and Walter H. Poesse formed the Domestic Electric Company of Cleveland, Ohio, to make motors. Domestic spun off part of its business to Lamb Electric, also founded that year,[60] but Domestic continued to operate successfully until 1929, when it would be acquired by Black & Decker.[61]

During this period, the process of manufacturing in America was being revolutionized by Philadelphia engineer Frederick Winslow Taylor (1856–1915), who in 1911 had written *Principles of Scientific Management* and founded the Taylor Society. Taylor's society established industrial engineering schools throughout the country, spawning efficiency experts with stopwatches to measure every step of the manufacturing process. This process, called "Taylorism," spurred manufacturing to improve its efficiency in a myriad of sophisticated, scientific ways, and influenced many manufacturers and followers.

Taylor followers Frank Gilbreth (1868–1924) and his wife Lillian (1878–1972) formed a management consulting firm based on Taylor principles. After Frank Sr.'s death, Lillian sought the one best way to perform household tasks, including the "work triangle" for linear kitchens. Their son, Frank Jr., would write a best-selling book in 1948, *Cheaper by the Dozen* (they had 12 children), which immortalized them both, and which was made into a popular film by the same name in 1950, starring Clifton Webb. A film of the same title starring Steve Martin, and its sequel, debuted in 2003 and 2005 respectively, but bore no resemblance to the original, except both featured a family with 12 children.

The industrial model based on Taylorism that amazed the world in 1913 was Henry Ford's new moving assembly line for his redesigned Model T, which within a year would be making over half (300,000) of all 543,679 cars produced annually in the U.S. The Ford assembly line's time to produce a Ford was reduced from 12.5 to 1.5 worker hours, which in turn reduced the price of a new Ford to consumers by 50 percent, down to an incredibly low $440. This efficient manufacturing process became known around the world as "Fordism."[62] Such a demonstration of modern industrial prowess was in sharp historical contrast to the July 1913 reunion of 53,407 surviving but aging Civil War veterans, both Union and Confederates, held in Gettysburg, Pennsylvania, in commemoration of the fiftieth anniversary of the famous battle there, reminding us of how rapidly industrialization had transformed a society of agriculture.

Along with industrialization, Taylorism also had a significant impact on domestic life. In 1913, another Taylor follower, Christine Frederick (1883–1970), respected household editor of *Ladies Home Journal*, a popular magazine for women, published a book, *The New Housekeeping: Efficiency Studies in Home Management*, based on her 1912 articles of the same title published in the magazine. As influential with housewives as Martha Stewart would be in the late twentieth century, Frederick developed her ideas and procedures in her small experimental kitchen in Greenlawn, Long Island, called the Applecroft Home Experiment Station. Frederick applied the same scientific principles to housekeeping that were being promoted in industry by Taylor.[63]

Frederick, through her articles and book, urged housewives to emulate these same scientific industrial principles in the home. She was advancing the interests of women, as had Ellen Richards and the Beecher sisters before her, by advising them on how to make their household tasks easier, more pleasant, and more efficient. Most of the book concerned kitchen layouts and the processes of food preparation, but among the household electric products she listed as recommendations were electric irons, electric washing machines, electric sewing machines, electric buffers and motors, electric vacuum cleaners, toasters, grills, percolators, electric fireless hot plates, stoves, fans, and ventilators. Her list of "Labour-Savers" definitely included vacuum cleaners, although she did not elaborate on their advantages or functions. Desiring to be up-to-date with modern lifestyles and the latest household equipment, many women responded to Frederick's scientific recommendations by buying efficient mechanical devices to perform manual tasks around the house, and vacuum cleaners were high on their list.[64]

At this time, vacuum cleaners were of two general types: manually operated, used where there was no electricity, and motor driven, which could only be used in urban areas that had electricity. They were both competing with traditional carpet sweepers, but many manufacturers had stopped making new manual cleaners after 1910, as electrification spread. Still, there were many on the market. Many department stores erected dramatic window and floor displays, featuring all types of cleaners. A sales article of the time in *The House Furnishing Review* urged department stores to sell a customer both an electric power vacuum *and* a carpet sweeper or manual vacuum.

> ... to use a heavy pulling electric power cleaner daily is not necessary, although it is quite necessary that such thorough house cleaning be affected at reasonable intervals, say a week or two weeks apart, and that the lesser force, exerted by the smaller hand types of machines, is quite sufficient to do the daily cleaning and as thoroughly as the occasion requires.[65]

There was a growing awareness in industry that design patents, on the books since 1842, were not effective in protecting the invention of unique product forms from being copied by competition. Such copying had become endemic. As a matter of fact, American art museums such as the Metropolitan Museum of Art in New York encouraged the copying of historical European designs with exhibitions intended to educate artists in the same way that artists traditionally learned by copying the masters. Artists who worked in industry were generally called industrial artists and their primary role was to provide surface decoration for textiles, pottery, and other manufactured goods, an activity called applied art.

By 1913, when over one million U.S. "utility" (functional) patents had been issued, there were only 45,000 design patents, which were numbered differently than utility patents. That year, U.S. commissioner of patents Edward B. Moore, presumably in response to competitive copying complaints in industry, had urged the modification of existing regulations to better protect the property of "industrial design, the distinguishing form of products that have marketable value."[66] Although this was the first formal mention of the more appropriate term "industrial design," it would not be used in industry for 20 more years, when manufacturers finally accepted industrial design as essential to the sales success of products.

Nevertheless, traditional art was changing dramatically in Europe. America was first made aware of European "modern art" by the International Exhibition of Modern Art, held in February 1913 at the 69th Infantry Regiment Armory at Lexington and 26th Streets in New York; the so-called Armory Show. It included the work of 300 modern artists, including impressionists Paul Cézanne, Vincent van Gogh, Edgar Degas, and Raoul Dufy; cubists Pablo Picasso and Fernand Leger; and the sensational "Nude Descending a Staircase" painting by Marcel Duchamp. Americans, accustomed to traditional realistic art, panned the show, and cartoonists had a field day. Even president Theodore Roosevelt commented, "That's not art!"[67] But within 15 years, product designs inspired by modern art would captivate Americans and force a change in the way consumer products looked.

America's elevation to world prominence through its technology provided award opportunities at international expositions. International awards were helping to publicize all kinds of consumer products, including vacuum cleaners. In 1914, Hoover was awarded the gold medal at the Taisho Exposition in Tokyo, and another gold medal at the Anglo-American Exposition in London.[68] At the 1915 San Francisco Panama-Pacific International Exhibition, a jury of electrical experts awarded Eureka the grand prize. Eureka cleaners were generally lighter, better designed, and with more innovative attachments, such as a hair dryer.

In 1915, the American School of Home Economics published *Household Engineering, Scientific Management in the Home*, "a correspondence course on the application of the principles of efficiency engineering and scientific management to the every day tasks of housekeeping," written by Christine Frederick, mentioned earlier, who was now identified as a "Household Efficiency Engineer." The book was highly popular and republished in 1919, 1920, and 1921, educating and influencing thousands of women in the growing home economics education movement that would be funded in colleges and high schools with the help of federal legislation in 1914 and 1917. The book, in over 500 pages, covers every possible aspect of household labor saving and home management, but of most interest to us here is her Chapter IV, "Methods of Cleaning," which gives us an idea of the nature of the vacuum cleaners market of the time.

Frederick characterized cleaning and dusting methods of the past as the "scattering," rather than the removal, of dust. "The new sanitary ideal today," she wrote, "has for its watchword 'absorption.' The broom is being largely replaced by the suction method of the modern vacuum cleaner, and the *dustless duster holds dirt as it cleans*. No one invention is as responsible for new cleaning methods as is the so-called vacuum cleaner." She then described the four broad types of vacuum cleaners:

> (1) Large portable vacuum cleaners, dustbag within the cleaner; hand or electric. (2) Small vacuum cleaners, dustbag outside the cleaner; electric. (3) Carpet sweeper, box model vacuum cleaner (manual type). (4) Stationary machines, located in basement, with pipes to all floors.

Frederick goes on to clarify the definitions, mechanical operation, and advantages of vacuum cleaners and suction cleaners for her students, in a manner similar to a write-up in today's *Consumer Reports* magazine.

> There is a slight distinction between vacuum cleaner and suction cleaner. In the vacuum cleaner, a diaphragm or rotary pump is used to create a partial vacuum in each stroke. The surrounding atmosphere then rushes in to fill the vacuum, bringing in dust. At the end of each stroke, when operated slowly, the vacuum model loses "pull" for an instant, but this is not noticeable at full speed. This type of cleaner can make a very strong pull.[69]

3. Electrics 1900–1920

In the suction type of cleaner using power, the suction is created by means of fans or turbines operated by a small motor encased in some portion of the cleaner. These fans in revolving cause an intake of air, bringing with the air dust, small fragments of paper, matches, etc. In this type of cleaner the "pull" is not very strong, but a large amount of air can be moved, permitting the use of a wide opening. In the suction type of cleaner operated by hand, the suction is created generally by a pair of bellows which, in being shut, caused a similar but less violent suction of air and intake of dust. In the hand type, however, the effort of the operator is needed to make the necessary power. In the suction type also there is a distinct "pull" for an instant, as mentioned above, at the end of each stroke.

The large portable vacuum cleaner (electric) is made with a powerful motor and is particularly suited to cleaning large houses or where there are many thick, all over rugs, carpets, and a quantity of draperies. This type of cleaner is always used with a hose attachment inserted in the intake opening. At the other end of the hose is what is called the nozzle opening, which is the actual part of the cleaner moved over any given surface. This nozzle tool may be a tool for the floor, for mattresses and tufted furniture, for draperies, etc., etc. The price of such a large portable cleaner is about $75.00 because of the cost of a powerful motor.

The large portable vacuum cleaner (hand) is operated by an air pump and lever, and preferably is worked by two persons, one to pump and one to move the hose attachment wherever necessary. In both of these portable type machines, the dustbag is within the body of the cleaner. This hand portable when operated by two people does almost as good work as any electric machine, but the labor is considerable. It comes fitted with all similar attachments, and costs from $20.00 to $30.00.

The small portable cleaner (electric) is made in many models and seems the best type suited for average use in homes wired for electricity. With this type, the dustbag is on the outside, the machine is light in weight, and can be operated on carpets and floor without a hose attachment. The hose is necessary here only to clean draperies, mattresses, etc.., and anything off the floor level. The advantage of such a machine are that it can easily be taken up and downstairs by a woman, that it rolls easily, and that it takes little storage when not in use. The price of such a cleaner is from $30.00 to $45.00. [An illustration of a 12-inch Hoover 1920 Model 105 suction sweeper is included in Frederick's book at this point.]

The sweeper type of cleaner is operated by hand. It consists of a box built like a carpet sweeper in which is contained a bellows. These bellows are operated whenever the cleaner is moved backward or forward over the carpet, the suction created drawing in the dust from the carpet. No attachments can be used with this type, which is strictly for floor use and is suitable only for carpets and large rugs. It does not clean bare floors well. This type is somewhat heavy and fatiguing to work in comparison to the usual carpet sweeper. In some models of this box type a sweeper just like a carpet sweeper is combined so that is swept at the same time that dust is sucked out of it. They cost from $5.00 to $12.00.

While the powerful electric machine and even the small, light machines suck dust from a given surface and pick up lint, matches, etc., most of them do not pick up threads, crumbs, or, in other words, brush the carpet at the same time that they clean it pneumatically. In order to provide for this picking up of threads, many cleaners, both large and small, are fitted with brush attachments either within the body of the cleaner or mounted without on a small platform. It is well to note this point in a cleaner before purchasing.[70]

Frederick's 1915 terminology is somewhat confusing to today's perceptions. What she sometimes calls "vacuum" types are actually manually operated (rotary pump, bellows, etc.) and what she calls "suction" types are actually electric, such as the Hoover example shown. And, of course, her "sweeper type" also mentions a bellows, which means it was manually operated. This indicates that at this late date, manual vacuums were still predominant in many homes. The reason for this probably was due to the higher cost of electric vacuums, their relative newness in the market (seven years), and the relatively small percentage of homes with electric power. Many households had probably acquired their manual vacs in the previous ten years, and were more satisfied with manual vacuum results, even with the extra physical effort, than with older and less effective carpet sweepers.

Frederick went on to advise students how to determine the type of cleaner needed for specific environments, describes the nature of "permanent vacuum systems" (built-in systems) for new homes ($200 and up, depending on the number of rooms), and tabulates the results of a scientific time comparison study of different methods of cleaning large and small rugs.

"Time Study" on 9 × 12 Rug by Different methods[71]:

	MINUTES
Broom sweeping (outside on porch)	10
(Moving to porch and replacing, 6 minutes)	
Hand vacuum sweeping	18
Electric vacuum sweeping	12

On Small 4 × 6 Rugs

	MINUTES
Broom	4
Hand vacuum	10
Electric	6

She concluded that while the electric vacuum cleaner offered no great advantage in time savings, it was *most* worthwhile on carpets, and its greatest advantages were that its thoroughness required less frequent cleaning, and that the effort to use it was much less than a broom or hand-operated (manual) vacuum cleaner, which she referred to as "hand vacuum sweeping" in her survey.[72] Certainly, this was no ringing endorsement of electric-powered suction vacuum cleaners in 1915. She seems to have almost equated manual vacuums with electric vacuums functionally. Later editions of her book by the American School of Home Economics did not seem to bother to update this particular section, perhaps due to distraction by the war.

On the other hand, we have to understand that in 1915, many specialists in vacuum systems disbelieved that a smaller and more dependable portable type was possible. In 1912, one manufacturer refused to show or demonstrate portable machines. Arthur Summerton, in his *Treatise of the Vacuum Cleaner* in 1912, stated: "We shall confine this treatise to the stationary (built in) system as we believe satisfactory results in cleaning cannot be expected from portable machines." He considered the portable vacuum cleaner "at the height of its career like the automobile." Since 1910, stationary systems (with suction pump and dust separator in the basement) were mainly used in residences.[73]

By 1915, World War I was well under way in Europe, a war America would finally enter in 1917. Because of the undesirable German connotation during the war, and despite its many occupants descended from German settlers, "New Berlin" was renamed "North Canton" by popular demand of its citizens. In 1915, Hoover had almost doubled its sale of vacuum cleaners over 1914 (11,756 over 7,415), so it decided to increase its capacity in 1916 by adding a three-story brick addition to its manufacturing plant and by doubling its working force. This more than doubled its 1916 production to 24,451 units. Hoover discontinued making horse collars in 1917, but its leather business survived during the war by making military goods, including leather straps for French 75 artillery, English eight-horse artillery harnesses with leather-covered cable traces, riding leggings, gun-slings for Russian rifles, and canvas water buckets and tarpaulins and French saddles for dump carts. But when the war was over on November 11, 1918, military leather contracts came to an abrupt halt. There was no peacetime market for any of these items, and the W.H. Hoover Company discontinued making leather goods, forever. Hoover was now in the vacuum cleaner business full time,[74] and that year, it had sold 79,381 of them.

After the war, America dominated the world, both in national prestige and mass-production manufacturing. It used its wartime capability to produce consumer products in high volume, and at lower and lower cost. The use of electricity became more widespread, prices of consumer products fell as more efficient mass-production techniques were introduced, and competition increased dramatically among U.S. manufacturers.

The situation was vastly different in Europe, with many countries devastated and their manufacturing capability at zero. In its process of recovery over the next five years, Europe would be greatly influenced by American business models such as Fordism and Taylorism, particularly in the new Soviet Union and Germany, but it would be some time before its manufacturing base could be restored. Unable to compete with the U.S. in manufacturing, Europe sought leadership in an arena where the U.S. was weakest — modern art and design.

At the turn of century, when America led the world in technology and innovation, Europe had dominated the world's decorative arts with Art Nouveau, the apogee of the Arts and Crafts movement, spearheaded by English designers in the late nineteenth century. But as early as 1903, German architect and head of schools of Arts and Crafts Hermann Muthesius (1861–1927) recognized that the complex shapes of traditional handcraft were not suited to being efficiently produced by machines. He envisioned a new style: a "machinenstil" (machine style) based on "form following function" (a phrase coined by Chicago skyscraper architect Louis Sullivan in 1896) to unite art and industry for German national economic advantage in trade.

In 1907, the same year James Murray Spangler was putting his first vacuum cleaner together, Muthesius founded the Deutsches Werkbund (German Union of Work), comprised of artists, architects, and businessmen. It was the first national effort to integrate industry with art and design. Immediately, Werkbund member architect Peter Behrens (1868–1940) was appointed as "artistic consultant" to Allgemeine Elektrictäts Gesellschaft (A.E.G.), the German equivalent of General Electric in the U.S. (which would not be concerned with product appearance until 1933). Behrens designed electric irons, electric kettles, advertisements, and buildings for A.E.G., demonstrating how corporate design could benefit sales. By 1910, Germany had overtaken England in innovation and originality in architecture, art education, crafts, and design.

In 1919, another architect member of the Deutsches Werkbund and a colleague and student of Behrens, Walter Gropius (1883–1969), became director of the Bauhaus, the first educational program of what we now call modern design, first in Weimar, and later in Dessau, Germany. Students of the Bauhaus pioneered in the design of mass-produced consumer products so that their forms visually represented their inherent machine origins, in other words, geometric, simple, and without surface decoration. Several German manufacturers, in the mid- to late 1920s, would use Bauhaus student designs to produce modest quantities of light fixtures and furniture.

This concept of utilizing artistic design in functional objects was not totally new. For centuries, artists had adorned the surface exterior of functional objects such as vases, lamps, sewing machines and stoves with representations of flowers, nature, classical images, or

Hoover Model 102 assembly line, ca. 1919. HHC.

stylish graphics. But never before had artists manipulated the form of the product itself to express visually the mechanical process by which it was made, or the function that it performed. Within ten years, this concept would revolutionize industry.

In the meantime, industry, particularly in America, followed the traditions of the past. Industrial artists worked mainly in the so-called art industries: glassware, ceramics, furniture, leatherwork, metalwork, lighting fixtures, rugs, carpets, silverware, textiles, and wallpaper. But this was only a small percentage of the expanding consumer market. Most new mass-produced products belonged to the so-called artless industries, which by its very title, discouraged artists from entering these industries. They included: household appliances (including vacuum cleaners), aluminum cookware, electrical machinery, cash registers, clocks, optical goods, fountain pens, scales, locomotives, and cars. In other words, these were anything traditionally more functional and mechanical than decorative. These were industries dominated by engineers. Industrial artists were neither interested nor welcome to participate.

U.S. industrial art received a further setback in 1917, when the Smith-Hughes Act established federal aid for vocational education, using the inappropriate term "industrial arts" for what had previously been called "manual training." Based on manual skills in a variety of hand, power, and machine tools in thousands of high schools, the training became known to many as "shop class." Industrial art for years thereafter acquired the perception of a lowly trade, as opposed to the high calling and prestige of fine art.

In 1917, appliance advertising was also becoming a fine art. In January of that year, the National Electric Light Association (N.E.L.A., founded 1885) committee on "Coordinate Advertising and Sales Campaigns" proposed a national program where manufacturers and sales organizations of electrical appliances would focus on a different specific appliance each month of the year, to maximize, coordinate and focus exposure of print ads, store promotions, streetcar ads, and door-to-door calls. March and September were proposed as the months for vacuum cleaners, coinciding with spring and fall housecleaning.[75] Many advertisers were selling the low cost of electricity. Torrington Electric Company was advertising its cleaners by citing the cost of electricity as only two cents per week, and that it used only one-third the electric current of a toaster.[76] Another vacuum manufacturer advertised, "There are a hundred brooms in every lamp socket, that will clean your home without Fuss-Dust-Muss."[77]

Some U.S. manufacturers took advantage of Europe's lack of manufacturing capabilities after the war. In 1919, The Hoover Suction Sweeper Company, Ltd., started business operations in Great Britain, and its first resale account was Selfridge's in London, selling its Model 102, made from June 1919 to July 1920. As early as 1917, the Hoover cleaner had been advertised as "The Only Sweeper Which Beats, Shakes, and Suction Cleans." By 1919, when a trademark for the slogan was applied for, the phrase was simplified to "It Beats As It Sweeps As It Cleans."[78] The beating described the basic principle of positive agitation used today by all vacuums, but at that time, it was erroneously believed by some that it might damage carpets, and Hoover's competition did all it could to encourage this misconception. United Electric Company of Canton, Ohio, claimed its Tuec home installation vacuums cleaned "without beating and pounding the carpet," and the P.A. Geier Company advertised that its Royal vacuums cleaned "by air alone." Eureka, by now making 2,000 cleaners per day, repeated that same phrase for its cleaners. In response to this negative publicity, Hoover pulled its "beating" ads for a few years but continued to perfect its concept of positive agitation.[79]

Vacuum cleaner manufacturing was advancing in Europe after the war, as well. Swedish entrepreneur Axel Leonard Wenner-Gren (1881–1961) was the founder of Elektrolux. Wenner-Gren was neither an inventor nor an engineer, but a sales genius. At age 21 in 1902, he had moved to Germany and took a position with a Swedish company called Separator, with offices in Berlin. After several years, he became the company's top traveling salesman, and in 1904 he left to become a salesman for agricultural engines.

According to Electrolux legend, in 1908, when Wenner-Gren was on business in Vienna, he saw a Santo *staubsauger* (German for vacuum cleaner) in a shop window and realized this was the ideal product he was looking for — one that was needed in every home. The Santo electric was made by the Keller Manufacturing Company in Philadelphia, starting in 1907, and was formally called the Keller-Santo. Wenner-Gren contacted Keller in Philadelphia, offering to act as their European representative. But Keller already had a representative, Herr Gustav Robert Phalen, with business interests in Vienna and Berlin. Wenner-Gren then began working for Phalen, heading up "Santo Staubsaug Apparate Gesellschaft" (Santo Vacuum Cleaner Appliance Business) in Berlin.

In 1912, Wenner-Gren returned to Sweden, where two companies, Artiebogaget (AB) Elektromekanistka and AB Lux, had started making copies of the stationary American Santo cleaner, called the Lux 1. Wenner-Gren was hired as an agent for Lux to work in Germany, the United Kingdom, and France, selling the Lux I, which used a universal motor made by AB Elektromekanistka.

The outbreak of World War I in 1914 halted vacuum cleaner exports, and Wenner-Gren contacted Elektromekanistka and Lux, proposing a new design with a universal motor, but found little interest. So he started his own company, Svenska Elektron AB, and began making his own design, which was quite successful. On October 30, 1916, he bought a major block of Elektromekanistka shares and became a board member. In early 1918, Wenner-Gren acquired shares of AB Lux, and cooperative arrangements were made. In 1919, the new company was rechristened AB Elektrolux, a combination of Elektromekanistka, Elektron, and Lux. Initially, AB Elektrolux was purely a sales company, whose mission was to launch a new Swedish vacuum cleaner manufactured by Lux on the world market. That year, AB Elektrolux introduced its Model 1 (made by Lux) vacuum cleaner in Sweden, Denmark, France, and the U.K. with turbine rather than bellows power, like the Santo.[80]

Back in the U.S., the Vacuum Cleaner Manufacturing Association in 1920 reported that there were 702,000 cleaners sold in 1919 (25 percent were Hoovers), and that during these early years of intensive competitive selling it became apparent that a common basis of comparative measurement was necessary to determine relative efficiency. A standards committee was appointed and work was started with the U.S. Bureau of Standards in Washington, D.C., with the result that a basis of performance was established and adopted for the work of the committee.[81]

Performance standards included a measurement of airflow in cubic feet per minute (CFM or ft^3/min) or in liters per second (l/s); or, alternatively, a measurement of suction, vacuum, or water lift, in inches of water, or pascals (Pa; 1,000 pascals are written as kPa). A typical domestic model of today has a suction of about negative 20 kPa, which means it can lower the pressure inside the hose from normal atmospheric pressure (about 100 kPa) by 20 kPa. The higher the suction rating, the more powerful the cleaner. One inch of water is equivalent to about 249 Pa, hence, the typical suction of vacuum cleaners is 80 inches (2,000 mm) of water.

Then there was also the principle of "sealed suction," the suction measured when there is no airflow through the suction motor, that is, by blocking the inlet of the vacuum cleaner while the motor was running at full speed. Manufacturer salesmen sometimes used this principle to demonstrate the power of their vacuum cleaner by picking up a 16-pound bowling ball. This is a good time to dispel the notion that this proved anything at all.

It was a totally meaningless public relations trick and had nothing to do with the cleaning performance or efficiency of a vacuum cleaner. With the example above, a cleaner with typical suction of 80 inches of water, simple physics can be used to explain the startling result. One cubic inch of water weighs 0.036 pounds, so the pressure for a vacuum cleaner with a suction of 80 inches of water would exert 2.88 pounds per square inch on a suction gauge at the sealed opening. A typical opening of a vacuum hose is about 1½ inches diameter, so this size opening would exert about 4.7 pounds of "pull." (I do hope I'm not boring you!) However, if one attached a wide, circular, inverted funnel-like device to the opening, enlarging it to about 20 square inches, and placed it on a bowling ball so that it sealed the airflow, it would exert over 57 pounds of pressure, easily lifting a small child, let alone a 16-pound bowling ball.

1919 Elektrolux Model 1.

The truth of the matter was that vacuum cleaner performance was, and is, a complicated balance of airflow, the size of the opening on the cleaning tool or nozzle, the design of the fan blades, the power of the motor driving the fan, and the mechanical action of brushes and beaters on the rug, among other engineering principles. All these factors needed to be balanced by engineers to manufacture a cleaner that was as lightweight as possible, and at a cost that was affordable. Finally, as Christine Frederick had explained in 1915, performance varied depending on the nature of the surface being cleaned.

Clearly, vacuum cleaners of 1920 did not have the airflow or suction that is common today, but these performance standards were becoming essential to the vacuum industry because of the growing market, and because of the increasing competition among more and more manufacturers. Claims of superior performance were the stock in trade of expanding armies of vacuum cleaner salesmen going door to door.

The coming decade would decide which manufacturers would survive, and which would fail, because there would soon be more vacuum cleaners manufactured than there were customers to buy them.

Chapter 4

Consolidation 1920–1940

The Roaring Twenties was the time many of its inhabitants regarded as the birth of the modern world. The decade began with the ratification of the Eighteenth Amendment, prohibiting the manufacturing, sale, or transportation of alcoholic beverages, and the Nineteenth Amendment, permitting women to vote (you win some, you lose some). Zeppelins were floating through the skies and commercial air service was in its infancy. For the first time in history, a majority of Americans lived in cities (51.4 percent). About half the country was electrified, with power reaching into many more homes as urban electric power systems expanded and became regional. Electric refrigerators, stoves, and vacuum cleaners were redefining household lifestyles. Automobiles became common, influencing where people could find work and taking people farther from home than they had ever been. Most households had telephones, and a new means of wireless communication and entertainment, radio, was the latest technological wonder that attracted consumers in a similar way as the internet would for us in the 1990s. Moving picture theaters provided entertainment. Life seemed good, and the vacuum cleaner industry initially prospered. The year 1920 was a banner one for vacuum cleaner sales. Unit sales were 50 percent higher than in 1919, at 1,024,167, and totaled $22.5 million in sales volume.[1]

The U.S. postwar economy, however, was not the best. The war had increased the national debt from $1 billion to $24 billion. During the depression of 1920–1921, demand fell, bringing business closures and rising unemployment. Many industries cut back production. In 1921, Hoover sales dropped from 281,889 in 1920 to 143,426, a fall of 50 percent. It would be 15 years before Hoover sales returned to 1920 levels. President Warren Harding, inaugurated in January 1921, promised to return the nation to "normalcy." This was accomplished by June 1922, as a result of lower taxes, a balanced budget, federal spending cut in half, and a reduction of debt. (Does this sound familiar?)

The decade of the 1920s saw the gradual transition of the vacuum cleaner business from its technical development stage into a highly competitive mass production and sales stage. The vacuum industry expanded along with industrial production in general, but it was even more accelerated by a dramatic reduction in the number of servants. Immigration quota laws of 1924, triggered by fears of criminal and defective immigrants, would allow no more than 2 percent representing each nationality.[2] This would reduce the influx of foreign-born young women, previously the largest segment of the servant population. At the same time, higher factory work pay for women without skills increased the cost of hiring servants. The wages of domestic servants during the decade increased by a factor of five, compared to middle-income wage increases which merely doubled. This significant trend encouraged many well-to-do housewives, who previously relied on servants, to take on housecleaning tasks, and vacuum cleaners became a logical choice for such upper class households. Still, because of their relatively high retail price, vacuum cleaners would be considered a luxury item until the 1950s, despite many advertisements of them as "electric servants."

A number of new vacuum cleaner manufacturers nevertheless entered that expanding luxury market. One was Air-Way. Daniel Benson Replogle (1863–1943) of Toledo, Ohio, was a vacuum cleaner inventor with many patents, including one he applied for in 1915 for a disposable vacuum bag, using suction to remove water from pulp to make a paper filter. He was interested in the health effects of vacuums, worrying about the dust people were breathing inside their homes, and the patent was his idea of filtering out the dust from the air stream before it left the cleaner. In 1917, while walking along a road, he was offered a ride by Pratt Tracy, who was on the board of several Toledo banks, as well as head of a wartime munitions manufacturing company. Pratt was looking for post-war consumer products along with his brother Clarence, and he liked both Replogle and his paper filter idea. Their meeting resulted in the formation of the Air-Way Electric Appliance Company in Toledo in 1919. That year, the company began to manufacture radios, the latest new household technology (and the inspiration of the company name), washing machines, and kitchen ranges.

In 1920, Air-Way introduced a vacuum cleaner called the Air-Way Sanitary System, designed by Replogle. Replogle had applied for his patent on February 26, 1919, which was granted on April 14, 1925, as U.S. Patent 1,533,271.[3] The vacuum cleaner was quite revolutionary—everything the Hoover was not. Sleek and lightweight (nine pounds, half that of the Hoover), the motor was mounted directly on the handle allowing it to be pivoted out of the way by twisting the handle, under and around obstacles, thus eliminating the "lawn mower" traditional operational method of competitive vacuum cleaners. A clear valve over the fan, called an indicator, allowed the suction to be diverted from the floor nozzle up the hollow handle, which could be used as a wand to vacuum cobwebs and baseboards without using attachments or hoses. All were highly innovative features, and its appearance was uniquely modern, in the sense that it was simple in form, visually expressive of its function, and dramatically different in appearance from its predecessors.

Wait, there's more! The cover of the indicator was clear celluloid, allowing the user to see the dirt flying toward the dust bag and thus to know when to stop. A feather renovating attachment allowed chicken feathers from pillows to be cleaned and blown directly into a pillowcase, and a moth control device pulverized moth crystals into a gas that could be blown into clothes closets. Self-adjusting wheels allowed the machine to go from bare floors to rugs automatically.

But Air-Way's secret weapon was the first patented, cellulose, 14-layer, disposable paper bag dust container, for which Replogle had to invent a machine to fabricate. The paper bag, inside a decorative cloth bag, served as a dust filter. After passing through it, the air was almost totally clean, and most dust remained in the bag. This innovation eliminated the housewife's highly undesirable task of shaking a cloth dust bag onto a newspaper, with dust flying all about. When the disposable paper bag became full, it was simply removed, deposited in the trash, and replaced with a new disposable paper bag, which Air-Way sold

"12 for a dollar." Housewives loved it more than sliced bread, which would not appear until 1928![4]

Another new manufacturer, Landers, Frary and Clark, entered the vacuum cleaner field in 1920 with a Universal cleaner boasting a friction-driven brush.[5] Founded in 1862 by George Landers to make household products, the company had adopted the trade name Universal for its products in the 1890s. Also in 1920, GE's Electric Vacuum Cleaner Company, which had acquired the Frantz Premier Vacuum Cleaner Company in 1915, established the Premier Vacuum Cleaner Company as a division and introduced the Frantz Premier Handy-Vac, a hand-held, hose-type vacuum, designed by the prolific inventor, James Kirby.[6]

In the early 1920s, Apex, headed by Clarence and Edward Frantz since 1912, used another inventor, Stanley McClatchie. In 1972, Clarence Frantz would recall that McClatchie "had designed an unusual cone-shaped vacuum cleaner which lay on the floor and rolled around easily to accommodate any position the user might assume with the hose to which it was attached." It looked like a child's extra large toy top and weighed only 8½ pounds. McClatchie's patent for this device (U.S. Patent 1,573,771) had been applied for on June 5, 1920, and was granted February 16, 1926.[7] The Frantz brothers agreed with McClatchie's suggestion that it could be tooled much cheaper in Germany, so manufacturing of the

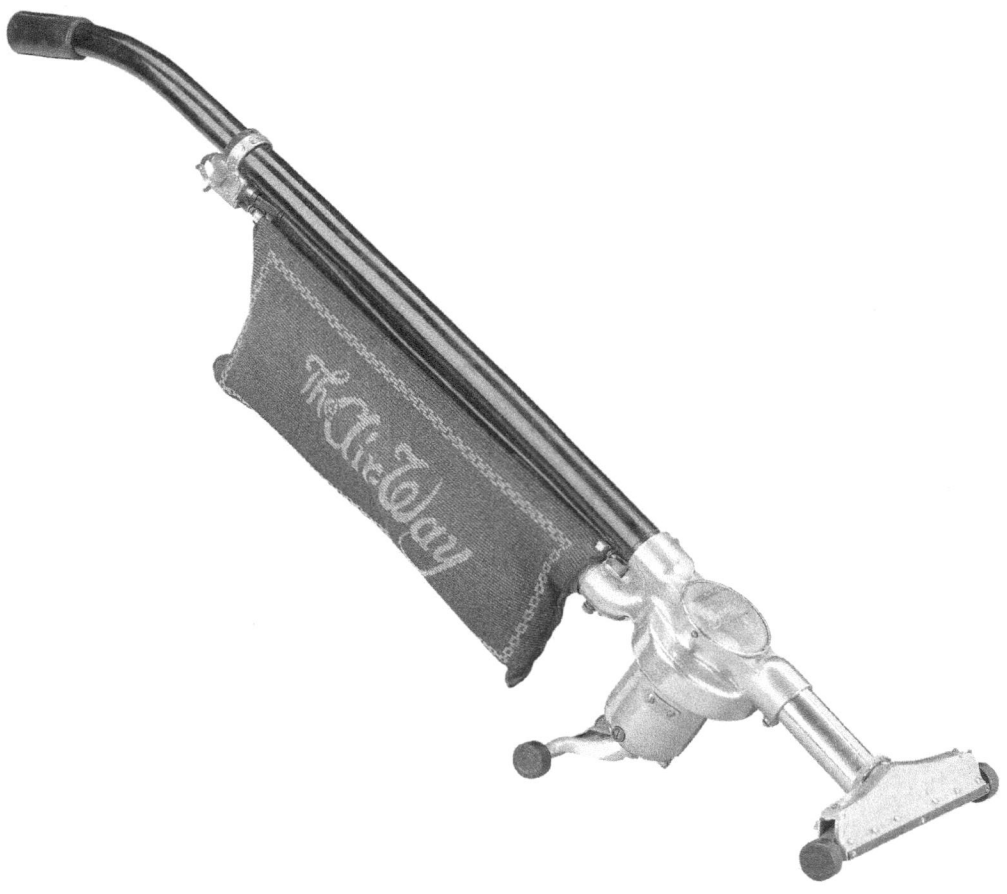

1920 Air-Way Sanitary System. Tacony.

Rotarex cleaner began in Stuttgart. The design was soon also made in Cleveland as the Apex Rotorex, but most of these were sold door-to-door by a Martin V. Kelley, of Renovator Incorporated in New York as "a new European type electrical purifying machine." The Rotorex would be discontinued when World War II broke out.[8]

It was then, for reasons other than his activities as an inventor, that McClatchie would be branded as a Nazi sympathizer. Born in 1893 as a member of a wealthy and well-known Southern California family, he had attended Harvard University and later, Leipzig University in Germany, where he was married in 1922. He had already begun a career as a mechanical engineer and inventor, including vacuum cleaners, with a number of patents assigned to Apex. He returned to Beverly Hills in 1932, but in 1936 attended the German Olympics in Berlin and was impressed with Hitler's consolidation of political power and the industrial, technical, and social revolution that had taken place in Germany under the Nazis. American athletes saluting the U.S. flag with the Nazi straight-arm salute also impressed him. He would then attract lasting notoriety by authoring a book, *Look to Germany—The Heart of Europe*, published in 1937 by Hitler's photographer, Heinrich Hoffmann, with 300 photos of the "new Germany" and with McClatchie's enthusiastic description of Nazism. When World War II began, English editions of McClatchie's book would be discarded as Nazi propaganda, and after the war, the book would literally be considered dangerous, causing postwar denazification committees to seek them out for destruction. But they are still available on the Internet. Freedom of the press, you know.[9]

Variations of different methods of cleaning sprang up everywhere in the early 1920s, including those that promised a profitable business in the home. In the April 1920 *Ladies Home Journal* appeared an advertisement by the Hamilton Beach Carpet Washer Company for a carpet washer that claimed to clean a 9 × 12 foot carpet in 30 minutes. How?

> With an action just like the human hand, two sponge rubber brushes rotated electrically 500 times a minute scrub the H-B compound deep down into the fibers of the rug. Mud and dirt and grime are instantly dissolved, suctioned back into the receiver-pan.

The machine was presented as "...an opportunity for wives to help their husbands and sons have a business of their own and financial independence," and the advertisement added that "...women, too, should seize this opportunity."[10] When Hamilton Beach entered the vacuum cleaner business in 1921, it had a typical problem faced by all new entries to the field. It was impossible to enter the business without becoming a licensee of the basic Kenney patents, and no more were being issued. Hamilton Beach solved the problem by acquiring the original Kenney license of the Richmond Radiator Company. By 1923, the Hamilton Beach Manufacturing Company of Racine, Wisconsin, would be acquired by the Scovill Manufacturing Company of Waterbury, Connecticut, and became a division of that company using the trademark Hamilton Beach for its products.

Selling cleaners in a luxury market was not easy. It could no longer be left to magazine or newspaper advertisements alone, the only mass communication means of the era. Radio transmission was just beginning and would not become influential until about 1926. Cleaners had to be sold one-by-one and door-to-door by aggressive salesmen, who demonstrated their product for women in their own homes. This followed the business plan precedent of the Fuller Brush Company, which by this time was a successful million-dollar-a-year business using door-to-door sales techniques. Vacuum salesmen often carried specially formulated "dirt" to place on carpets so that the full power and cleaning effectiveness of their product would be made obvious. They were personable, aggressive fellows who excelled in customer relations, and in their ability to sell products. Most vacuum cleaner companies had a huge number of such salesmen, which they trained and rewarded handsomely with generous commissions.

Typical of this practice was Hoover, which in 1921 held its first international convention in North Canton, attended by 300 salesmen from around the world. In 1920, Hoover had begun international expansion by manufacturing cleaners in Windsor, Ontario, Canada for sale in both England and Canada. That year, Hoover was making 700 units per day, while Eureka was making 2,000. Competition was intense. It was at the 1921 sales convention that the Hoover Song, based on the company's somewhat controversial slogan, first debuted, to the tune of the well-known "Field Artillery March":

> All the dirt, all the grit,
> Hoover gets it, every bit,
> For it beats, as it sweeps, as it cleans.
> It deserves all its fame,
> As it backs up every claim,
> For it beats, as it sweeps, as it cleans.
> Oh it's Hi, Hi, Hee!
> The kinds of dirt are three,
> We'll tell the world just what it means,
> Bing! Bing! Bing!
> Spring or Fall, the Hoover gets them all.
> For it beats, as it sweeps, as it cleans.[11]

The song was also used in Hoover advertising and, later, radio, but letters would come in from all over the country complaining that a commercial company should not use the patriotic "Field Artillery March" for such a purpose, and the company promptly ceased public use of the song. But it would not die, and it would still open and close Hoover international conventions well into the modern era. It also appeared on all employee service pins as "IBAISAIC."

In 1922, with W.H. "Boss" Hoover's eldest son, Herbert W. "H.W." Hoover (1877–1954) as president, the name of The Hoover Suction Sweeper Company was shortened to The Hoover Company. Hoover was by far the industry leader in vacuum cleaner sales, with annual sales of nearly a quarter of a million units. In 1923, Hoover introduced its Model 541, which reduced weight by several pounds by using a die cast aluminum main housing. The price was $65.[12] On February 20 of that year, the millionth Hoover cleaner rolled off the assembly line. In 1924 a new, four-story, brick office building would be built facing the town square in North Canton — the building that still stands today.

Up to this time, upright vacuum cleaner users (except those using Air-Way with replaceable bags) had to shake the collected dirt out of what usually

Herbert W. "H.W." Hoover, Sr. (1877–1954). HHC.

Hoover Suction Sweeper Company service truck, ca. 1919. HHC.

was a cloth bag. The only opening was at the bottom, where the bag attached to the cleaner, and thus was relatively small in diameter, making it difficult to shake out the compacted dirt. Royal solved the problem by making a cloth bag in 1923 that opened wide at the top, making it easy to shake out the contents from the top.[13] Still, it was a messy job, scattering dust clouds that women detested.

The proliferation of vacuum cleaners and other household products in the early 1920s awakened the federal government to the industry. On January 4, 1922, Senate Resolution No. 127 directed the Federal Trade Commission to investigate the house furnishings industry for causes of factory, wholesale, and retail price increases — specifically, competitive conditions, prices and profits. Three reports were issued. The third, Volume III, on kitchen furnishings and domestic appliances, issued October 6, 1924, included vacuum cleaners (Chapter II) and examined in detail the activity of the Vacuum Cleaner Manufacturers Association (VCMA) since its formation in 1913. It concluded that indeed, in 1920, vacuum cleaner industry profits averaged 44.1 percent return on investment, the highest in the home goods industry, with furniture second at 28.2 percent. Also, prices to retailers averaged 25.75 percent above prices to wholesalers. No federal trade restrictions appear to have been placed on the industry, but trade associations were chastened and began to become more formalized and organized.

On December 5, 1919, the VCMA met in Cleveland to reorganize, adopt a constitution, and create an executive committee. On October 6, 1922 at the semiannual VCMA meeting in Cleveland, the following officers were elected: Chairman, H.W. Hoover, Hoover Suction Sweeper Company; Vice Chairman, G.L. Smith, B.F. Sturtevant Company; Secretary-Treasurer, C.G. Frantz, the Apex Electrical Manufacturing Company. The executive committee consisted of Chairman Fred Wardell, Eureka Vacuum Cleaner Company; Philip A. Geier, the P.A. Geier Manufacturing Company; and H.H. Wright, M.S. Wright Company. Clarence G. Frantz, secretary-treasurer of the group, would still be serving in that same capacity in 1972, 50 years later. The association was still restricted to licensees of Kenney patents, which, in fact, would expire on March 19, 1924. Current concerns included patents, the gathering of statistics, cooperative advertising, and standardization of parts. The VCMA in 1922 initiated a massive national advertising campaign in newspapers across the country promoting the slogan, "Banish Dirt and Dust the Electric Vacuum Cleaner Way," but after some objections from members, the executive committee discontinued it.[14]

In Europe in 1923, the first international exhibition of household appliances was held in Paris. It was the first time household appliances alone were the subject of an international exhibition; all previous exhibitions had included a broad range of products and industries. The annual exhibition would a few years later be called the Salon des Arts Ménages (Salon of Domestic Science), "domestic science" being what most Americans understood as "home economics." Paris would soon have a powerful impact on American industry, as we shall soon see.

Another European country, Sweden, was also having an impact on the vacuum cleaner industry in Europe. In 1921 and 1922 the Elektrolux company introduced its Model V, a horizontal, cylindrical tank-type cleaner with a pistol grip on one end to push or pull it over the carpet, sliding on removable wire runners. International demand for the new Elektrolux cleaners was tremendous, and they would soon appear in the U.S market.[15]

In the last chapter we noted the emergence of the Ford Model T in 1913, in which Ford revolutionized the automobile market by cutting the price in half and dramatically increasing production speed. Now, in 1923, a similar event occurred in the vacuum cleaner market. Eureka introduced its Model 9 upright cleaner at nearly half the price of the Hoover. It was also 6 pounds lighter, at 11 pounds, and had the same motor horsepower. It also had a front-mounted bag that did not strike furniture and spew dust, as did the rear-mounted

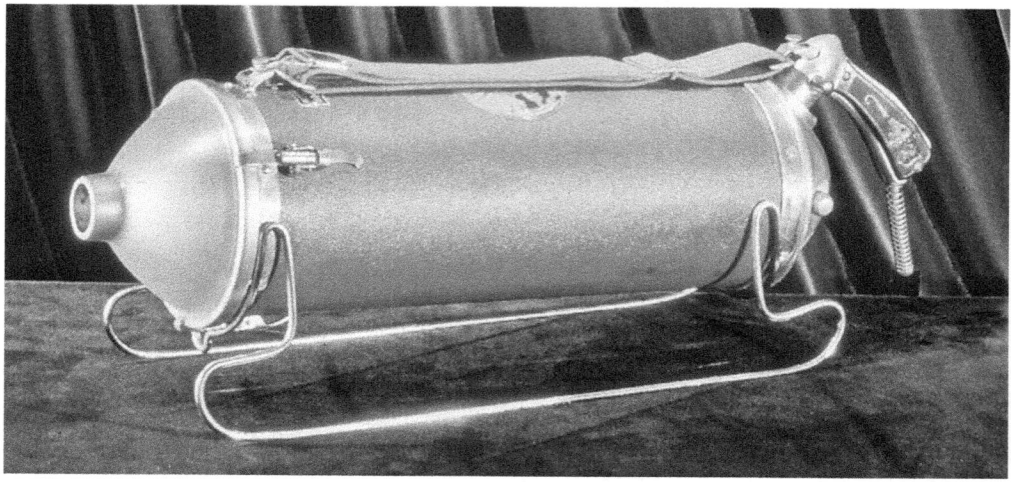

1924 Electrolux Model V, imported to the U.S. IHA/Lifshey.

bag of the Hoover and other uprights. The Model 9 was the best-selling single model Eureka ever made. Over the next three years, it would sell 1 million cleaners (that's almost 1,300 per day), and capture a large percentage of Hoover's market share, which sold only 713,073 cleaners during that same period. Eureka partly gained this clear advantage because, unlike most of Hoover's competitors, it did not wait until the original Hoover patents expired in 1926 to introduce new models.[16]

The 1920s are often considered good times, and in fact, they were better than the previous generation in many ways. But in fact, the sale of vacuum cleaners was made difficult because many consumers were still economically challenged. An exhaustive study was conducted in Muncie, Indiana, by Helen and Robert Lynd in 1924, and published in 1929 as *Middletown, A Study in American Culture*. Middletown, a fictitious place, was actually Muncie, Indiana. The study found that an annual salary of $1,920 was required to maintain a family of five, and half the households in town did not meet that standard of health and decency. None of the working-class wives of Muncie reported spending less than four hours a day on cooking, cleaning, and laundering, and most reported seven or more hours a day, seven days a week. Moreover, 55 out of 124 working-class wives worked for wages, many citing as the reason that "it takes the work of two to keep a family nowadays." In 1926, in Zanesville, Ohio, 70 percent of all families earned less than $2,000, and only 52 percent had vacuum cleaners.[17]

In many households, it was the automobile that was competing for any extra cash. Cars, often just called "machines" at the time, had become the latest hot technology to possess in the 1920s, just as computers would be in the 1990s. Whereas there were only 5.5 million cars registered in the U.S. in 1918, by 1924 there were 15.5 million—a tripling in six years. Of 123 working families in Muncie in 1924, 60 had cars. In 1924, you could buy a Studebaker touring car for $975, while a Hoover Model 103 upright cleaner was only $52.50. But cars typically were bought on time, with payments of $35 per month, about one-fifth of a good monthly salary of $165. General Motors had begun offering car purchases through installment plans in 1919, and Eureka followed this automotive lead, financing vacuum cleaner payments at $4 per month.

Counting depreciation, repairs, fuel, and loan payments, the cost of owning a car was about $100 per month, or $1,200 per year, over half of a generous annual salary. But the appeal was powerful. Many mortgaged their homes to buy a car, which, although a necessity for some, for most was recreational, and much more exciting than sitting on the front porch or waiting for the streetcar. People wanted to travel to see the sights and visit friends and family. Plus, a car sitting in the driveway or garage was also a visible mark of social prestige and financial success. By the end of the decade, more than half of all American families owned a car.

These factors obviously had a negative impact on vacuum cleaner sales. There was not much visible prestige for a vacuum cleaner sitting in the closet, and for many, less costly carpet sweepers seemed to work good enough. Although the cash purchase of a vacuum cleaner was not much more than one monthly car payment, for many, the choice was obviously for a car. Reduced vacuum sales closed down many new manufacturers and brought Regina near bankruptcy, although later in the decade it would become a viable floor care company.

The first significant European vacuum cleaner import to the U.S. made its appearance in 1924. Sales representative Gustaf Sahlin brought to America the Electrolux Model V (5), versions of which had been produced in Sweden and sold in Europe since 1919. He opened the first Electrolux sales and service office in New York. Axel Wenner-Gren had secured the rights to use the name "Electrolux" (with a "c") in the U.S., instead of the Swedish "Elektrolux" (with a "k"). Up to this time, most vacuum cleaners in the U.S. were of the upright style. The Electrolux was the first tank-type cleaner in the American market to become popular. The Model V had an internal cloth bag with a sanitary air filter placed before entry to the motor, and thus was the second dustless cleaner on the market (the first was Air-Way) and was advertised as a combined "Cleaner and Air Purifier." It used a flexible metal hose and initially had removable wire runners to allow the unit to be pulled across carpets, but when the Model 11 was introduced in 1927, the runners were permanently fixed. The Model 11 also used a rubber hose with cloth braiding over the rubber to improve durability.[18] These models were quite underpowered, and most people bought them not for carpet cleaning, but for dusting above the floor to clean draperies or furniture. Further, the Model V was only sold in limited areas, such as New York, where it was a highly popular choice for small apartment dwellers with few rugs. The market would not take the Electrolux seriously until its Model 12 in 1933, when tank and canister type cleaners would become more popular.

The primary difference between tank cleaners and canister cleaners was the orientation of the motor, fan, and bag assemblies during operation. In tank cleaners, sometimes called cylinder cleaners, these components were arranged in a horizontal manner, resulting in a long, horizontal cylinder form that was essentially transported on its side, often with skids, around the room. In canister cleaners, these components were arranged vertically, resulting in a vertical, cylindrical form, usually more squat or potlike than tank cleaners, and which were made mobile with wheels or casters on the bottom. Some designs allowed the cleaner to be used either horizontally or vertically, further confusing the terminology.

New vacuum cleaner manufacturing companies continued to enter the market. In 1925 Scott & Fetzer entered the vacuum cleaner business with an improved design by Jim Kirby, vacuum cleaner inventor since 1906 who served as Scott & Fetzer's wartime production foreman.

The Scott & Fetzer Company had been organized as a machine tool company in 1914 with George H. Scott as president and general manager, and Carl S. Fetzer as vice-president and treasurer. It was incorporated in Cleveland, Ohio, in 1917 making tools and dies.[19]

In 1912, Domestic had marketed a cleaner Kirby designed in 1910, to be used to serve housewives in rural communities without electric power, which was still about 55 percent of the population. The design was dubbed the Grasshopper, which it resembled, with its horizontal cylindrical drum as the body, a floor nozzle in the front as the head, wheel struts as the legs, and the handle as the tail. There was a cloth bag filter in the front portion of the cylinder and a cylindrical bellows in the rear of the cylindrical drum attached to wheels on extended legs. When a shovel-type handle fastened to the top of the cylinder was pushed forward and downward, the nozzle slid forward across the rug, while the wheel-mounted rear end of the bellows remained stationary. Thus, the bellows expanded, sucking in air and dust from the rug. At the end of the forward stroke, the handle was lifted, permitting the bellows to contract, and the rear end moved forward on its wheels.[20]

Kirby had developed an improved version of the Grasshopper in 1919, also a nonelectric, called the Vacuette, which looked like a typical upright vacuum cleaner. Its fan was powered by the wheels through a roller-type clutch driving a wheel and worm. Each time the machine was pushed forward, the clutch wound up a spring, which drove the worm wheel, which in turn powered the fan. When pulled backward, the clutch released enabling the fan to continue spinning by itself. It was also the forerunner of all later multi-attachment vacuums, and featured a removable floor nozzle and handle. Kirby would remain with Scott & Fetzer for 10 years, designing vacuum cleaners after its initial Vacuette Electric of 1925, but the Kirby name would not be used on a Scott & Fetzer vacuum cleaner until 1935.

Kirby became almost as famous in the washing machine industry, since he had started a new relationship with another Cleveland business in 1916, the Laun-dry-ette Manufacturing Company. The Laun-dry-ette, designed by Kirby, was the granddaddy of today's spinner-agitator automatic washers. It combined, for the first time, washing and extracting in a single tub, making it unnecessary for the housewife to touch the clothes until after they had been damp dried. Things went along prosperously until the company went bankrupt in 1926.[21]

Meanwhile, events totally unrelated to vacuum cleaners, but important to design, were occurring in Europe. In the Soviet Union after the 1917 revolution, the Communist Party placed all factories under government control to implement Vladimir Lenin's (1870–1924) directive of adopting the Taylor system of industrialization. Leon Trotsky (1879–1940) believed that to serve the revolution, art had to be on the cutting edge of industrial modernism. Accordingly, artist Aleksandr Rodchenko (1891–1956) initiated an art movement called Constructivism in 1921, calling for artists to "go into the factories, where the real body of life is made." Although Constructivism's geometric designs of furniture and porcelain soon faded in popularity, the movement had an impact on the Bauhaus in Germany, where by 1924, at the height of the school's international influence, student designs of modernist furniture, pottery, and lighting fixtures were being manufactured in limited quantities.

The Bauhaus development of a new "machine style" had by this time migrated to France. A French proponent of modern style was architect Charles Edouard Jenneret Gris (1887–1965, better known as Le Corbusier), who had worked in 1910 for Peter Behrens in Berlin. Le Corbusier's books, *L'Esprit Nouveau* (The New Spirit, 1920), and *Vers une Architecture* (Towards a New Architecture, 1923), described his theory, which connected architecture with mass-production standards, practices, and forms; in particular, his architectural designs were inspired by symbols of postwar technology: automobiles, airplanes, and ocean liners.

On the heels of these developments, in 1925, *The Exposition Internationale des Arts Décoratifs et Industriels Modernes* (International Exposition of Modern Decorative and Industrial Arts) was held in Paris. France wanted to recapture its prewar international dominance in style with modern design. Organizers intended to set new standards and required that participants display only "those exhibits that were not dependent on the art of the past," a deliberate departure from past international expositions.[22]

The French dominated the exposition in which 26 countries participated, but the U.S. did not, because it felt it had no innovative "modern" examples to show, which of course, was true. President Herbert Hoover did, however, after appeals by artistic and cultural organizations, send a commission to report on the event, and its report pointedly warned the U.S. Department of Commerce that the modern movement in applied arts would soon reach America.

This exposition, its name shortened to Art Deco by many, would indeed be highly influential in the U.S. in the latter part of the decade, particularly in promoting new, more modern styles for consumer products, in what would become known in the U.S. as the Machine Age. Until the 1930s, mechanical engineers designed the appearance of most U.S. consumer products, except automobile bodies that were custom styled for wealthy owners. Most ordinary consumers cared only about the performance and function of what they bought, primarily because they were the first mass-produced mechanical products they had ever purchased, and they were delighted with anything that saved labor or provided convenience. This would soon change, after European styles became popular.

In the U.S. vacuum cleaner industry, following Hoover sales methods, Air-Way began selling its cleaners door-to-door in 1925, and sales soared when people saw it demonstrated. Previously, Air-Way vacuums were sold along with its washing machines, ranges, and radios in retail outlets. At this time, The Hoover Company was still a leader in the industry, with annual sales of 229,005 units,[23] and with highly successful patents going back to 1908. But these would expire in 1926, and many upright cleaner manufacturers would begin copying Hoover's unique patented feature of a motor-driven agitator and brush.

Only one company had copied this feature before the Hoover patents expired: Pneu-Vac introduced its Sweeper-Vac with a motor-driven brush, although it didn't have any front wheels. The nozzle sat directly on the floor, and the drive belt ran underneath the cleaner from the rear at the bottom of the motor to the front. A lever turned a worm gear on and off so that the brush would stop or start at the control of the user. Hoover, of course, sued Pneu-Vac, but Pneu-Vac sold out to Western Electric, which continued to make the cleaner until the lawsuit was resolved. Hoover also sued Air-Way, which had a smooth, polished beater bar mounted on rubber instead of on the agitator itself, but the courts said it was the same thing, and Air-Way had to stop making it.[24] Other manufacturers avoided the Hoover patent by slight modifications. GE's 1924 Premier Deluxe cleaner used a revolving "brush" driven by the motor, but it used rubber fingers so it wasn't technically a brush. Of course, in 1926, when the Hoover patents expired, many manufacturers introduced their own versions of motor-driven brushes. GE then put a motor-driven real brush on its Premier Duplex, Hamilton Beach followed suit on its AV-1 with a power-driven brush, and Apex introduced a motor-driven brush. Royal did not have a revolving brush until 1930.

A new, patented improvement was needed by Hoover to stay ahead of these competitors, especially Eureka, breathing down its back with its highly successful Model 9, which had captured one-third of the vacuum market. Hoover's answer in 1926 was the Model 700, called the Greater Hoover by salesmen and customers alike, because it was superior to the model it replaced, the 1923 Model 541, and was the last major evolution in upright technology. Some still call it the finest upright cleaner ever made. Recognized by its distinctive orange triangle label on its polished metal body, it was priced at $75 without

1926 Hoover Model 700, the first featuring a motor-driven agitator with replaceable brushes and beater bars. HHC.

tools and had a 17-inch nozzle. It was produced until 1929 and featured the first motor-driven rotating agitator with the first replaceable brushes combined with rigid metal beater bars—the patented implementation of its unique "positive agitation."[25]

The beater bars, designed by Hoover chief engineer D.G. Smellie,[26] gently vibrated the carpet to loosen dirt and would never wear out, as would the brushes alone. This innovative sales feature was highly successful, greatly increased dirt removal, and gave new meaning to the Hoover slogan "it beats, as it sweeps, as it cleans." Hoover's Model 543 was the economy version of the 700.[27]

Other features on the 700 included a power switch conveniently located on the new pistol-grip handle rather than on the motor, the first rubber-covered electrical cord, a motor 25 percent more powerful than its predecessor, and a new washable cloth called Norca (for North Canton) for the dust bag. The 700 was the most expensive cleaner you could buy at $87.50 with attachments. Two million 700s were sold over the next three years, counting both American and Canadian models, compared to the Eureka Model 9's one million sales from 1923 to 1926. But they both were competing fiercely.[28]

Although vacuum cleaners, the best antidote to dirty carpets, have been our primary focus, there was usually something under and beside the carpets that also needed care. These were the bare hardwood floors, which generally, before wall-to-wall carpeting became popular, extended beyond the carpet around the periphery of the room. In many cases, rooms remained uncarpeted entirely. These floors were usually varnished and were cleaned and dusted fairly easily with a damp or dry mop, but over time succumbed to wear and tear, and unless protected, required costly replacement.

In 1888, Samuel Curtis Johnson (1833–1919), of Racine, Wisconsin, had developed a paste wax that protected the life of the wood and imparted a lustrous shine to hardwood floors, but which required periodic polishing and rewaxing. In the nineteenth century this was an onerous task that required kneepads and exhaustive hand labor. Some inventive souls made the task easier by fixing a brick, or piece of wood the size of a brick, to a piece of carpeting and pushed the resulting polishing device with a long handle. This saved the knees and increased the reach, but was still tough going for those large Victorian mansions with miles of wood floors.

The history of mechanical floor polishers is much murkier than for vacuum cleaners, although they seem to have appeared at about the same time, certainly by 1912. They used brushes made of tampico and bassine, vegetable fibers used for centuries for floor scrubbing and polishing. The first machines were known as divided weight machines, with the bulk of the weight on rear wheels that remained on the floor during operation. Carnuba wax was applied to the floor before polishing. Soon, manufacturers realized that the more pressure on the brush, the better the action. This led to swing machines, which centered weight on the brush, and the rear wheels lifted off the floor during operation.

In about 1920, a number of patents were granted for floor polishers, some of which were still manually operated, and others, electrical. But it is not clear whether any of these were put into production, or whether they were commercial or domestic. In 1920, the General Signal Corporation applied for a patent under the Regina name, and the Regina Company re-entered the floor care market with a dual-brush floor electric polisher. The General Signal Corporation would be in business until 1993. Another electric home floor polisher was the one introduced in 1926 by S.C. Johnson. Johnson's son, who then headed the firm, had commissioned a neighbor to design an electric home floor polisher. The neighbor happened to be none other than Chester A. Beach, coinventor of the Universal motor of 1909.[29]

Some manufacturers offered optional floor polishing attachments for their vacuum cleaners in the 1920s, such as Hoover's Model 700. In 1928, unit sales of floor polishers would be 59,000, and in 1929, 53,000. In the Depression, sales would drop to 32,000. At prices of about $78, floor polishers were in 1931 where vacuum cleaners had been 15 years earlier.[30] But inexpensive household floor polishers would not become very popular in the consumer market until the 1960s.

The year 1927 was a memorable one in America by any standard. Charles Lindbergh flew solo across the Atlantic, astounding the world; the New York Yankees won 110 games that season and captured the World Series with superstars Babe Ruth and Lou Gehrig; motion pictures introduced the wonder of sound in the Al Jolson film *The Jazz Singer*; and although only a few knew of it at the time, Philo Farnsworth (1906–1971) invented electronic television.

It was quite a year for the vacuum cleaner industry, too. Eighty percent of all affluent households in 36 cities had vacuum cleaners as well as washing machines. There were 41 companies making cleaners, offering a total of 77 different models. This was almost as many as automobile manufacturers, of which there were 43 at the time. Many vacuum companies had huge sales organizations like Hoover, which employed 4,000 salesmen.

Many vacuum manufacturers in 1927 were familiar names encountered in earlier chapters: Air-Way, Apex, Clements, Duntley, Eureka, General Electric, Hamilton Beach, Hoover, Invincible, Landers, Frary & Clark, Premier, Regina, Royal, Scott & Fetzer, Spencer, and Sturtevant. Others were new names, such as Allen & Billmyre, Arco, Electrical Household Utilities Corp., Edison, Electric Vacuum Cleaner, Federal, Graybar, Kent, Metal Specialties, Morrow, National Stamping and Electric, Norma, Standard, Sunshine, Super Service, Thor, Torrington, United Electric, Vose, White Cross, Wise-McClung, and M.S. Wright.

Upright cleaners dominated the scene, constituting about half of the field, and ranging in price from $34 to $87, the Hoover Model 700 being the top price (with tools), and the Birtman Model G the lowest at $29.50. Prices of some familiar names included Regina, $74.75; Clements Cadillac Model 92, $65; Hamilton Beach Model 1, $62.50; Scott & Fetzer (electric) Vacuette, $59.50; Air-Way Model E, $58.60; Royal Model J, $57; Apex Model A-3, $55; Eureka Model 9, $54.50 (with tools); Landers, Frary & Clark Model E, $49.50; General Electric Model 69, $49; and Premier Model 57, $45. Some were priced with and without attachments, indicating that some buyers were not interested in cleaning anything but carpets. The battle between Hoover and Eureka raged on. In 1927, Eureka would produce 270,563 cleaners, and Hoover, 259,114. In 1928, Eureka would respond to Hoover's 700 with its Model 10, which had a larger motor and an adjustable nozzle.

Lightness of vacuum cleaners was an important criterion for easy consumer use, and uprights ranged from 10 to 17 pounds, with Air-Way Model E the lightest and the Hoover Model 700 the heaviest. Eureka, Royal, Regina, Scott & Fetzer, and Apex all weighed in between 10½ and 11½ pounds. What about power? Some brands, such as Air-Way, Apex, Hoover, Premier, used ⅛-horsepower motors, and some, such as Hamilton Beach, Eureka, Royal, and Birtman, used larger, ⅕-horsepower motors. Clements used a powerful, ½-horsepower motor.

There was a number of what we would consider heavy-duty commercial cleaners that ranged in price from $150 to $590 and which weighed from 80 to 2,700 pounds. Motors ranged from 1.5 horsepower to 10 horsepower. Those we would classify as domestic canister cleaners were still very large and unwieldy. There was the Duntley, weighing 37 pounds, which looked a bit like the classic 12-gallon Shop-Vac cleaner; the Invincible Universal model at 70 pounds; the Kent Vacuna at 60 pounds ($235); and the Super Service Model G at 28 pounds. Canister-type cleaners were not as thorough in carpet cleaning as uprights, but offered user flexibility to clean above floor level and on stairs, as well as on carpets.

Several innovative hand-held vacs were in the 1927 mix, offering low cost and easy handling. The Day-Fan Handy-Vac weighed 3 pounds with a $\frac{1}{16}$-horsepower motor and sold for $18.75; the Presto-Vac Junior by Metal Specialties Manufacturing Company weighed 4½ pounds, sold for $25, and had a $\frac{1}{30}$ horsepower motor; and the O.K. Vacuum Brush Model A weighed 3½ pounds and cost $19.75. The Apex Rotorex, an innovative canister type described earlier, weighed only 8½ pounds and cost $24.50.[31]

The foregoing information is from a special 1927 issue of the *Electrical Record* magazine, so it is not surprising that only American-made cleaners are listed. But at this time, the Electrolux tank cleaners had been sold in the U.S. for three years and were still made in Sweden. They were very lightweight and easy to handle, and Electrolux claimed that by 1927, it had captured a third of the canister cleaner market. The Model 12 in 1930 would be the last Electrolux model to be made in Sweden, and after that, they would all be made in the U.S.

Several new vacuum cleaner companies entering the market in 1927 included the Breuer Electric Manufacturing Company in Chicago and the Westinghouse Electric & Manufacturing Company, which made vacuum cleaners in East Springfield, Ohio, and had its sales office in Mansfield, Ohio. In 1928, the Singer Manufacturing Company (of Singer sewing machine fame) would enter the field with vacuum cleaners made by its subsidiary organization, the Diehl Manufacturing Company in Elizabethtown, New Jersey.

Another famous name reluctantly entered the vacuum cleaner field in the late 1920s. As you may recall, Anna Bissell had taken over the company in 1889 after her husband, Melville, passed away. Bissell was the unchallenged leader in carpet sweepers and had built an extensive worldwide business. When household electrification swept the country and paved the way for hundreds of new electric appliances, Bissell had remained confident that the public would not overcome its fear of this strange new power source for many years. Anna thought of electric vacuum cleaners as monstrosities that could damage expensive rugs, and sometimes could even cause fires as a result of shorted out wires. Besides, Bissell had a well-established position in retail stores, in contrast to the aggressive and annoying door-to-door salesmen selling vacuum cleaners.

But by the late 1920s, vacuum cleaners had increased in both quality and popularity, and when displayed in retail settings with carpet sweepers, the Bissells looked ancient by comparison. Bissell decided that to avoid losing its place in the market, it had to introduce its own electric vacuum cleaner. Like others, it was loud, clumsy, and kicked up much dust, but had motorized brushes and a fan for creating suction. Bissell, of course, continued to make Bissell carpet sweepers, improving them with better bearings and a handle that adjusted the sweeping pressure on the brushes. In its 1928 design, after Melville R. Bissell, Jr., had taken over the company from his mother, the Bissell sweeper automatically adjusted the height of the brushes to different surfaces.[32]

In 1927, a number of national events triggered a movement that would within a few years impact all appliance manufacturers, including those in the vacuum industry. A number of influential department stores such as Macy's, Franklin Simon, Wanamaker, and Marshall Field held exhibitions of European products from the 1925 Art Deco Exposition in Paris. Typical was Macy's Art in Trade exhibition in New York in May 1927, which attracted 40,000 people and was hyped extensively on radio. The Metropolitan Museum of Art had also assembled an exhibition of Swedish industrial arts and sent it on tour to member institutions. The European designs shown were modernistic, dramatic, sleek and exciting, designed by leading French, Scandinavian, and German designers. European design had been refined since the end of World War I by educators and architects, who based their product designs on artistic and intellectual theories, rather than on functional mechanics, as in America, and these designs were promoted by European governments seeking new international markets, the largest of which, of course, was the U.S. By comparison, these designs made most American manufactured products look pathetically obsolete, crude, and clumsy; seemingly of the nineteenth, rather than the twentieth century. U.S. consumers, wanting to advance along with the latest European styles, began to demand more modern-looking designs.

Public appeals such as this, in those times of record U.S. sales and unprecedented production levels, fell on the deaf ears of most manufacturers. But not on those of some forward-thinking advertising men. In the August 1927 issue of *Atlantic Monthly*, Ernest Elmo Caulkins (1868–1964), head of the Calkins-Holden Advertising Agency and called by some "the dean of advertising men," wrote an influential article, "Beauty, the New Business Tool," which strongly urged manufacturers to incorporate modern design into their products.

Some did. In the automobile industry that year, the homely Ford Model T, which had dominated the industry since 1913, was dethroned in 1927 by a more stylish and colorful General Motors Chevrolet, which captured 45 percent of the market and forced Henry Ford to consider a redesign of his masterpiece. General Motors hired California custom automobile designer Harley Earl (1893–1969) in 1927 to establish an "Art and Color Section" at GM that by 1929 had 100 stylists working on new modern automotive designs.

Professional practitioners also were paying attention. A few creative, forward-thinking individuals in such diverse fields as advertising art, set design, engineering, exhibit design, sculpture, art, ceramics, architecture, education, fashion illustration, and publishing, some of whom had attended the Art Deco exhibition in Paris and "saw the future," began to implement professional practices in what was then called "industrial art" by the public. But some of these practitioners called themselves "industrial designers."

Among these pioneers were Walter Dorwin Teague (1883–1960), an advertising artist who established a New York industrial design firm in 1927 and was designing cameras for Eastman Kodak by 1928; Norman Bel Geddes (1893–1958), who left a highly successful New York career as a set designer in 1927 to design cars for Graham-Paige Motor Company; Raymond Loewy (1893–1986), a fashion designer who would open his office in 1929 designing duplicating machines; Egmont Arens (1888–1966), who, starting in 1929, headed the industrial styling department of advertising firm Calkins & Holden; and Henry Dreyfuss (1904–1972), another New York set designer who would also open his office in 1929 to design telephone hand sets for Bell Telephone. These and other early designers focused on the external appearance of products for consumer appeal, not on the internal mechanical workings, and already in 1927, some had organized into the American Union of Decorative Artists and Craftsmen (AUDAC).

AUDAC's designers of modern styles were intensely concerned with the proliferation of design piracy, the industry practice of illegally copying the external appearance of successful products. Museums actually encouraged industry to do so, by exhibiting examples intended to educate manufacturers in more "artistic" European design. There was little industry interest in the importance of the design patent's "distinguishing form of products that have marketable value" until now, when designers themselves became aware of the potential protective value of design patents, recognized their "marketable value," and had formed an organization to voice their concern. AUDAC generously pledged "to cooperate with manufacturers and the public in the placing of American arts and crafts on a basis of honesty, dignity, and merit." But they also began filing design patents to protect their work.

What exactly is an industrial designer? The English term still leaves many people confused as to what industrial designers do. After all, there are many designers in industry: mechanical designers, electrical designers, software designers, and industrial engineers (who design production lines). One of the founding industrial designers, Harold Van Doren (1895–1957), offered this definition in 1940:

> Industrial design is concerned with three-dimensional products or machines, made only by modern production methods as distinguished from traditional handcraft methods. Its purpose is to enhance their desirability in the eyes of the purchaser through increased convenience and better adaptability of form to function; through a shrewd knowledge of consumer psychology; and through the aesthetic appeal of form, color, and texture.[33]

The Germans have a more precise term: *Formgebung* (form giving). The result of such activity is, in German, called *Gute form* (good form) because industrial designers are concerned with the form of an industrial product, not just the decoration or embellishment of its surfaces. Industrial designers rely on their artistic sense of shapes, similar to a sculptor, in creating abstract forms that are more pleasing or evocative to the eye than others. They know what emotions shapes can communicate to observers, either consciously or unconsciously: a sense of order, elegance, beauty, newness, simplicity, clarity of function, ease of use, comfort, richness, humor, or high drama, to name a few. Industrial designers see form as a way of visually convincing customers that a product will perform well and satisfy their physical and emotional needs, but the bottom line, from a business standpoint, is that the industrial designer strives to make the product look different from competition, look fresh and new, and look exciting and dramatic to consumers.[34]

Promotion of modern design continued unabated in 1928, with an exhibition called Twentieth Century Taste by the B. Altman department store and Macy's dramatic International Exposition of Art in Industry. But industry, riding the crest of high demand and maximum production, was in no mood to make changes in tooling. Most consumer mass-produced products, including their appearance, had been designed for many decades by engineers, who by now numbered about 230,000, were trained in management, headed many production facilities, and who were reluctant to give up their corporate authority, even in part, to "artists" who knew nothing of manufacture, business, or mechanical design.

By way of contrast to industrial designers, engineers generally saw product forms not in a visual, artistic sense, but as individual functional components of a mechanical device, each component shaped as it should be to function best, and requiring no further external change. Therein lies the historical conflict between engineers and industrial designers. Manufacturers in 1928 either assumed that engineering expertise also extended into the realm of aesthetics, or, based on their reliance on door-to-door selling, did not consider aesthetics an important part of the sale.

By 1929, the year the Museum of Modern Art was founded, even business leaders in the American Management Association (AMA) became concerned about the problem shared by many manufacturers: overcapacity. There were more goods than customers. There was general agreement that artistic values had become essential to sales. The AMA responded to the challenge by conducting seminars called "How the Manufacturer Copes with Fashion,

Style, and Art," and "The Renaissance of Art in American Business." The appearance of products was becoming an issue of fashion, the farthest thought from the minds of vacuum cleaner manufacturers. At the AMA annual convention that year in Detroit, the keynote speaker, E. Grosvenor Plowman, advisor on merchandising problems to the Associated Industries of Massachusetts, stated that the modernistic style was the "off-spring of the jazz age," and that

> the modern style will remain as long as the machine in its present form is the characteristic of world production, [changing] its outward form from year to year, just as women's dress styles change, [and that] manufacturers would have to learn to balance faddism and permanence.[35]

In other words, he was considering modern design as a passing fashion or fad that might in the future go away. In further discussion of the subject at the convention, someone stated that although a simple radio cost only $50 to make, in an artistic cabinet it brings $150 or $750 from the well-to-do buyer. It certainly sounded like an argument to use style for more profit.

This phenomenon was not merely a matter of style, however. Industrial designers were also demonstrating the ability to reduce costs and to dramatically increase sales, qualities that manufacturers could not ignore. A man from KitchenAid cited another example during discussion and drove home this point. He reported that Egmont Arens's 1928 redesign of the original 1919 KitchenAid stand mixer had cut its weight in half and reduced its cost, as well as improved its appearance. Sales had increased by 100 percent.[36] (It is relevant to note that Arens's subsequent redesign of this as Model K in 1937 is still being produced today, 75 years later, with very little appearance change, and is still the most popular in the stand mixer category.)

The date of this AMA convention was none other than Tuesday, October 29, 1929, a date memorable to many as Black Tuesday, when the stock market crashed and initiated the Great Depression, which would last for the next ten years.[37] The stock market lost one-fifth of its value. Consumer spending dropped from 7.4 to 1.4 percent.

Meanwhile, back in the vacuum cleaner industry, which was still totally oblivious to the emerging trend of modern design, "sanitation fever" was at its height. The public was highly concerned with invisible germs and disease, as we are today, but things were different then. In the 1920s, life expectancy was only about 54 years. Diseases such as tuberculosis, diphtheria, coronary disease, typhoid fever, pneumonia and typhus were all quite common. There were no antibiotics in 1929; penicillin was just being tested in clinical trials, as was angioplasty for coronary disease. Blue Cross was in its infancy in Texas only. National health organizations of the day were waging campaigns urging personal hygiene, cleanliness, and sanitation systems for disease prevention, not unlike today's campaigns against smoking, fast food, salt, sugar, environmental damage, and obesity. A number of vacuum cleaner manufacturers attempted to address these concerns of health.

Sanitation Systems, Inc., a Chicago-based distributor of home cleaning systems for Cleveland's Scott & Fetzer, had been founded in 1928 by Martin J. Callahan, his son, and associate Ray Owen. Within two years, it was the largest distributorship in Chicago. The three men contracted with Cleveland's P.A. Geier to make an upright vacuum cleaner for them to sell under the name of Health-Mor, which was accomplished by putting the Health-Mor name on Geier's existing Royal Super upright cleaner design.[38] This was called a "purifier model" and was equipped with a "germ-killing" chemical chamber through which the air had to pass. In 1929, Royal also produced a purifier model with a chemical chamber that contained formaldehyde for killing germs. Health-Mor in 1930 would change its name to Health-Mor Sanitation Systems, Inc.[39]

Air-Way's success with its patented disposable bag, which also claimed to address health concerns, finally drove Hoover to put a disposable bag called the Hygienisac on its Model 725 in 1929, and Air-Way promptly sued — successfully.[40] At the same time, Hoover also offered a "mothimizer" accessory, dispensing air over paradichlorobenzene crystals to rid clothes closets of moths.[41] That year, Hoover also brought out its Model 200 Duster, a model 700 motor on its side with a dust bag and a set of skids, used exclusively with cleaning tools. It sold for $29.75, but was so unpopular that only 9,000 were made. In 1930, Hoover would introduce its first hand-held cleaner at $16.50, the Model 100 Dustette at 4½ pounds, for which it received the Good Housekeeping Seal of Approval.[42] Universal also debuted a hand-held vac, Model E-115, for car cleaning.[43]

Despite the health claims of disposable bags, the feature required users to continuously buy replacement bags at additional cost. At least one inventor had the idea of a cleaner that required no emptying of dirty bags, nor even the purchase of costly replacement paper bags. In fact, it would require no bag. In 1922, John W. Newcombe had invented and patented a mechanical separator, a device separating dust and dirt from the air stream, without a filter or bag, but did little with the concept until he joined with financier Leslie H. Green to develop the Newcombe Bagless, a hand-held cleaner that needed no bag at all. It was introduced in 1928 and worked so well that an upright version soon followed (one might say it was the first Dyson!). Newcombe and Green founded Rexair ("King of the Air"— the logo was a flying lion) Company in Troy, Michigan, in 1929. After the Depression began in 1929, Green bought the patent rights from Newcombe and hired engineer Clarence A. Brock in 1930 to refine the design. Green and Brock would work to develop a patent, which would be applied for in 1935, and would introduce a unique canister, which will be described later.[44]

After the stock market crash of 1929, and over the next three years, global economies, including the U.S., ground to a halt. General Motors stock fell from $73 to $8, and auto sales declined 75 percent by 1932. The Radio Corporation of America (RCA) stock dropped from $549 to almost zero. These were the "high tech" stocks of the time. In the U.S., manufacturing and sales fell by 50 percent and unemployment soared to 23.5 percent. By the end of 1930, vacuum cleaner sales had slipped back from 1,312,000 units per year in 1929 to 840,000, a 16 percent decline. Without jobs, people could not afford to buy much of anything, let alone vacuum cleaners. In many cases, they could not even afford electricity.

Power companies, then referred to as central stations, realized that their profits were hurt by plummeting sales in electrical appliances. Electricity was relatively cheap at $1.50 per month for just lighting, but an essential electric stove and refrigerator boosted utility costs to $7 per month, so adding yet more electrical appliances such as vacuum cleaners was a tough sell. Power companies normally sold electricity service with teams of sales representatives in various departments, such as "appliances," with the objective of helping a subscriber of electric service to "get the most out of the two wires that lead into his house." In Depression times, such efforts were expanded to become massive sales campaigns for new appliances. In 1930, Middle West Utilities Company in Chicago temporarily involved hundreds of employees on their staff to sell vacuum cleaners or other electrical appliances in sales campaigns via telephone calls to acquaintances, offering appliances at considerably reduced cost.[45] The New York Edison Company from 1930 to 1931 ran three similar two-month successful campaigns in New York and the Bronx, where apartment living was dominant, selling $1 million worth of vacuum cleaners (Premier/GE and Eureka), twice what vacuum dealers sold in the same period.[46] The cleaners sold were probably the Premier Junior ($37.50), the Premier Spic-Span hand-held cleaner, and the Eureka Model 10.

Manufacturers became desperate and grasped at any way to attract customers to their products. To inspire sales, Detroit introduced the first streamlined production cars that were distinctly different than their 1920s predecessors. Hearing news of dramatic sales successes by industrial designers, manufacturers sought them to make their products more appealing to buyers. But there were probably no more than 20 industrial design practitioners at the time in the country, and they already had more work than they could handle, as described earlier.

Unable to find available practitioners, some manufacturers went to academia to find designers. In 1930, Westinghouse hired Pittsburgh art and design educator Donald Dohner (1892–1943) to design everything Westinghouse made, from locomotives to ashtrays. In 1931, Montgomery Ward established a Bureau of Design, staffing it with display designers. Over the next few years, many major manufacturers, such as Toledo Scale, Sears Roebuck, DeVilbiss, Herman Miller, Philco, and Alcoa, would all engage or hire industrial designers to incorporate modern design into their products. General Electric would establish its own internal industrial design department in 1933, as would RCA.

In 1931, three new Royal cleaners were introduced: the Royal Purifier, which had a chemical chamber for sterilizing germ-laden dirt, using "Royal germicidal crystals" ($65); the Royal Super ($48.50); and the Royal Princess ($38.50).[47] Hoover launched its Model 750 ($79.50) in December 1931, the first with a two-speed motor and the first cleaner with a headlight (called a Hedlite by Hoover and added in July 1932 as a $5 option), its purpose presumably being to illuminate dark corners for better cleaning, but, as in many such features, more to provide a competitive selling point. Hoover also added the Hedlite to its Model 900 commercial cleaner.[48] Adding a headlight posed a serious technical problem. Standard 110-volt house current would quickly burn out a small bulb, which would overheat. Hoover's solution was to use the field coils of the motor as a transformer to step down the 110-volt AC house current to eight volts DC.[49] In 1932, Hoover opened its English manufacturing plant in Perivale, near London, and with virtually no competition, within a few years would dominate the British market. It was there that "hoovering" became a verb.

Birtman Electric Company, which had been making Bee vacuum cleaners since its incorporation in 1909, had begun to manufacture various electrical products for Sears Roebuck under the Kenmore brand name in 1927. Now, in 1932, Birtman began to manufacture Kenmore brand vacuum cleaners for Sears, a valued and profitable supplier relationship that would continue for 25 years into the mid 1950s.

Hoover introduced its Model 800 in 1933 with the built-in headlight as standard, but renamed it a Dirt Finder. The 800 had a handle made of Duralumin, a lightweight aircraft metal,[50] a double ball bearing motor, an "aromadore" (which added a pleasant scent to the room), and quadruple beater bars on the agitator. Hoover was so proud of the 800 that it had one in a glass case for many years at the entrance to its factory in North Canton.[51]

Other companies were adopting the Hoover-type power-driven brush rolls. Eureka in 1933 introduced its Model G with a motor-driven brush, as well as its hand-held, easy-to-carry Model H, similar to hand-held models that had been introduced in 1932 by Hamilton Beach and A.C. Gilbert. Air-Way introduced its twin-motor Model 35 Chief in 1935, making it the first vacuum cleaner with a power nozzle. It featured a backwards-revolving brush that did not grab the carpet fringe (an annoying tendency of most uprights) and rubber

1933 Eureka Junior Model H hand-held vacuum. MIAD.

beater bars with steel tips. Air-Way's DirtMasteR boasted the same features. Hoover responded by suing Air-Way over its copy of the beater bar. Other cleaners, such as all Apex models, and GE's Model 111 and its Premier Grand, also featured the backwards-revolving brush.[52] Air-Way's 1936 Super Chief added a headlight and a switch to shut off the brush roll, making it the first motor-driven upright that could vacuum bare floors without damage.[53] Clements' Cadillac model and Birtman's Bee Vac also added motor-driven brushes. This feature soon became an accepted standard on all upright cleaners.

By March 1933, when Franklin D. Roosevelt was inaugurated as president, there was a bank panic with 389 failures and many in the process of closing, U.S. industrial production and sales were only half of what they had been in 1929, many companies were going out of business, and two million people were homeless. On top of this, wages had fallen by 20 percent. But spirits were lifted that summer by the Century of Progress Exposition in Chicago, which during the summers of 1933 and 1934 attracted 48 million visitors to a showcase of new technology, transportation, architecture and modern design, further influencing the public toward modernity. It was also a showcase for industrial designers, who designed exhibits, streamlined trains, cars, radios, and a range of consumer products.

One of the most prominent industrial designers, Norman Bel Geddes, made a prescient prediction in 1932:

> In the perspective of fifty years hence, the historian will detect in the decade of 1930–1940 a period of tremendous significance. He will see it as a period of criticism, unrest, and dissatisfaction to the point of disillusion — when new aims were being sought and new beginnings were astir. Doubtless he will ponder that, in the midst of a world-wide melancholy owing to an economic depression, a new age dawned with invigorating conceptions, and the horizon lifted.[54]

In February 1934, an influential feature article in *Fortune* magazine identified the ten most well-known industrial designers, described their work for dozens of companies, and praised their designs for lowering costs and increasing sales anywhere from 25 to 250 percent. Among these were designers who would soon be designing vacuum cleaners: Henry Dreyfuss, Raymond Loewy, and Lurelle Guild (1898–1985), as well as others mentioned earlier. Despite the enormously high fees of these prominent designers, manufacturers rushed to cash in on what appeared to be a sure bet for higher sales results in a depressed market when people were just not spending. Industrial designers were suddenly in extremely high demand and were becoming nationally famous, thanks to *Fortune* magazine. The somewhat cryptic title of the article, "Both Fish and Fowl," correctly alluded to the fact that industrial designers, after centuries of the perceived separation of "artists" and "businessmen," were now functioning as both.[55]

The article, which was the first of major media to use the term "industrial design" instead of the traditional "industrial art," also inspired academia to educate young designers in what some design educators called the wave of the future. That same year, Carnegie Institute of Technology (now Carnegie Mellon University) initiated the first degreed educational program in industrial design, establishing a pattern that would be followed by dozens of other schools over the next 10 years. Westinghouse industrial designer Donald Dohner, one of the top ten in the 1934 *Fortune* article, headed the Carnegie program and the next year would initiate a similar program at Pratt Institute in Brooklyn. The first industrial design graduate, from Carnegie Tech in 1936, was a female émigré from the Philippines, Maud Bowers, who graduated alphabetically in her class of five: two females and three males.[56]

The *Fortune* article made many manufacturers aware of industrial design for the first time, and drawn by the possibility of sales increases and reduced costs, they rushed to find the nearest industrial design consultants. Even the vacuum cleaner industry jumped on the modern design bandwagon. Montgomery Ward's new Bureau of Design had just hired Chicago industrial designer Dave Chapman (1909–1978) to head its product design department. Chapman was well aware of the emergence of industrial design in industry, which prompted him to enter the field (he had formerly been an architect). As head of Ward's product design, he was working with Apex to convert its 1934[57] Which Way upright vacuum cleaner model into an MW (Montgomery Ward) brand model and to incorporate Art Deco styling.

The Which Way, made by Apex, was called that because it had a swiveling pivot connection directly above the handle bail, allowing the user, with a twist of the wrist, to change the direction of the machine, a feature adopted 70 years later on the Dyson Ball models.[58] The new MW would also incorporate a headlight, closely following Hoover's innovative feature of 1932.

The decision to incorporate modern design into the new MW model may or may not have been influenced by Chapman, but in a 1971 interview, Clarence Frantz of Apex would recall that in 1933, he had realized that appearance had become a stronger factor in sales, particularly in automobiles, and that vacuum cleaners "surely did look angular and mechanical with much to be desired in having an attractive product," and that "it was about that time that products of a very wide type were given some modification to attract the home-type people who were in no way connected with mechanical matters."[59]

So Frantz said he had hired prominent automotive "stylist" (as Detroit's industrial designers were called well into the 1990s) George W. Walker (1896–1993) to "style an Apex vacuum cleaner," which was probably the one Apex made for Montgomery Ward's 1934 MW, and which had an Art Deco headlight grill and an MW nameplate, both with a strong automotive look.[60] Walker had attended the Cleveland Institute of Art and worked as an automotive designer since 1928 with Peerless, General Motors, and Graham Paige until the latter failed in 1929. He had just opened his own industrial design firm in Detroit in 1934. The 1934 MW was a somewhat simplified version of Walker's Design Patent 94,370, filed on May 1, 1934, and granted January 15, 1935,[61] but still recognizable. It was probably the first vacuum cleaner in the industry styled by an industrial designer, rather than an engineer, but there would soon be many more. Walker specialized in automotive design and later (in 1955) would become Ford's vice president of styling.

4. Consolidation 1920–1940

Hoover was probably the next to adopt modern design. Hoover's annual sales had fallen from 242,280 units in 1929 to 156,494 in 1933, a decline of 35 percent and lower than it had been since 1919.[62] Something needed to be done to increase sales, and the 1934 *Fortune* article may have inspired Hoover's 1934 hiring of Henry Dreyfuss, one of the new profession's luminaries, to work on the design of its cleaners, at an annual retainer fee of $25,000,[63] when most corporate executives were making a mere $5,000 per year. In 2011 dollars, Dreyfuss's retainer would be over $400,000. In 1933, Dreyfuss had designed GE's new "flat-top" refrigerator, hiding the previously exposed refrigeration unit (the 1927 Monitor Top design by engineers) by enclosing it within the cabinet, and the Sears Toperator washing machine, both highly successful in the market place.

For these early industrial designers, getting into the business was like shooting sitting ducks in a pond. Every product in sight was ugly, ungainly, and obsolete in style, designed by engineers who were totally focused on functional performance, but oblivious to the new modern design trends and totally unaware of the public desire for more attractive appearance. It was an incredibly lucrative business, and it was no surprise that scores of artistically trained professionals entered the new profession due to the extraordinary demand.

Dreyfuss would design many of Hoover's cleaners during the decade. First was the Model 825, made from 1935 to 1936, which had a headlight that also dramatically lit up the Hoover logo on top of the motor. A rectangular nameplate on top of the motor flowed like a waterfall down to the chassis, stopping at the furniture guard. The Model 475 was an economy version of the 825. Dreyfuss also worked on the Model 125 Dustette (1935), and the Model 300 (1935–1939). Next was the Model 150, selling for $79.50, which was produced from August 1936 to October 1939 and

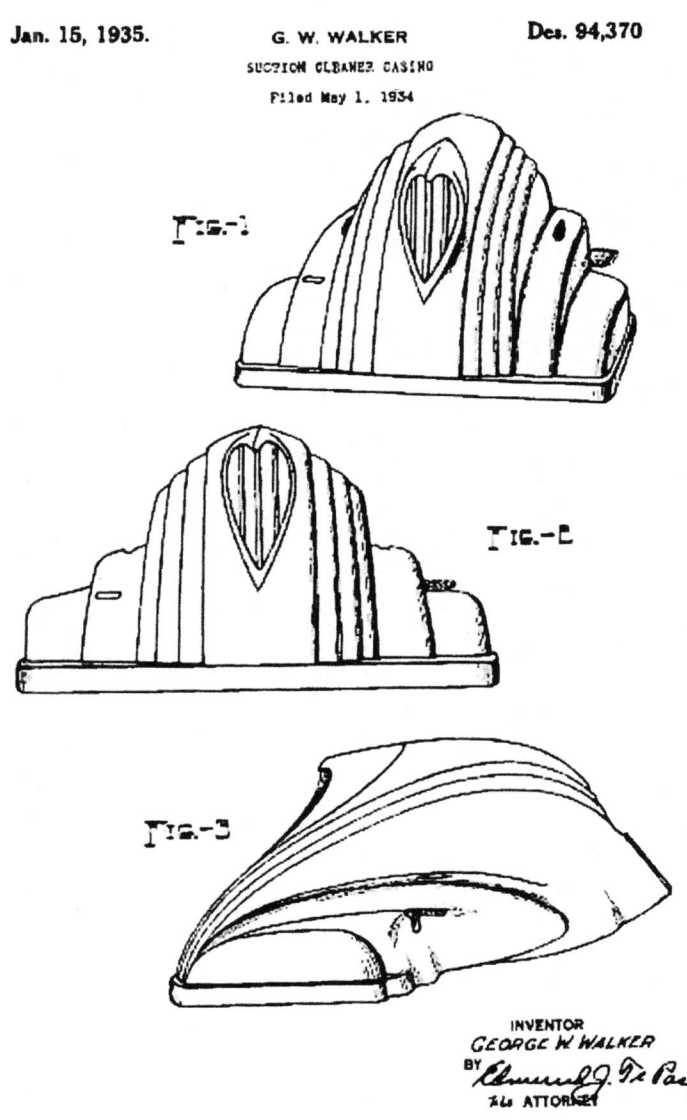

George Walker's 1935 design patent D94,370, for Montgomery Ward's 1934 MW cleaner. U.S. Pat. Off.

was the first basically new Hoover model in 10 years. Features included a semiautomatic nozzle height adjustor, a time-to-empty bag signal, a 12-inch nozzle, and a new style converter for hose and attachments.[64] The 150 was the first electrical appliance to use magnesium, a third lighter than aluminum, for its die cast body, which reduced the weight from 18 to about 16 pounds, and it was the first to use Bakelite plastic for the motor hood. Model 25 (1937–1938), at $65, was an economy model of the 150.[65] By 1936, Hoover's annual sales had doubled since 1934, to 314,225 units,[66] but it was still battling its key rival, Eureka. By 1936, Eureka estimated that 25 percent of all electric vacuum cleaners in use in the U.S. and Canada were theirs.[67]

There were, not surprisingly, still many manual vacuum cleaners being used in rural areas. Before this time, a small but growing number of farms had installed small windpowered electric plants, but by 1935, only 11 percent of American farms had electricity, and in Mississippi, less than 1 percent. In 1935, the Rural Electrification Administration (REA) was formed as part of Roosevelt's New Deal. Under REA, nonprofit rural cooperatives were organized to build power lines and distribute electricity, financed by long-term federal loans at 3 percent interest rates. In the south, the power came from the Tennessee Valley Authority; in the Northwest, from Bonneville, Grand Coulee, and Boulder (later renamed Hoover) Dams. By 1941, 50 percent of the nation's farms would be electrified, and finally, by 1970, about 98 percent. Nothing has done more to eliminate rural poverty than bringing electricity to remote locations. It sure didn't hurt vacuum cleaner sales, either.

Neither did the advent of plastics, which made vacuum cleaners lighter and less expensive. The earliest was Bakelite, a thermosetting resin of phenol and formaldehyde, reinforced with finely ground sawdust or flour, which was invented in 1907 by U.S. chemist Dr. Leo Hendrik Baekeland (1863–1944), initiating the era of plastics. It had previously been used only in electrical components, but became available for appliances, if only in very dark, conservative colors. In 1926, the B.F. Good-

1936 Hoover Model 150, the first with a plastic motor hood. HHC.

rich Company had developed polyvinyl chloride, commonly called PVC, which was a flexible and inexpensive substitute for rubber.

The 1930s introduced a cornucopia of new plastic materials that were used by many manufacturers, including those of vacuum cleaners. Neoprene and fiberglass for insulation were developed in 1931. In 1933, British chemists developed the first man-made polymer, polyethylene. Nylon was developed by DuPont in 1934, as was urea formaldehyde, called Plaskon, which achieved brighter colors impossible with Bakelite. Polyester would be developed in 1936, and in 1937, the American Cyanamid Company would develop melamine formaldehyde, called melamine, which would feature color stability and abrasion resistance. Acrylic, a transparent material, would be marketed in 1936 as Plexiglas and Lucite, and Dow would first market polystyrene in 1937. Teflon would be discovered in 1938, and Dacron in 1941.

Although women were historically presumed by manufacturers to be the primary users of vacuum cleaners, it was a woman who first reminded manufacturers to pay more attention to the needs of women, because they constituted a large, influential segment of consumers. The problem was that men, who up to this time dominated engineering and industrial design, had always designed most consumer household products, including vacuum cleaners. Often, men used their own physical dimensions and sensitivities, which were often quite different than women, to determine the height of handles, the circumference of handgrips, and the weight of the product, as well as the general appearance of products. It was the first female industrial designer, Belle Kogan (1902–2000), who would strongly urge manufacturers to design products to satisfy their female users. In a 1935 article in a plastics magazine, she warned that

> in considering the marketing and merchandising of any commodity, the modern manufacturer is confronted with the problem of pleasing the American housewife ... the feminine viewpoint is one to be studied, to be understood, to be coddled.... Thirty million women — all potential customers — constituting practically the entire buying structure of the nation — comprise a force, which cannot or should not be disregarded. The tastes of the American woman, her reaction to color and form, are of vital importance to the manufacturer.[68]

Cost reduction was an obvious strategy for all manufacturers during the Depression. Because of a shortage of consumer cash during the Depression, all manufacturers were desperately trying to lower their costs to entice frugal consumers with lower prices. Hoover made cheaper motors by using sleeve bearings instead of ball bearings and reduced the blades on the fan from twelve to six to produce "economy" models 425 (1932–1933) and 475 (1935–1936). Eureka painted its Model 9 gray instead of polishing the metal brightly and called it an "economy" model.[69] Consumers were very responsive to any options that lowered cost. Canister and tank-style cleaners, sometimes called "hose-type" cleaners because they included a long hose to reach above-floor dirt, were beginning to become more popular. They were less costly to make and buy than uprights, although they did not clean carpets as well. Regina made its first canister in 1935.

That same year, Scott & Fetzer introduced its first in a long line of cleaners to be named in honor of its original inventor, James B. Kirby, calling it the Kirby Model C. By that time, Kirby himself was 51, and he had in 1930 developed a new feature on the electric Vacuette for Scott & Fetzer: the Sani Em-tor, which, claimed Kirby, was "a device which simplified emptying a cloth cleaner bag without fuss or muss of flying dust and even without removing the bag from the cleaner."[70] The device consisted of an aluminum hopper attached to the exhaust part of the fan case. The cloth bag was attached to the hopper and was emptied by detaching the bag support chain at the top and shaking the bag into the aluminum hopper, which was then emptied and replaced.[71] Kirby would continue to work with Scott & Fetzer exclusively on vacuum cleaners, and for Apex exclusively on washing machines, until well after World War II, and would work on his own various inventions even after his retirement to Florida in 1961.

Despite their popularity, the main problem with disposable bags, even though they eliminated messy emptying of cloth bags, was that they clogged quickly with dust, thus reducing suction, and had to be replaced with clean bags. Some manufacturers were trying to compete with the highly successful Air-Way disposable bag by making canisters that needed no bag at all. This not only promised to maintain constant suction, but to save considerable cost to consumers, the latter extremely important in the Depression.

James B. Kirby (1884–1971). IHA/Lifshey.

As described earlier, one of the earliest manufacturers of bagless cleaners was Rexair, which had introduced the first bagless cleaner in 1928. Rexair engineer Clarence A. Brock had immediately discarded the upright version and worked to develop a canister cleaner based on the Newcombe Bagless cleaner. He applied for a patent (2,188,031) on December 18, 1935, which would be granted January 23, 1940.[72] This canister vacuum had two sections: the motor in one side and the dirt container in the other. At the top was the separator, which directed the incoming air into a centrifugal motion, and which separated the dirt and dust into the dirt container, removable for emptying. Had this design been used, it would have worked perfectly. But in 1936, Leslie Green hired Russ Hill from Air-Way as sales manager; he advised revisions to the Brock design, combining the separator and dirt container, and placing the motor on top, making it one of the industry's first canister-type portable vacuums. Unfortunately, the relocation of the separator caused it to fail to separate the fine dust properly, and customers expressed their dissatisfaction by returning many of the first units. Concerned, Rexair's board demanded a revised design, resulting in the 1937 introduction of Rexair's Series A, which had a unique water filtration system, trapping the dust lifted from carpets by passing it through a dirt container half-filled with water. Rexair's new slogan became "Wet Dust Can't Fly."

Rexair was the first cleaner since James Kirby's 1906 attempt to use water as a filtration system to remove dirt, in lieu of a bag or filter, although the use of water as a dust filter can be traced back to the Leaycraft patent for a manual suction cleaner in 1883, mentioned earlier. The Rexair guaranteed no loss of suction, similar to the Dyson "cyclonic" system of 1979, and it is still the only cleaner on the market that does not clog with dirt. Because of its ability to eliminate airborne dust in the home, life was made easier for allergy and hay fever sufferers, and the medical profession recognized Rexair as a breakthrough product. It was the first combination vacuum cleaner/air purifier and the first that separated water from the airflow, making it the first wet/dry vacuum using a water filtration system, which would lead to its being renamed the Rainbow in 1955. Between 1937 and 1940, Rexair would sell nearly 200,000 Series A, followed by Model B and Model C cleaners, which differed only in color and were mounted on wheeled dollies to move around the room.[73]

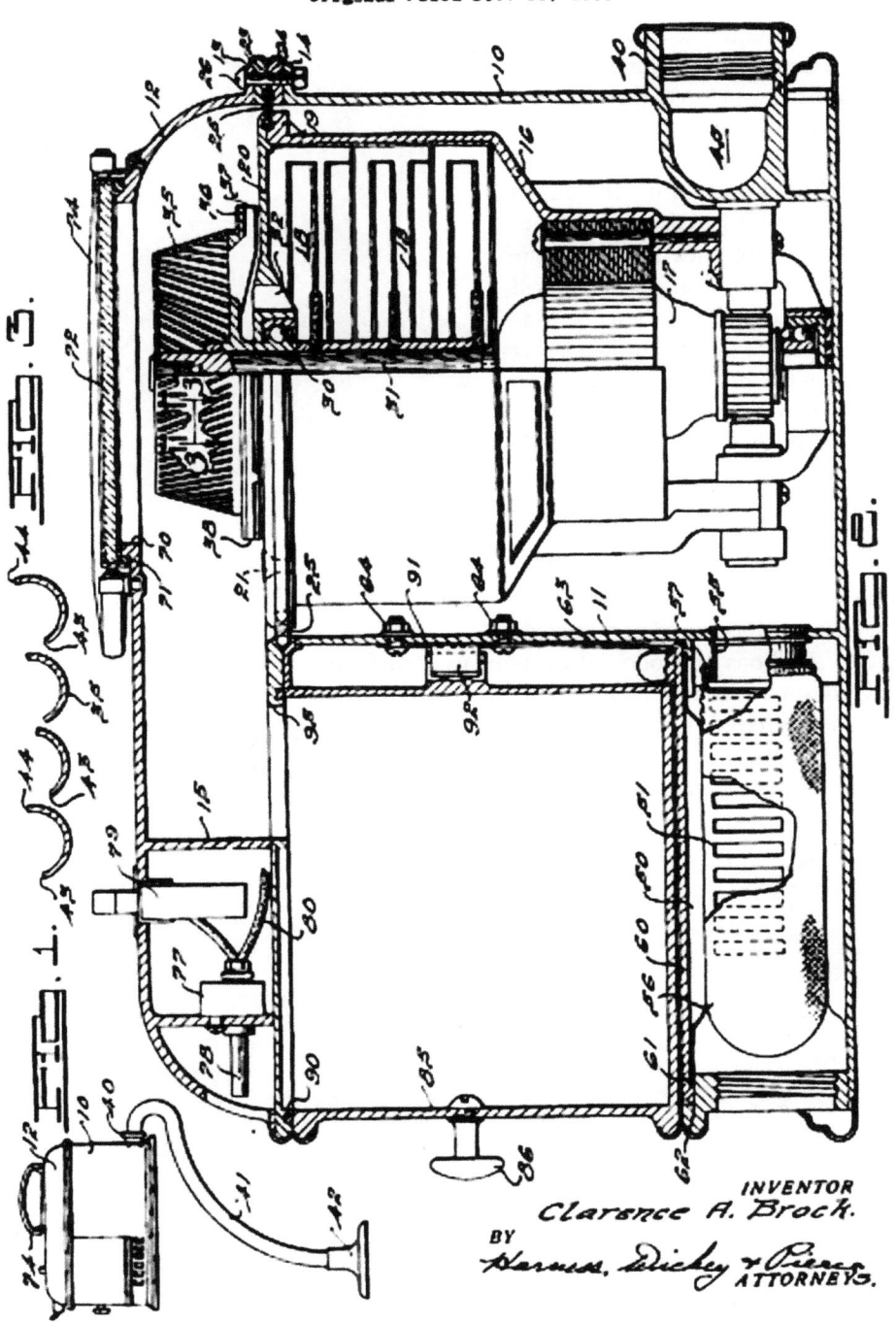

Clarence Brock's 1940 patent 2,188,031 for a centrifugal separator canister. U.S. Pat. Off.

In 1933, Electrolux had decided to open an American factory in Old Greenwich, Connecticut, because in 1932, a boat from Sweden loaded with Electrolux Model 12 vacuums had sunk in the Atlantic. As mentioned earlier, many banks were closed that year, so the Electrolux could not get money to build its factory and had to accept "moratorium checks" until the banks opened. Electrolux had developed a special model for America known as the Model 12A (A for American) and used motors from the White Sewing Machine Company, a division of Electrolux that Sweden had established in the U.S. in 1933. The company purchased cords, switches, and dust bags from outside vendors. In 1937, Electrolux, now uniquely made in America, introduced its Model XXX (Model 30) tank cleaner, designed by prominent industrial designer Lurelle Guild in a modern style described in design patent 106,662.[74] Millions of Model 30 vacuums were sold from 1937 through 1954, the longest run of any single vacuum cleaner ever, at a price of $69.95. This design became a classic, and one is displayed in the Smithsonian Institution. In 1949, Guild would also design a floor polisher attachment for the Electrolux (design patent 158,743).[75]

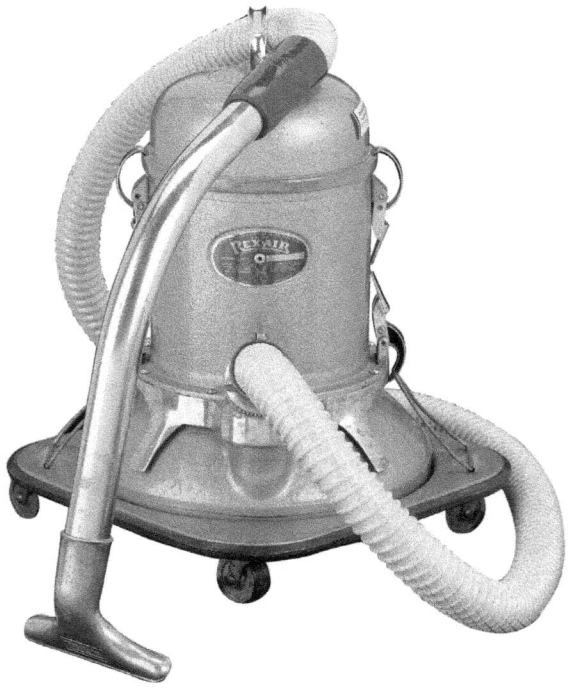

Rexair Model C, ca. 1939–40. Tacony.

During the Depression, of course, vacuum cleaner sales were extremely weak, since

Electrolux Model XXX (model 30). MIAD.

few people had money to spend on expensive electric vacuums. Instead, many opted for a less expensive Bissell carpet sweeper. As many vacuum cleaner companies went out of business, Bissell was the beneficiary. It, too, decided to leave the vacuum cleaner business and discontinued its manufacture of vacuums. Instead, Bissell believed that its sweepers had a unique segment of the market. While electric vacuums could be used for heavy duty cleaning, the sweeper would be favored for quick and easy touch-ups, in the same way that a broom might be used. This, indeed, would become the typical strategy of carpet sweeper manufacturers. The E.R. Wagner Manufacturing Company, maker of baby carriages, go-carts, shock absorbers, and hinges since 1899, came to the same conclusion and introduced its first carpet sweeper in 1932.[76]

By the middle of the decade, advertising began to flood the mass media as manufacturers sought to increase sales. Consumers were faced with the problem of separating advertising hype, of which there was much, from facts, and of distinguishing good products from bad. To address this problem, Consumers Union, a nonprofit organization, was formed in 1936 to comparatively test products and report performance and prices directly to consumers in its magazine, *Consumer Reports*.

Hollywood films of the 1930s popularized the modern look in both extravaganzas and in the staged interiors of well-to-do apartments and homes, which often displayed modern décor and accessories. People wanted to emulate this modern look in products they bought. More and more manufacturers turned to consultants or hired industrial designers to modernize their products, and the vacuum cleaner industry was no exception. The Singer Company in Elizabeth, New Jersey, without an industrial design staff of its own, sought the assistance of General Electric's new Appearance Design Division, established in 1933 by Ray Patten (1897–1948) in Bridgeport, Connecticut. The result was its R-1 vacuum cleaner, designed in 1936 by Malcolm S. Park (design patent 99,723, filed April 2, 1936, and granted May 19,

1936 Singer Model R-1.

1936) featuring a cord reel mounted on the handle.[77] It was essentially GE's model 111, or its Premier Grand, with an Art Deco–styled hood.[78] Sears Roebuck established its design department in 1934, headed by John Richard "Jack" Morgan (1903–1986), who designed its upright cleaner made by Birtman Electric Company, as shown in design patent 105,495, filed May 28, 1937, and granted on August 3, 1937.[79] It looked pretty much like Hoover's Model 150 with a headlight, shown earlier.

In other instances, engineering provided new mechanical features to increase sales. C.H. Sparklin, a 1924 graduate of Kansas University, had been considered a genius as a kid in Hiawatha, Kansas, because he had built a movie projector, a roller coaster, a sail wagon, and a glider (which crashed him into a cornfield from a height of 30 feet). As an employee at Birtman Electric Company in Chicago, he began pondering the general vacuum cleaner problem known as the "seal," the correct distance of a cleaner nozzle above the carpet surface. If the nozzle was too close, it stopped the passage of air and wore off the nap of the carpet, and if too far, the air was dissipated and did not pick up the dirt efficiently. Sparklin's solution, which required complex mathematical calculations, was to allow two or more wheels to sink into the carpet to lower the nozzle. By employing wide and narrow wheels, he caused some to ride over and others to sink in, thus achieving the optimum seal. This simple invention, which Sparklin patented on September 29, 1936, greatly improved the efficiency of vacuum cleaners.[80]

Consumer cost of vacuums continued to be a major problem. In 1937, a Rexair cost $90, an Air-Way $99, and a Hoover Model 150, $95. One could buy an Electrolux for $69.95 on installment for $10 down and $6 per month with a $4 interest charge.[81] These prices seem reasonable today, but when translated into 2011 dollars, a Rexair would cost $1,396; an Air-Way, $1,536; a Hoover 150, $1,474; and an Electrolux, $1,086. Given these prices, it is easy to understand why, in a deep recession, selling vacuum cleaners was extremely difficult, and why they were still considered a luxury item.

To make matters worse, in 1937 there was a "recession within the Depression." In the six-month period between Labor Day of 1937 and March 1938, four million more jobs were lost, an increase of 45 percent over 1936. It was in 1937 that the P.A. Geier Company introduced its hand-held cleaner, the Royal Prince; Apex debuted its first tank cleaner, the first square cylinder, designed by industrial designer Dave Chapman of Chicago; Royal introduced its first tank cleaner[82]; and Eureka debuted its Challenger upright, which, according to its owners manual, was "designed by famous stylists" (actually George Walker, who had his office in Detroit) and was made in various colors until 1941.[83] Hoover also made a cleaner called the Norca (for North Canton) Model 1 in 1937 that had no beater bars (to save metal cost) and used only brushes. Hoover sold it over the counter at retail stores for $39.75.[84] By mid-1938, there was a recovery in the recession, as the GDP began an upward climb that would continue into the war years. Hoover introduced Model 26 in 1938 at $68, one of which in 1939 became the five millionth Hoover produced, and introduced an economy model, the 305, at $52.[85] In 1939, Hoover's annual sales soared to 335,586, 33 percent higher than before the Depression.[86]

Aircraft design, which had become dramatically sleeker and faster during the 1930s, was beginning to influence product designs such as automobiles. By the end of the decade, many of the car designs we now regard as classics had been made. The practice trickled down to consumer household products; the Wagner Company, which had entered the carpet sweeper market in 1932, debuted a new modern design in 1938 that looked like a streamlined section of an aircraft wing (design patent 111,033 by E.R. Smith).[87]

In 1938, Air-Way competed with the popular Electrolux by debuting its cylindrical, purple and chrome Model 55A Sanitizor canister cleaner, which looked somewhat like an Electrolux tank-type cleaner standing on end with a wide, flared base for stability. It was the first hose-type (canister) cleaner to use a disposable bag, Air-Way's unique patented feature. The hose was over eight feet long, and when stored, coiled around the top half of the cylinder, retained in place by a chromed metal basket. When the hose was attached to the top of the cleaner, it could swivel, so that with the cleaner stationary in the center of a room, one could clean the entire room without moving the cleaner, which incidentally, had no wheels, but metal runners. Inside the cleaner was a GE motor with double fans that provided more suction than any other hose-type cleaner. There was a filter after the motor to eliminate carbon dust from the exhaust. The bag was in a cage with 36,000 holes, to equalize pressure over the

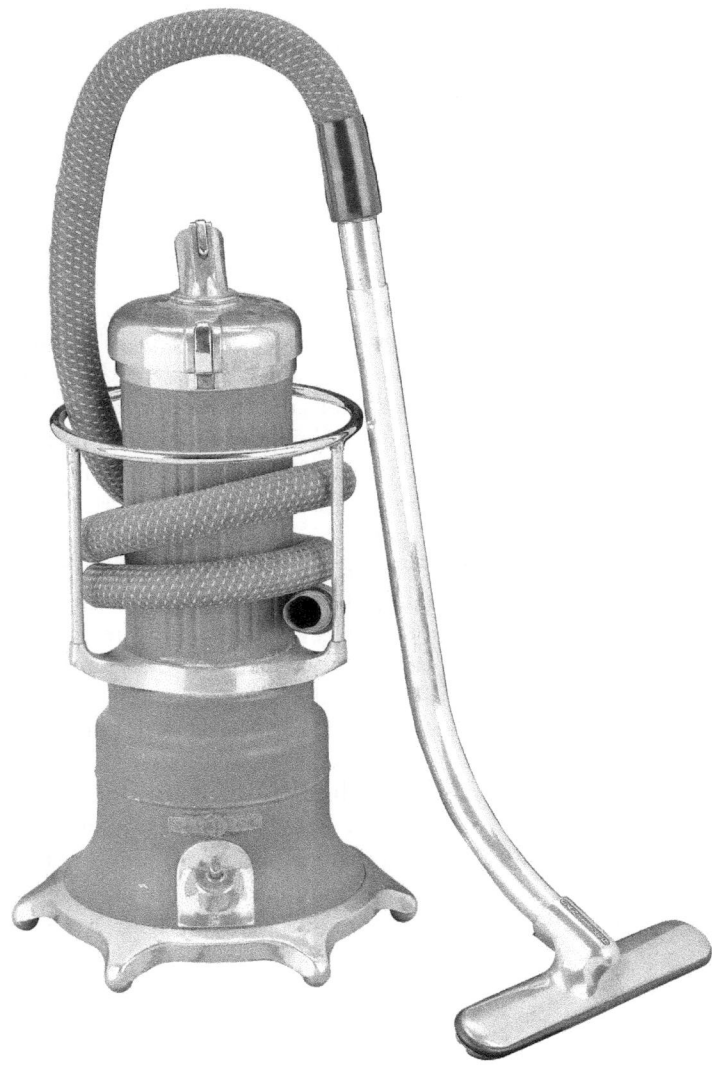

1938 Model 55A Air-Way Sanitizor. Tacony.

entire bag surface and allow the bag to be filled completely. When full, the bag could be closed with a bag seal Air-Way provided, so that no dust or germs could escape.

Air-Way's sales pitch was "Bringing Sanitation to the Nation."[88] The paper bag which Air-Way provided with their cleaners through the years was so effective, one salesman in Milwaukee, Don Clark, often gave this dramatic demonstration: after demonstrating the cleaner in a prospective client's home, he would remove the bag from the machine and lick the outside of it, proving just how well it filtered and acted as a barrier. Many times this clinched the sale.[89] Although Air-Way continued to make uprights until World War II, it would sell very few of its 1935 DirtMasteR and its 1938 Zephyr upright models, which, although better performers, were quite expensive.[90]

Electrolux struck back against the 1937 Apex tank with its 1939 square design, rather than its traditional round cross section tank cleaner. Famed industrial designer Raymond Loewy, who had just designed the streamlined S1 locomotive for Pennsylvania Railroad that would star in the 1939 World's Fair, designed Electrolux's highly streamlined cleaner. Loewy was the most flamboyant industrial designer, and already the most widely known.

In Europe, bad economic times were made worse by political turmoil, persecution, and Hitler's war preparations. These were tragic years in Germany, after Hitler rose to power in 1933. From 1937 to 1938, Hitler's progressively repressive National Socialist (Nazi) regime caused many liberal European intellectuals to immigrate to America, including many scientists, educators, and architects. These included architect/designer/educators such as Walter Gropius, Ludwig Mies van der Rohe (1886–1969), László Moholy-Nagy (1895–1946), Marcel Breuer (1902–1981), and Josef Albers (1888–1976), many of whom were faculty members or directors at the German Bauhaus, closed by Hitler in 1933. These émigrés were welcomed into universities and professions to continue the advancement of architecture and industrial design in the U.S., and they all became highly influential in these fields. In 1938, the Museum of Modern Art mounted a highly publicized exhibition, Bauhaus 1919–1928, which lionized the Bauhaus and committed the U.S. to a historical and qualitative standard of modern design based on the "best of Europe."

The New York World's Fair, optimistically entitled The World of Tomorrow, opened in April 1939 in Flushing Meadows, New York. Walter Dorwin Teague, by now the undisputed dean of industrial design, was not only one of the eight directors of the fair, but his firm also designed exhibits for the National Cash Register Company, Ford, Kodak, and U.S. Steel. All the leading industrial designers, including Henry Dreyfuss, Norman Bel Geddes, John Vassos, Raymond Loewy, and Russel Wright (1904–1976) designed major exhibits at the Fair.

New technologies exhibited at the fair included electronic television, aviation, super highways, new plastics such as nylon, the first fluorescent light bulbs, color photography, air conditioning, and mechanical robots, all pointing to a glorious future. It is likely that a number of vacuum cleaner manufacturers had exhibits. General Electric and Westinghouse both had major exhibits, promoting their appliances. The Production and Distribution Zone and the Community Interest Zone featured a large number of manufacturing companies showing the most modern conveniences and consumer products for the home. The fair, attended by 200,000 people, had been conceived during the Depression to create an event so fantastic that it would lift the country out of its doldrums, and it seemed to accomplish this, based on all the publicity and excitement, although unemployment was still at 17 percent.

By March 1939, Germany had already invaded Czechoslovakia, and by September, Poland. The fair closed in October under a cloud of approaching war in Europe. When the fair reopened in April 1940, the theme had changed from The World of Tomorrow to For Peace and Freedom. Soon, the Depression was essentially over, as orders for military lend-lease equipment for England, under attack by Germany in July, boosted the economy, and jobs became more readily available. But unemployment was still at 14.4 percent, and since 1929, tax rates had gradually more than doubled over the decade from 4 to 10 percent.

In November 1939 a canister cleaner known as the Filter Queen 200 was introduced by Health-Mor and made by the P.A. Geier Company.[91] Edward H. Yonkers, Jr., of Illinois had invented the concept (patent 2,198,568, filed on September 8, 1937, and granted April 23, 1940).[92]

Yonkers was a 1924 engineering graduate of Dartmouth. Helping his wife about the house, he became annoyed by the way tank-type cleaners jammed with dirt and thought

4. Consolidation 1920–1940

the trailing bag was unwieldy. He discovered that by injecting air into the middle of the cleaner and making it spin, the force of gravity would cause the dirt to drop to the bottom, and the air could be taken out through a paper filter at the top. Health-Mor bought the patent from Yonkers in 1940 because it wanted a bagless canister cleaner similar to Rexair, which had a patent on a bagless canister, but without a filter before the motor. Yonkers circumvented this with a disposable filter before the airflow entered the motor.

The Filter Queen was perfected by a team of P.A. Geier engineers, including Max Fairaizi, Ted Fistek, and Gene Martinec, and was powered by a ⅔ horsepower motor, which turned two rotating impellors over four inches in diameter at the rate of 13,000 to 15,000 rpm. This unique, patented technology used a cone-shaped filter that used centrifugal force to fling dirt toward the outside walls of a container; it was the same principle that Dyson would much later reintroduce and call "cyclonic" action.[93] At this time, canister cleaners were only a niche market, but soon they would become a staple of the industry.

1940 Yonkers bagless cleaner, patent 2,198,568. U.S. Pat. Off.

Within a year, the Filter Queen was challenged by a competitor called Silver King, the name itself encouraging an obvious comparison but with a touch of superiority (the name could only have been more obvious if it were Filter King). Silver King was founded by the Thompson family in Illinois in 1940 to make commercial cleaners, but in the 1950s it would enter the domestic market with its Lightweight All Metal Silver King Professional wet & dry canister, which is still being produced today by Silver King International in Colorado Springs at a rather steep price (over $3,200).

With the Depression fading, consumers were more inclined to spend. Eighty percent of U.S. households were wired for electricity, now including many rural areas, and their spending priorities were clear: 79 percent owned electric irons, 66 percent owned a car, 52 percent had electric washing machines and refrigerators, and 42 percent had vacuum cleaners.

In 1940, Hoover introduced its Model 60 with a lower, more angular, knife-edged motor hood designed by Henry Dreyfuss. Model 60 featured a headlight, an automatic nozzle adjustment, a handle counterbalance spring, and a fingertip handle control. It was the first with an exhaust muffler to reduce noise, the first to use nylon bristles, and the first with helical brushes on a single bar agitator. At $82, the Model 60 would be made until the war started.[94] Eureka debuted its Model D-171 in 1940, styled by industrial designer George Walker of Detroit; it won a prize for beauty.[95]

Vacuum cleaner sales were soon back up to one million units annually, the same as before the Depression.[96] Still, unemployment was at 9.6 percent, which in 2011 would be called a recession. But the problem now became a shortage of household goods. Manufacturers could not keep up with increased demand for consumer products when the Defense Department began to offer lucrative defense contracts for essential military supplies and equipment for lend-lease to England. Because consumer demand was so great, inventories of consumer goods in retail stores and warehouses quickly disappeared as production facilities were converted to fill government orders.

For consumers, the worst was yet to come, for after the December 1941 attack on Pearl Harbor by the Japanese and a declaration of war by president Franklin Roosevelt, most manufacturers of consumer goods converted completely to war production. All those prewar vacuum cleaners were going to have to last five more years.

CHAPTER 5

Postwar 1940–1970

After World War II was declared in December 1941, many vacuum cleaner companies stopped or slowed making consumer goods and converted their facilities to the manufacture of war materials. Many employees enlisted or were drafted into the armed services, their places filled in many instances by women on the home front. Eureka won five "E" (for efficiency) awards, conferred by the federal government for outstanding war effort, for its service in making oil burners, oil gear units for Swedish aircraft guns, aircraft landing gears, telescope sights for tanks, gas masks, and carburetors for B-29 bombers. Regina made bomb fuses. Rexair produced radar cases, rocket launchers, and ship partitions and linings. The P.A. Geier Company, maker of Royal vacuums, made smoke bombs, aircraft fittings, and tank transmissions. Electrolux made munitions and air cleaners for Swedish forces. GE made electric motors, generators, radar equipment, aircraft gun turrets, howitzers, radio transmitters, and turboprop aircraft engines, as well as other types of war equipment. Bissell built a variety of light industrial implements. Hoover manufactured incendiary bomb parts, shot bags, helmet liners, life belt inflators, parachutes for fragmentation bombs, turret motors for bombers, and air blower units, as well as 25 million highly secretive and well-guarded variable time proximity fuses, described by some as being "as difficult as compressing the components of an aircraft engine into the shape of an ice cream cone,"[1] and that some considered second only to the Manhattan Project in importance to Allied victory. Hoover won every possible production award from the government.

During the war, in 1943, Hoover stock was offered to the public for the first time, but family members were still in charge and did what families do. In a war-related incident of corporate generosity, H.W. Hoover, who was in England at the time of the invasion of Poland in 1939, offered safe haven to children, many of whom were employees at Hoover's English facilities. Seventy-eight children between the ages of 5 and 16 made the trip to North Canton and stayed in private homes for the duration of the war. After the war, the children returned, a few at a time, to their homes in England.[2]

Many industrial designers also served in the war effort. Henry Dreyfuss, Walter Dorwin Teague, and Raymond Loewy were called upon to design strategy rooms for the Joint Chiefs of Staff, featuring a world map on a curved surface 12 feet high and 25 feet wide. Dreyfuss designed and built four rotating globes, 13 feet in diameter, one each for Roosevelt, Churchill, Stalin, and the Joint Chiefs of Staff. Other industrial designers used their skills while in the military to design camouflage, war-related scale models, instrument colors, and training devices, or, like so many other Americans, engaged in military combat roles. John Vassos (1898–1985), head of industrial design at RCA, worked for the OSS (which would later become the CIA) and parachuted into Greece to organize the underground. After 1943, when the end of the war was in sight, many designers who stayed on the home front worked with manufacturers to develop new postwar products.[3]

After the war, the vacuum cleaner market returned with an explosion of pent-up consumer demand never before seen. Returning servicemen desperately needed and wanted homes, cars, furniture and appliances. Manufacturing of all these commodities had stopped during the war, including home construction.

There was such an acute shortage of housing after the war that low-cost housing developments began springing up all around the country, such as Levittown on Long Island, Bucks County in Pennsylvania, and Delanco Township in New Jersey. Houses in these planned communities offered ideal new owner packages that included a yard, shrubbery, landscaping, a garage, shutters, wall-to-wall carpeting, and built in appliances. Crews of masons, carpenters, plumbers, electricians, and painters followed each other down streets, completing 30 homes per day. From 1945 to 1950, annual housing starts grew from 200,000 to 1,154,000.

Many civilians had been financially distressed since the Depression started 15 years before, but the economy had been booming during the war and provided excellent income for millions of defense workers, who now wanted to spend it to improve their standard of living.[4] Manufacturers rushed to production with prewar or quickly designed new products. Electrolux got a jump on the entire industry during the war by offering potential customers a certificate as a $25 deposit toward the purchase of the first new vacuum cleaner after the war ended, giving the company a huge financial "war chest."

After the war, GE dropped its Electric Vacuum Cleaner Company name and its Frantz Premier brand name and began using General Electric (GE) as its brand name. GE's first postwar cleaner was the Model 807 upright with a two-speed motor.[5] Tank-type cleaners had became extremely popular and were imitated by many competing manufacturers, including GE, which entered the fray with its Model 811 tank cleaner.

Most of the new homes being built were furnished with wall-to-wall carpets that seemed to beg for cleaning but could not be taken out of the house and beaten on an outdoor clothesline. In fact, clotheslines were disappearing, thanks to automatic washing machines and dryers becoming available and popular. By 1945, 48.2 percent, or 13.7 million homes, already had a vacuum, and millions of new homeowners were demanding more. Vacuum cleaners finally became an essential household appliance, rather than an expensive luxury. The vacuum industry virtually tripled in scale from 1.3 million full-sized units in 1940 to 3.5 million units in 1950. Twenty-one vacuum cleaner manufacturers competed for market share. During this period, household tradition still regarded Saturday as housecleaning and washing day, when carpets were thoroughly vacuumed, furniture cleaned and waxed, silverware polished, lawns mowed, and cars washed. This was a cultural carryover from prewar traditions of frugality and hard work by those middle-aged people who had come of age in Depression times.

During the war, Regina, one of the oldest names in the industry, underwent a major transformation. Prior to 1941, most of its cleaners were sold door to door, in the tradition of the industry. After the war, Regina converted its distribution to trade and retail channels. Business increased dramatically in 1945, when Regina introduced its best-known product, the Electrikbroom. One-third the weight of a traditional vacuum, the Electrikbroom had been under development since 1934, but it had taken ten years to solve engineering and financial problems and to acquire licensing rights. Regina had acquired such rights from the Quadrex Company before the war, where engineer Ward Leathers had designed it.[6] The Electrikbroom, initiated a new category of vacuums, called stick vacs because of their slim profiles, light weight, handiness of operation, and compactness of storage (just like a broom). They were offered with a variety of colored and decorative exterior bags and would remain popular for years, but Regina would fold in 2000.

Not all vacuum manufacturers fared well during the war. Eureka had been in the red since 1937, averaging annual losses of $199,000, and in 1939, after Wardell had lost enthusiasm for running the company, Henry W. Burritt assumed its presidency and wanted to make major changes. Burritt discontinued door-to-door salesmen in 1940, but in 1941, the company lost $500,000 on sales of $5 million and had fallen to about 7 percent of the industry total. In June 1945 Eureka merged with Williams Oil-O-Matic, a company known for its heating business, changed its name to Eureka-Williams, and relocated from Detroit to Bloomington, Illinois.

By 1947 Eureka had built an organization of 5,500 vacuum cleaner dealers and distributors and had realized profits of $1 million, but still lagged consistently behind Hoover. In 1952, oil burner manufacturing would be switched to Sweden, and in 1953, the company would be purchased by the Henney Motor Company of Freeport, Illinois, and controlled by C. Russell Feldmann. The company would focus on Wooden Lungs, for treatment of polio, and air-conditioning equipment until 1957. It would then diversify into school furniture (1960), would develop the Henney Kilowatt, a battery-operated car (1961), and later, would develop thermal batteries (1968).

However, in 1960, Eu-

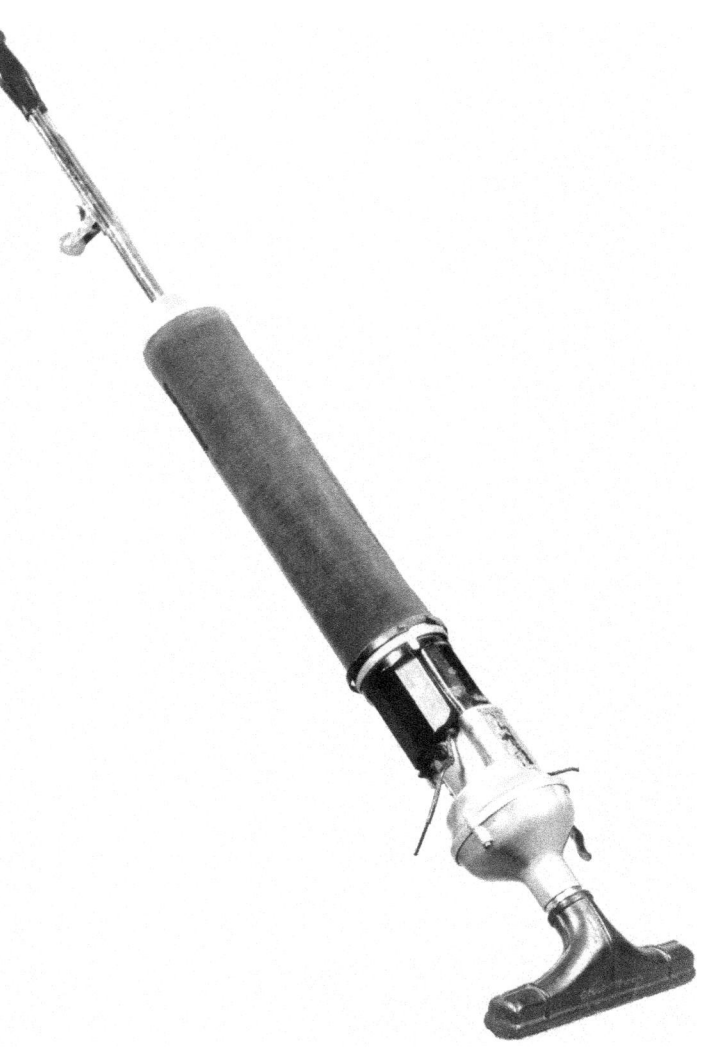

1945 Regina Electrikbroom. IHA/Lifshey.

reka-Williams would merge with National Union Electric Corporation, to which all its heating operations were transferred. This would give additional manufacturing capacity to Eureka, allowing it to again concentrate on household floor care equipment. By 1971, Eureka-Williams would constitute 40 to 50 percent of National Union's sales and profits, and National Union would report that vacuum cleaner volume had climbed for the twelfth consecutive year. Feldmann would take Eureka private and would make it a division of National Union. Eureka would then develop a complete 65-product line of floor care products, including canisters, uprights, hand cleaners, and commercial models and polisher-scrubbers.[7]

Industrial designers were again in high demand after the war. In 1940, Harold Van Doren had written a comprehensive, 388-page textbook, *Industrial Design, a Practical Guide*, which described "how to become an industrial designer" in a step-by-step process. Both industrial design organizations, the American Designers Institute (ADI, founded 1938), and the Society of Industrial Designers (SID, founded 1944), developed prototype curriculum for higher education. Under the postwar GI Bill, hundreds of returning servicemen and women, as well as young high school graduates, took up the profession at dozens of schools, which now offered degrees in the new career. Many manufacturers, as they pumped out new products to meet the unprecedented consumer demand, invested in industrial design to improve sales through design appeal, using new modern shapes unlike any of those before the war.

In 1946, Singer introduced the most modern-looking upright vacuum cleaner to date, the Magic Carpet S-2, followed by the S-3 ($79.75). Designed by famed industrial designer Raymond Loewy, who by then had a staff of 200, the Magic Carpet was a total redesign of Singer's 1936 R-1 upright, with a low motor housing no higher than a cigarette pack, accomplished by turning the motor on its side. The Magic Carpet could clean under furniture with ease. Instead of the usual broom handle, it had two tapering, thin, chromed metal struts extending downward from a hand grip, widening to embrace a plastic cord reel housing above the motor hood. The cleaner could be hung on a closet wall, saving storage space. Instead of the brush roll being mounted directly to the nozzle housing, it was suspended by two side arms attached near the motor, allowing it to float freely and adjusting automatically to most carpet heights. The front wheels could also be adjusted to standard or thick

1946 Singer Magic Carpet.

piles. As absolutely modern as it was, the overall design, particularly the split handle, suggested an uncanny resemblance to the 1869 Whirlwind (see group photographs of vacuums at the end of chapter 2).

Henry Dreyfuss designed all Hoover postwar products. These included its 1946 upright Model 61 ($89), and its 1950 Model 62, both of which were modifications of its 1940 Model 60 and featured an added foot pedal to release the handle from an upright position and a foot pedal for carpet height adjustment. An economy upright Model 28 with a teardrop, streamlined motor hood and headlight was also introduced. In 1947, Hoover introduced its first tank cleaner, to compete with the popular Electrolux Model XXX (Model 30), the 1937 model that was still going strong after the war. Hoover's tank cleaner was the Model 50 Aero-Dyne (meaning "air power," $79.50), which was lighter and less expensive than the Model XXX and featured an "ejector" that emptied the bag while the cleaner sat open on a sheet of newspaper. In 1948, Hoover debuted a small apartment-size Junior upright Model 115, designed with a ten-inch nozzle for England.[8]

Several new vacuum manufacturers entered the postwar scene. In 1946, the Interstate Engineering Corporation (I.E.C.) introduced its Compact Model C1 canister cleaner. I.E.C. was founded in El Segundo, California, in 1937 to design and manufacture aircraft for the U.S. government and had made the cleaners to vacuum Howard Hughes Corporation aircraft during the war. The Compact, an aerodynamic cross between a canister and tank cleaner, had two rear wheels and two front skids, and a hose with several attachments so that custodians could easily clean under aircraft seats and in overhead bin compartments. It worked so well that I.E.C. decided to sell it to the public, door-to-door. In 1949, an I.E.C. look-alike called the Revelation ($119.95) would debut, to be sold only in retail stores until 1954. The exterior design would remain essentially unchanged to the present day.[9]

Another new entry to the market was by Alex Lewyt (1909–1988), who since 1927 had been president of the Brooklyn-based Lewyt Corporation, founded in 1882, and was a World

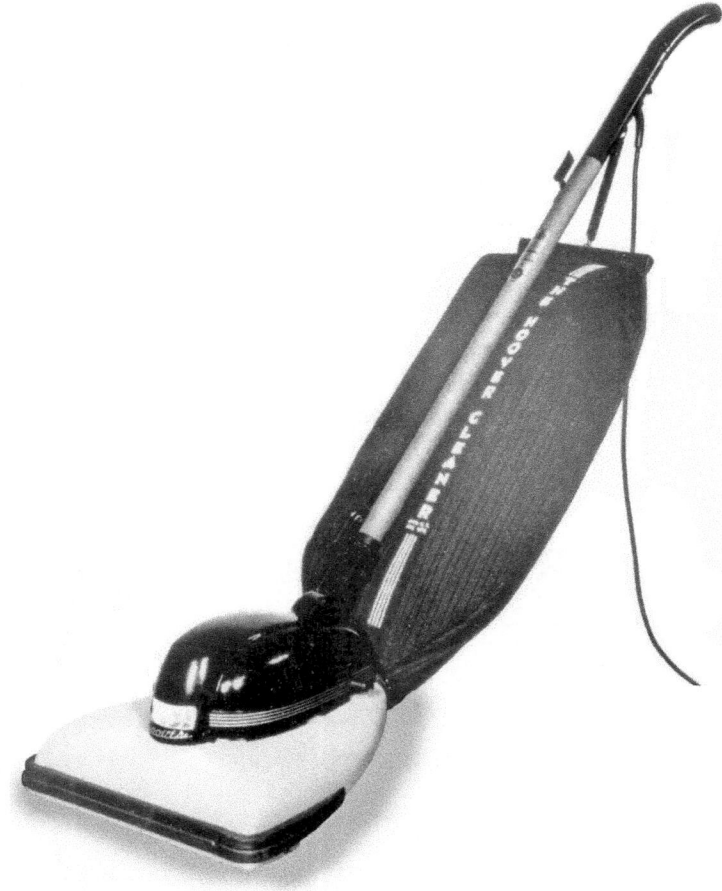

1946 Hoover Model 28. HHC.

War II manufacturer of industrial components. Lewyt had never made consumer products, but during the war produced a cleaning device for the Navy. One day in 1945 Alex overheard one of the women working on the assembly line say, "Wouldn't it be nice to have a cleaner like this around the house?" Alex had been thinking about postwar possibilities, and this convinced him. He entered the vacuum cleaner field in 1947 with a cylindrical canister with gray hammer-tone finish and no wheels (a wheeled base would be added later by consumer demand). Model 44 weighed 16 pounds and had no dust bag. Lewyt had replaced the messy dust bag with a dust separator, paper filter, and cloth filter ("No bag to empty!"), the system that had been invented by Ed Yonkers in 1937 and introduced by Health-Mor in 1939 as the Filter Queen 200. This would later get Lewyt into serious trouble.

By 1949, *Kiplinger* magazine stated: "The startling newcomer is the Lewyt Corporation.... Lewyt rhymes with 'Do it.'" In a paper presented to an American Management Association meeting in 1950, Lewyt explained why he chose a vacuum cleaner.

> We selected it first of all because it represents a big business. This year something like three million vacuum cleaners will be sold. In terms of dollar volume the vacuum cleaner is the third most important electrical appliance. We looked at the market and found that there is only a saturation of about 50 percent in the vacuum cleaner field.... [W]e felt that the lack of complete merchandising ... was the industry's Achilles heel and our one possibility of cracking the market.[10]

Lewyt mounted an aggressive $1.5 million advertising campaign using radio, television, newspapers, magazines and billboards before a single Lewyt cleaner was in the field, but in the first year of operation he sold 100,000 cleaners bearing his name at $89.95 each and amassed a $3 million backlog of retail orders. To help his 6,000 franchised dealers sell more Lewyts, he introduced demonstration displays within stores called The Market Place, with carpets and artificial dirt, and a 10-second demonstration of Lewyt cleaners called the Junior Demo.[11]

Increased competition after the war caused problems for some of the older manufacturers. In 1947, P.A. Geier was struggling, reluctant to invest in research and development. Three of its key engineers, Gene Martinec, Max Fairaizi, and Ted Fistek, who had engineered Health-Mor's 1939 Filter Queen 200, and their secretary, Annette Clark, left the company to form an independent firm called Team (later Jem) Development. By 1948, Jem developed three prototype Model 350 Filter Queen vacuums, and P.A. Geier manufactured and sold

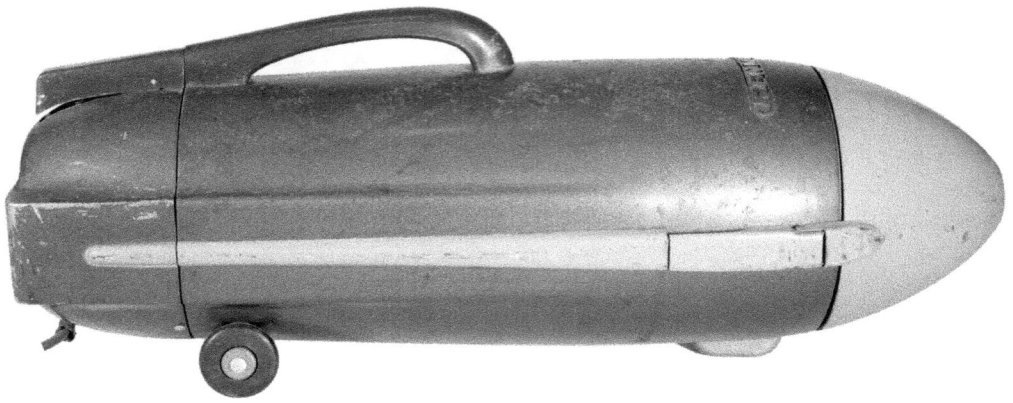

1948 Kenmore Commander or Bullet Model 116.722.1. MIAD.

them for $119.45 each. P.A. Geier encountered serious financial problems in the early 1950s and broke its long-standing contract with Health-Mor. Health-Mor undertook its own manufacturing operation, maintaining its corporate headquarters in Chicago, and launched its first self-produced vacuum cleaner in 1952. In a few years, Health-Mor introduced the first painted and chrome models with automatic cord reels. Health-Mor had a direct sales scheme and high prices that were about 20 times the average retail vacuum cleaner, which consigned it to a niche, but lucrative, market.[12]

Kenmore had been the brand name under which Sears Roebuck sold appliances made by a variety of independent manufacturers since 1927. The first vacuum cleaners sold under the Kenmore name appeared in 1932. In 1948, Kenmore introduced an unusual tank-type vacuum cleaner made by the Birtman Electric Company called the Commander Model 116.722.1. The Commander was called the Bullet cleaner because of its uncanny resemblance to a 1000-pound naval artillery shell, a shape no doubt inspired by World War II, and which made the Electrolux look like it was standing still. But the Commander wasn't much different than the Montgomery Ward Supreme tank cleaner of 1945, designed by George W. Walker, which looked, for all the world, like a torpedo.[13] These designs emulated the military and aircraft imagery that was common after the war, typified by GM's 1948 Cadillac tailfins inspired by the twin tails of the P-38 fighter plane, or the 1950 Studebaker with its aircraft nose spinner, a design claimed by Robert E. Bourke (1916–1996) of Raymond Loewy Associates. Birtman's Commander would be sold until the mid–1950s in a sequence of colors.

In 1942, Air-Way's patents on disposable bags, which enabled Air-Way to become one of the few vacuum manufacturers to make it through the Depression in the black, had finally expired, and most major manufacturers adopted this highly popular feature after the war. Air-Way came out with its classic Sanitizor Model 66 ($94.50) in 1949,[14] designed by consultant industrial designer Gordon Florian (1909–1984). It was a blue tank cleaner style, but rather than a cylinder shape, it had a square cross section body. Attached to each of the four edges of the square body was a chromed wire skid, parallel to and extending the full length of the body. In a vertical position, looking a bit like a rocket on a launching pad, the four skids extended away from the body by about three inches, thus giving the unit significant vertical stability. In this mode of operation, a hemispherical, swiveling hose connector on the top, looking like a nose cone and surrounded by a bright red ring, enabled the hose to be turned in any direction without tipping over the body. The four skids also enabled the unit to be laid on any of its four sides and pulled across the floor horizontally, like a normal tank cleaner.[15] Air-Way would follow its Model 66 with a green Model 77 in 1954 with a metal Atacha-Carrier for tools, and a Model 88 in 1958 in two-tone turquoise.[16]

Another postwar competitor was the Fairfax canister made in Fairfax, Virginia. When it was first introduced is not known, but by 1948 it was an imposing, sturdy, chrome-plated cylindrical canister topped with a chrome hemisphere with red trim. It traveled on casters and resembled the R2-D2 robot character from *Star Wars* films. The Fairfax survives to the present day with only minor design changes, made by Eco Products. Fairly costly (today's price: nearly $2,400), the Fairfax competed in 1948 with Kirby, Silver King, Health-Mor's Filter Queen, Rexair, and other niche manufacturers, mostly door-to-door.

For many manufacturers, door-to-door salesmen were becoming a prewar anachronism. By 1950, Hoover decided to sell "special" products through retail stores, rather than its own salesmen. The first item was a 1947 Hoover Model 010 dry iron with a large "pancake" temperature dial, followed by a 1952 Model 011 steam/dry iron, both designed by Dreyfuss. Next, Hoover began selling refurbished uprights in retail stores, as well as a 1953 Hoover Holiday tank cleaner made in England, which was, of course, inspired by the enormous postwar success of Electrolux. Hoover's own door-to-door salesmen revolted; they were not pleased with this turn of events that took away their commissions.[17] They were also having trouble getting into the new suburban homes for demonstrations, where increasingly, such salesmen were considered a nuisance. The fact was that with the advent of shopping centers, greater mobility by automobile, and an explosion in advertising via television, radio, magazines and newspapers, consumers were attracted directly to retailers for comparative shopping.

Comparative shopping was also encouraged by nonprofit organizations such as Consumers Union, founded in 1936. In 1951, Consumers Union reviewed vacuum cleaners in its July *Consumer Reports* magazine, and after testing 36 models, they found three uprights and two tanks to be outstanding in rug cleaning efficiency. The uprights were the Apex 5400 ($79.95, "the best"), the Eureka 250 ($84.95), and the Wards catalog no. 85B439M ($51.95). The tanks were the Hamilton Beach 26 ($74.95, "the best," although compared to the performance of "the best" upright, it was only fair) and the Electrolux XXX ($77.50, "Second-best"). There were only three canisters among the 21 tanks tested, illustrating the high popularity of tank cleaners compared to canisters. Detailed descriptions of all 36 cleaners reviewed can be found on the Web.[18]

To establish more dramatic product identity at the point of sale, many manufacturers updated their trademarks to a more modern, postwar style. The now-familiar Hoover circular trademark made its first prominent product appearance on the Model 42 tank cleaner in 1952 (in blue, not red), but had been used in advertisements and on the Dustette bag since 1950. It would become red on Hoover's Model 55 in 1953.[19] Designed by Dreyfuss, the horizontal line of the Hoover "H" spanned the diameter of the circle, not exactly horizontally, but at an upward angle of precisely 18 degrees. Under the "H" line and parallel to it, were the capital letters "OOVER." The trademark would be modified slightly in the 1960s, when many corporations updated their trademarks to a more simplified, modern style.

Electrolux had only two postwar rivals, Rexair, with its water filtration system, and Air-Way, with its disposable bags. But Electrolux's prewar appearance design was beginning to fall behind competitors. In 1948, Electrolux had engaged Swedish industrial designer Sixten Sason (1912–1967), aircraft and automotive designer for Saab, to design a whole new series of vacuums in the streamlined style pioneered in the U.S. Most notable of these was the 1957 Z 70, made and sold in Sweden.[20]

Selling replacement bags had

1948 Fairfax Fax-o-Matic canister. MIAD.

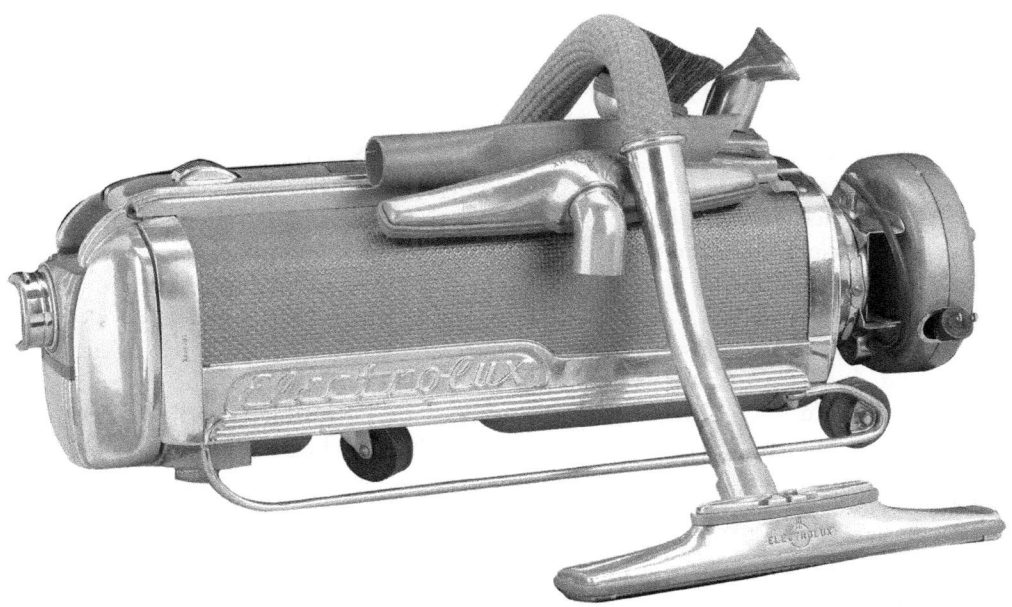

1955 Electrolux Model LX1 (61). Tacony.

become a profitable and infinite income stream for all manufacturers that used them. Electrolux not only adopted them, but figured out a way to sell even more bags, as well as maintain suction efficiency. What resulted after six years of development was "the vacuum cleaner with a brain," the "automatic" Model LX (60) Electrolux, introduced in 1952 for $109.97. When the pores of the disposable bag were full of dirt, the LX automatically stopped the motor, opened the front cover, and ejected the filled bag from the cleaner. Presto! Insert new bag!

The effective Electrolux sales pitch was "If it's running, it's cleaning — when it stops cleaning it will stop running. Electrolux LX guarantees you a cleaner home *automatically*." The LX was a high-quality product, weighing 24 pounds (28 pounds with the industry's first automatic cord winder attachment, an optional accessory). The hose was eight and a half feet long, and a new attachment, the Tuftor, was offered for cleaning long-pile cotton and shag rugs, which were becoming popular. Later, Regina and other manufacturers produced "shag rakes," accessories that fluffed up shag rugs when they became trampled with footprints.

There was, however, an immediate problem with the Electrolux bag ejection system. Although 18 replaceable bags were provided with the cleaner, they were quickly used up because the single-ply bags immediately clogged with dust. There was no way to delay the automatic ejection of bags, and customers returned their cleaners, objecting to the idea of having to buy so many new bags. A multi-ply bag that clogged less was soon offered as an alternative, but the real solution was the development of a new control dial that users could advance or delay the ejection of bags, introduced on later LX models. In 1955, Electrolux would debut its most advanced version of the Automatic Electrolux, the LXI (Roman numeral 61). The LXI featured runners and three wheels, one of which swiveled, making it easy to move over rugs and floors. It also had a new swing away cord rewinder and new colors.[21] The bag ejection system would prove to be somewhat problematic, but, to this day, when an Electrolux bag is full, on its top of the line models, the machine will shut itself off.[22]

With the expiration of the Air-Way disposable bag patent, all vacuum cleaner manufacturers adopted disposable, replaceable, paper bags. The bag's original intent, of course, was to eliminate the messiness of emptying cloth bags onto newspapers for disposal of dirt, a process that was not only distasteful to predominantly female users, but which scattered dust and dirt back into the air. Upright cleaners using disposable bags enclosed them in soft decorative cloth or plastic bags for appearance purposes, accessible by zipper or snaps. Once women used disposable bags, they would never again return to the old-style permanent cloth type.

But beyond the obvious benefit and popularity with users, disposable bags became a profit windfall for manufacturers. They had discovered the beauty of the Gillette "razor blade principle" of eternal replacement income. King Camp Gillette (1855–1932) had founded the American Safety Razor Company in 1902 and began selling safety razors in 1903 for $5 (about $114 in 2011 dollars). This was half a workingman's weekly pay, yet they sold by the millions. His invention was a thin, hardened steel, replaceable razor blade. In 1904, Gillette sold 90,884 razors and 123,648 blades. By 1915, razor sales reached 450,000 units but blade sales exceeded 70 million.

Thus was established the so-called razor and blades business model, the fruits of which vacuum cleaner manufacturers were now enjoying with the sale of replaceable bags. To maximize such sales, manufacturers advised customers to "buy only genuine (insert manufacturer's name) bags for best performance," even though there sprang up dozens of independent competitive bag makers, derisively labeled Bojack by the Hoover Company (Hoover's term for any competitor). Hoover had introduced disposable bags on its Model 63 upright ($116.95) in 1953. In fact, many of the competitive bags were in fact inferior in quality and performance. But disposable bags were here for the rest of the century and beyond, to the benefit of users, and of course, for eternal extra profit for vacuum cleaner companies. They were inexpensive to make and provided a very high profit margin.

Some of the most effective disposable bags in the future would be made of HEPA filters. HEPA is an acronym for High Efficiency Particulate Air. The original HEPA filter was designed in the 1940s and used in the Manhattan Project to prevent the spread of airborne radioactive contaminants. They were introduced to the commercial market in the 1950s under a registered trademark. It is the best known filter for removing at least 99.97 percent of particulates from the air, such as dust, animal dander, smoke, mold and other allergens that are 0.3 microns or larger, thus improving air quality. HEPA is also a general term for all highly efficient filters, which are available in various gradations of 85 to 99.9999 percent efficiency. However, it would be some years until they would be generally used in household vacuum cleaners.

In the carpet sweeper world, Melville Bissell III, a nephew of Melville, Jr., took over leadership of the company in 1953. He immediately engaged Harley Earl Associates (HEA), industrial designers, to design a new sleek Bissell Sweepmaster carpet sweeper. Designed by James G. Balmer, Jr., Frederick W. Hertzler, and Carl B. Denny of HEA, it was introduced in 1954 and won a national design award in 1955.

Unlike his uncle in the Depression, Melville III did not feel the Bissell line should be restricted to carpet sweepers, but should expand into diversified floor care. Bissell would continue to avoid vacuum cleaners, but initiated development of a wet-cleaning carpet machine. The new product, called the Bissell Shampoomaster, was a nonelectric device that

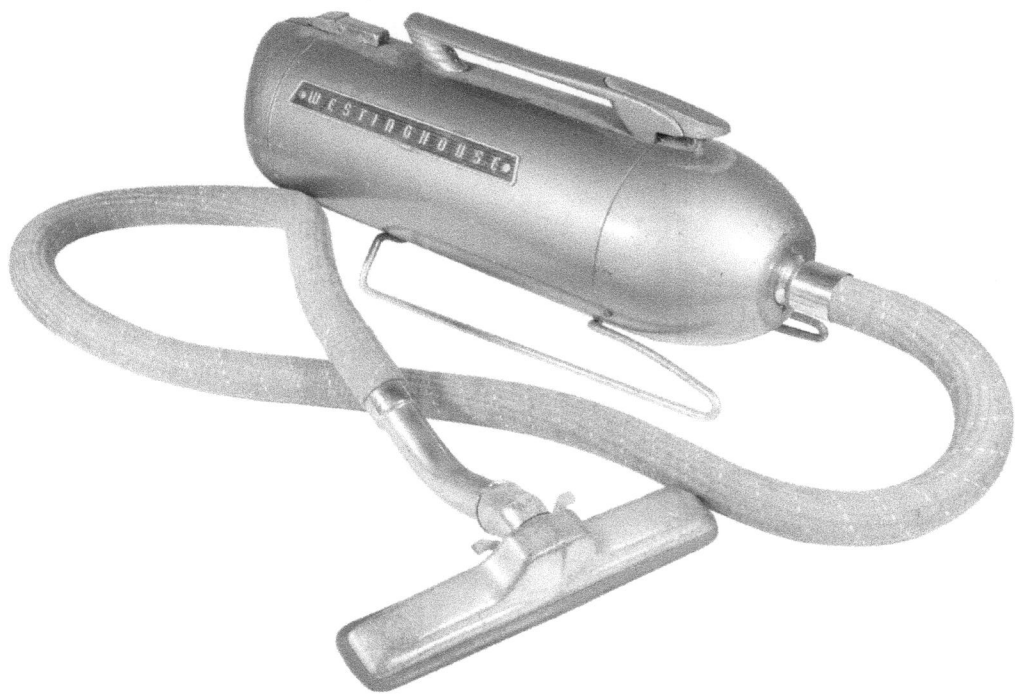

1953 Westinghouse Rocket tank cleaner. MIAD.

used only water and a detergent. During its manufacture between 1957 and 1967, Bissell promoted it even more than its traditional carpet sweepers.[23]

In 1954, the Consumer Union (CU) again tested vacuum cleaners for rug-cleaning efficiency and reported results in its magazine, *Consumer Reports*. Forty-three vacuum cleaners were tested. Among uprights, Ward's Supreme 85A493M was rated as the "best buy" at $64.95. In cylinders and canisters, "best buys" included the Eureka Roto-Matic 800A ($69.95) and the Wizard 830 ($69.95, made by Western Auto Supply Company in Kansas City, Missouri). Fifteen canisters and sixteen cylinders (tanks) were reviewed, compared to only three canisters in CU's 1951 review, reflecting the dramatic upsurge of canisters in the market. The tanks included a 1953 tank cleaner by Westinghouse Electric Appliance Division in Mansfield, Ohio, designed by Raoul Lambert, head of Westinghouse industrial design staff. It was advertised "as modern as a rocket," which it did in fact resemble ($89.95 with 12 attachments). Lightweight cleaners reviewed included the Regina Electrikbroom L ($49.95) and the Westinghouse Porta-Vac T-12 ($49.95). CU tested these in a separate category and considered them only "acceptable for limited service." Descriptions of all 43 cleaners reviewed can be found on the Web.[24]

Canister cleaners were rapidly increasing in popularity because of their adaptability to both floor and above-floor cleaning. In 1955, Hoover introduced its first canister cleaner, Model 82. It had a unique spherical shape with a horizontal vinyl furniture guard around its middle, suggestive of a ring at the equator of a planet, such as Saturn, thus inspiring its name, the Constellation. Claiming to be the quietest on the market, its appearance was designed by Henry Dreyfuss (design patent 175,210),[25] and it came with an extra-long and stretchable Ultra-flex hose that could reach up or down a flight of stairs. It had a swivel

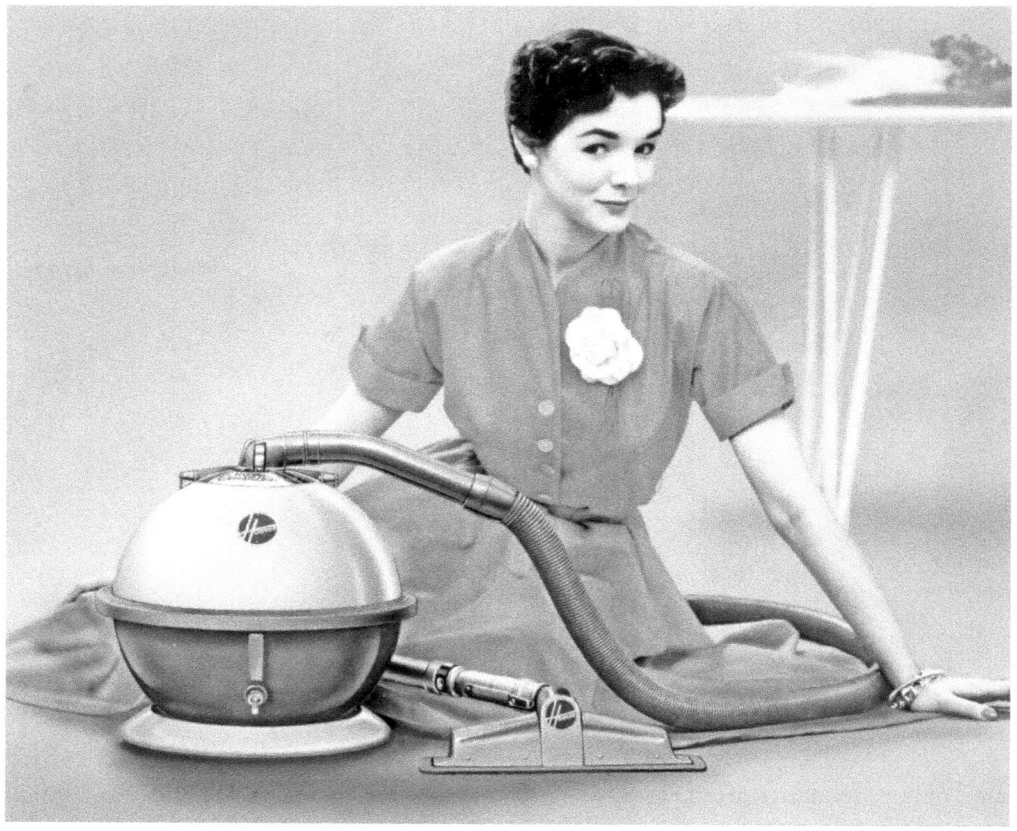

1955 Hoover Model 82 Constellation. HHC.

attachment to the cleaner that enabled the user to walk around the cleaner and reach an entire room with the hose. It had no wheels, but a five-caster dolly that latched to the cleaner was available as an accessory for easy transport.[26]

The following year's Constellation, Model 84, included a feature called Air-ride, which floated the cleaner on its own exhaust, functioning like a hovercraft on carpet or hard floors, an innovation that was patented by Hoover engineers G.P. Daiger, Dale C. Gerber, and Werner G. Seck (U.S. Patent 2,781,103).[27] Constellations would be changed and updated over the years until discontinued in 1975.

Inevitably, increased postwar competition resulted in more corporate problems of older manufacturers. The P.A. Geier Company had been making Royal vacuum cleaners since 1905, but in 1953, it suffered bankruptcy, a hostile takeover by the Walter E. Schott Organization, and was renamed the Royal Appliance Manufacturing Company. In 1954, the company would be bought by its own employees, headed by Stan Erbor (1901–1981), and headquartered in Cleveland, Ohio.[28] Its cleaner on the market that year was the Royal 801 ($79.50). In other corporate moves, GE sold its industrial vacuum cleaner business to the MultiClean Company in March 1953, but continued to make consumer vacs.[29]

Rexair had no such problems. It had been highly successful with its 1936 Series A cleaner with water filtration of dirt. Subsequent models B and C ($121.50) were functionally and in appearance the same, but in 1955, Rexair launched an improved version that became its flagship product, its Model D, the first to use the term Rainbow, the name by which we know it today and which described its unique water filtration system, a system that became exclusive to Rexair. Leslie Green had worked with T. Russ Hill and his key salesman, J.V. Sanders, to perfect the design. It was twice as powerful as its predecessors and was restyled with modern postwar design, in gold with brown accents. It was sold in that color until 1961, when the color would be changed to chrome and copper, and it would be called the Chrome Dome.[30]

Alex Lewyt had gotten in trouble when he copied the "cyclone" bagless system from Health-Mor's Filter Queen 200 in 1947. Health-Mor had sued Lewyt for patent infringement, and Health-Mor not only won, but collected enough money from Lewyt to open its own factory in 1955 to make the Golden Monarch Model 500 Filter Queen.[31] Alex Lewyt would acquire lasting fame that year for what some call the "worst prediction of all time" in the *New York Times*: "Nuclear-powered vacuum cleaners will probably be a reality in ten years." It didn't happen.[32] Lewyt's last design, initiated by Model 77 in 1955, was called the Big Wheel canister because it had two large wheels (about 6 inch diameter) with a cube-like body between them in a hammer-tone gray finish. It featured a Power Dial to vary suction and a power nozzle. In 1956, a green hammer-tone finish would replace the gray on Model 88.

Vacuum cleaner sales in 1955 increased by an impressive 25.28 percent over 1954, with sales of 3.3 million units. Canisters were gaining, accounting for 60 percent of sales, while uprights and tank cleaners were both in the 20 to 40 percent bracket. New features were a key factor, with emphasis on mobility and color styling, as units could be found in every color of the rainbow. Sixty-four percent of vacuum cleaner sales were replacements, indicating that market saturation was increasing. However, it was the best year ever for floor polishers, with unit volume advancing by 15.4 percent. This was attributed to the building boom, more fashionable bare floors,[33] and vinyl flooring, which had replaced linoleum for kitchens and was now used for most high-traffic areas of new homes. Vacuum cleaner manufacturers were quick to respond to the fast growing opportunity. Some, like GE, had marketed polishers after the war, certainly by 1951. Hoover offered polisher attachments for its Model 28 in the late 1940s, but made its first two-brush polisher in 1950 (Model 21). GE offered its two-brush FP-1 in 1957. The GE series would expand with improved models through FP-10 in the early 1960s, and to the FP-13 and FP-14 in the latter 1960s.[34]

In the ten years since the war, the appearance design of many consumer products, including vacuum cleaners, had become nearly as important as their functional attributes (engineers probably would disagree). Few manufacturers of consumer products would introduce anything without the collaborative participation of an industrial designer, either on staff or as a consultant. Sales and competitive designs demanded appearance design excellence. This had been due largely to the rise of the industrial design profession, public desire for the new and the modern, and since 1941, the national promotion of Good Design, as defined by the Museum of Modern Art and its highly promoted Good Design shows across the country from 1950 to 1955. Consumers were made aware of good design, and demanded it. Industrial designers knew this and capitalized on it.

In 1944, the Society of Industrial Designers (SID) had been founded by the 15 most prominent industrial designers, which included the "big three," Walter Dorwin Teague, Raymond Loewy, and Henry Dreyfuss. All three wrote popular books on design, lectured nationally and internationally, and were responsible for many of the most successful postwar designs. In 1947, Dreyfuss followed Loewy as president of SID, which by then had regional chapters in New York, Chicago, Cleveland and Los Angeles. On October 31, 1949, Loewy graced the cover of *Time* magazine, and in 1951, Dreyfuss was on the cover of *Forbes* magazine. Industrial designers had become famous international personalities.[35]

In 1955, Dreyfuss published *Designing for People*, a book that promoted not only industrial design in general, but also his new concept of "human factors" (later called "human engineering," and later, ergonomics), by proposing:

> It must be borne in mind that the object being worked on is going to be ridden in, sat upon, looked at, talked into, activated, operated, or in some way used by people individually or en masse. If the point of contact between the product and people becomes a point of friction, then the designer has failed. If on the other hand, people are made safer, more comfortable, more desirous of purchase, more efficient — or just plain happier — by contact with the product, then the designer has succeeded.[36]

Dreyfuss illustrated his book with blueprint style diagrams and charts of males and females in standing, seated, and reaching modes, with each diagram annotated with average anatomical dimensions derived from comprehensive and precise measurements of thousands of military personnel, accumulated during the war when Dreyfuss designed military equipment. The dimensions included maximum (95th percentile) and minimum (2½ percentile) dimensions of each "average" dimension. He described how he used these charts to insure that designs fit the size, weight, and physical capabilities of users. His charts and measurements, and later iterations of the concept, became the bible for industrial designers, and expanded and updated versions are used to the present day.[37]

Henry Dreyfuss Associates (HDA) in New York, as we know, had designed most of the new Hoover cleaners since 1935. Considered as one of the Hoover family, Henry was welcomed as such with luxury accommodations, fine dining, and corporate deference during his professional consulting visits to North Canton. However, the Hoover family had been going through trying times since 1948, when H.W. retired from the presidency due to ill health and was succeeded by his brother, Frank G. Hoover, also in poor health. In 1951, John Frank "Jeff" Hattersley became president. On January 1, 1954, after a two-year family battle during which several family members left the company, Herbert W. Hoover, Jr. (1918–1997), H.W.'s eldest son, who had become vice president of sales in 1952, was elected president. In September 1954, H.W., Sr., passed away, and that same year in December, Frank G. also died. That year, Hoover profits were a mere three percent of sales, and its market share was down to nine percent, a pitiful performance.

Henry Dreyfuss (1904–1972). Photograph ca. 1955.

Those difficult financial times may have inspired Hoover to question the exorbitant retainer fees it was paying Dreyfuss. Perhaps more significantly, many in the Engineering Division, who worked directly with Dreyfuss during his visits, were complaining that he had become arrogant and inflexible to engineering concerns regarding cost and manufacturing capabilities, refusing to modify any of his proposals even slightly. There were stories of his angry tantrums when a design detail was challenged by engineers. To many, his fame seemed to have gone to his head. Imagine how engineers felt after they had designed a new cleaner: Dreyfuss came in and proposed the same cleaner with a different external shape, and then claimed to the world that he had designed it. Due to this, and with his senior Hoover family member supporters now gone, Henry found himself getting a pink slip in 1955, despite his international fame and reputation. Hoover, more specifically its engineering division, decided it could meet its own industrial design needs with an internal staff.

Herbert W. Hoover, Jr. (1918–1997). Photograph ca. 1955. HHC.

Hoover factory and offices, North Canton, Ohio, 1949.

By this time, a new generation of industrial designers, formally educated in postwar academia, had expanded the field and provided design services that were much more affordable than the Dreyfuss organization, or any other famous pioneer of the profession. But at Hoover, in the manner of the family tradition, the industrial design responsibilities in 1955 fell to 50-year-old E. Russell Swann (1905–1958), a Hoover employee since 1925, who in 1942, when most younger men were in the armed forces, had been sent to work in Dreyfuss's office in New York for four years to be trained as an industrial design apprentice. When he returned to the Hoover Company's engineering division in 1946, he established and staffed a small model shop, building appearance models of Dreyfuss designs for review when Henry visited. Swamped by new postwar products by 1956, Swann hired a 25-year-old industrial design graduate of Carnegie Tech to serve as his design assistant at a salary of $450 per month. This is how I became a Hoover employee for the next 16 years, and after Swann's death, headed the industrial design section in the engineering division for the last 14 of them.

The engineering division also hired a new consultant. Since the war, color had become an essential tool in the merchandising of all consumer products, including many vacuum cleaners. In some cases, color provided consumers with alternative decorative choices, such as at General Electric, where major kitchen appliances were available in pink, turquoise, and yellow, as well as white. In other instances, color became the primary way to distinguish multiple models, which differed only in minor or invisible features from each other. This was a major competitive advantage for vacuum cleaner manufacturers with a growing line of products. So it was logical for Hoover in 1956 to engage as consultant Faber Birren (1900–1988), a well-known New York color consultant who directed *House & Garden* magazine's color program. The program annually identified the most popular colors for home interiors, fashion, and consumer products. Birren soon created a virtual Easter egg basket of Hoover products, using the latest fashionable *House & Garden* colors.

At that time, the Hoover engineering division was comprised of 123 employees, consisting of 38 graduate engineers and scientists; 49 designers, draftsmen, staff coordinators and laboratory assistants (including two industrial designers); 10 clerical and library assistants; and 26 shop mechanics.[38]

Engineers dominated the company, as they had from the beginning, which was typical for most manufacturing companies. All employees, including hourly workers and salaried professionals, had a personal time card in a slotted metal panel, with which they punched in and out on a time clock each morning, lunchtime, and evening. The engineering division, headed by G. Pierce Daiger, vice president of engineering with Dale Gerber as chief engineer, was located in four floors of the north wing of the brick manufacturing plant and offices built in 1924 with over 200,000 square feet, 16-foot ceilings, and facing the town square in North Canton. Daiger, a quiet but determined defender of engineering authority over industrial design, had probably been the primary source of Dreyfuss's departure. He regarded engineering as the keystone of Hoover's historical competitive success and regarded industrial designers as artists applying cosmetics on engineering designs. Industrial design was not recognized as a new technical discipline that could make Hoover products highly competitive. The comparative numbers of staff above tell the story of priorities.

Division executive offices, engineering design, a library, package design, blueprint machine, and a drafting room were on the fourth floor. Except for the executive offices, all work spaces were of the open "bull-pen" 1924 style, where each employee was visible to all others, including a few sea-green, removable metal wall cubicles with glass above the waist level for department heads. Large, six-foot wide, green-painted, oak drafting tables with drafting machines were provided for designers and draftsmen; and gray steel office desks and file cabinets for engineers. In the next wing on the fourth floor was what vacuum collectors later called the "vault," a huge storage facility with metal shelving, filled with antique models, competitive models, development models, prototype models, and production models, a sort of three-dimensional engineering archive (see group photographs of vacuums at the end of chapter 2). Lower floors included test labs, research, a conference room, an experimental kitchen, the patent department, and on the ground floor, engineering shops, where prototype models were built. On any of these engineering floors, one could walk directly into the manufacturing facilities in the adjacent wing to observe various production lines in action, or across to further wings, to departments or divisions of finance, billing, accounting, service, marketing, or, further south, corporate executive offices, an art department, and a large conference room.

North Canton was a conservative, traditional, church-going, Midwestern small town with a population of about 5,000. It was not at all unusual for the Hoover Company to provide all employees with Thanksgiving turkeys, or for a boss to show up at the doorstep of his employee on Christmas morning with several aged prime steaks. Employee picnics were held in Hoover Park every summer. The working atmosphere was collegial, friendly, and cooperative among employees ranging in age from 18 to retirement, many of whom were neighbors. In the legacy of its founder, W.H. "Boss" Hoover, the Hoover Company was, in fact and practice, a family of people, many of whom had worked at Hoover their entire careers. This was not unusual in the 1950s; rather, it was typical of many well-known companies.

Among these employees there were many talented people. One example was Louis Segesman, an engineering mechanical designer at Hoover since the 1920s, who in 1956 was still making mechanical drawings for development models that were absolute works of art. He meticulously detailed each screw and each tuft of bristles on an agitator brush roll, the latter in perfect perspective. Dozens of cross sections and views were arranged with perfect spacing between them and inscribed with careful, artistic, hand lettering to describe components and details, all with consistent line weight, uniformity, and clarity.

Another was Charlie Strausser, head of the drafting department and also mayor of North Canton. Many Hoover employees held voluntary leadership positions in various community and church organizations. Charlie had adopted the World War II aircraft manufacturing process of "lofting" to precisely specify complex nongeometric shapes, for example, those that Dreyfuss had introduced on motor hoods. The lofting process took a plaster cast of an industrial designer's clay or wood model of an irregular shape and sliced it in parallel sections about a half-inch apart, in both directions, with a band saw. These slices were then traced precisely on production drawings, so that toolmakers could duplicate the specified shape perfectly in the steel molds. Before this procedure, draftsmen could define a shape only with straight lines and compass radii that could be replicated in tooling, resulting in simple, geometric, but necessarily mechanical-looking forms.

Jack Duff headed the research department and was the source of many new concept developments. He managed a divisional design sketch program that reviewed dozens of ideas submitted by many engineering employees each month. The winner of the "sketch of the month" had his or her photo posted at the fourth floor entrance to Engineering, and received a $25 bonus. Employee inventors named on functional patents received a bonus of $50, and for design patents, $25.

The development process at Hoover was a complex, detailed, and traditional four-year cycle from concept to production, starting with rough mechanical and appearance concept models. Designs were constantly refined with engineering test models and then prototype models before preparing final specifications so that the production division could initiate tooling and production. These engineering models were periodically reviewed and critiqued by executives in the production and marketing divisions. A parallel review process for industrial design occurred with a series of appearance models and sketches. These procedures were typical of many manufacturing organizations, including those making vacuum cleaners.

There were many more vacuum cleaner manufacturers like Hoover, all diligently developing new models and features in similar environments and with similar procedures. In 1956, Electrolux introduced its Automatic Model E tank cleaner, which was half the weight and half the manufacturing cost of its previous LXI. This was accomplished with the use of lighter-weight plastics rather than metal and the use of "flip-over" attachments that combined two tools into one. The model retained its refined automatic stop-and-open action. A similar version, the Automatic Model F, would come out in 1957.[39]

Some established vacuum manufacturers were encountering problems caused by heightened postwar competition. Apex Electrical Manufacturing Company, a powerful force in the industry since 1913, had cut production, gone to a four-day week, and laid off 700 workers in December 1948 and January 1949, due to a government curb on installment buying, and in 1956, Apex was absorbed into the White Sewing Machine Company.[40] The last vacuum cleaner with the Apex name was probably its canister cleaner with a polished chrome top designed by Chicago industrial designer Dave Chapman, the same who had worked with Apex as head of Montgomery Ward's design group in 1934, but since 1936, head of his own office in Chicago, Dave Chapman, Inc. By 1964, Apex's parent company would change its name to White Consolidated Industries (WCI).

Hoover introduced its Model 65 Convertible upright cleaner in 1957. When the converter on the end of the hose was clicked in place on the base of the upright, the motor automatically advanced to a higher speed, which provided an extra burst of power for straight suction cleaning using the attachments on the end of the hose. The new motor hood was lower than previous models, making it easier to clean under furniture. The Convertible would remain a classic on the market for 30 years, with periodic changes in color, motor hood design, and model numbers.[41]

After the war, many major appliance companies, such as General Electric, RCA Whirlpool, Frigidaire, and Kelvinator, had promoted new product ideas in futuristic, traveling concept kitchens at home and abroad. Whirlpool Corporation had acquired RCA air conditioner and cooking range lines in 1955, and would use the RCA Whirlpool brand name on a full line of major appliances until 1966. RCA Whirlpool's Miracle Kitchen, introduced to 15 million television viewers in 1957, included a working, robotic vacuum cleaner that prowled around the kitchen, vacuuming up crumbs and debris. Designed by Sundberg-Ferar, a Detroit industrial design firm, the cleaner was never produced in quantity, but it inspired many competitive vacuum cleaner manufacturers to initiate similar research projects. The RCA Whirlpool kitchen also would appear in the U.S. trade exhibition in Moscow in 1959, where vice president Richard Nixon and the Soviet Union's premier Nikita Krushchev met in their so-called kitchen debate of democracy versus communism. The Cold War was in full swing.

In 1957, Whirlpool dropped the RCA from its name for small appliances and entered the vacuum cleaner business by merging with the Birtman Electric Company. Whirlpool began manufacturing vacuum cleaners for Sears under the Kenmore label, using Lamb motors (see below). It also sought to sell vacuum cleaners under the name of Whirlpool for five years. Actually, the Whirlpool acquisition of Birtman was the result of a hostile takeover engineered by Sears. Birtman, Sears's former supplier of

1957 Hoover Model 65 Convertible (Model 564 version shown).

vacuums, built their own motors, which cost more than Lamb motors. Sears had demanded a lower motor price, which Birtman was unable to provide. Subsequently, by plan, Sears employees bought up all available shares of Birtman, then turned them over to Sears, which caused Birtman to go out of business, and allowed Whirlpool to take over making Kenmore vacuum cleaners for Sears.[42]

The less expensive motor that triggered this event was the Lamb motor. Ohio industrialist, broadcasting executive, and labor lawyer Edward Lamb (1901–1987) owned the Lamb Electric Motor Company, a leading manufacturer of small electric appliance motors founded in 1915, with plants in Kent and Cambridge, both in Ohio. Many vacuum cleaner and floor care companies used Lamb motors over the years. In 1955, Ametek had bought Lamb Electric, which in 1961 would be renamed Ametek Lamb Electric Division. In 1958, Edward Lamb in turn bought Air-Way and ceased its production in Toledo. He then engaged Eureka to redesign and manufacture the Air-Way Sanitizor. Wheels were added in lieu of the previous metal runners, and a more powerful Lamb motor was added. The result was the Air-Way GermMaster that dispensed glycol through the exhaust, "sanitizing" the air in a room.[43]

The launching of Sputnik by the Soviet Union in 1958 ended the post–World War II era and initiated the space age, which would have a significant impact on American attitudes and priorities. While scientists struggled to compete in space technology, corporations would seek increased profitability through expansion of markets and new product lines. The world had suddenly become smaller, and there was a new challenge of international competition for global markets, because of the recovery of European economies. This inspired Hoover to adopt a major policy change: expanding its line to include all types of portable appliances.[44]

Other companies were thinking along similar lines in order to diversify their product lines and to increase their market share. A basic law of mass-production is that the more volume produced, the lower the costs of production, and the more competitive the sales price can be. Cost competitiveness became the order of the day for all manufacturers, and economy models had the upper hand. Another basic law of mass production is that the more products you make, the more you have to sell, and selling rests heavily on design, marketing, distribution, and promotion.

The vacuum cleaner industry was suffering from its own success. Vacuum cleaners had been made so durable since the 1930s, and they functioned so effectively for so many years, that many people had no reason to replace them. Since the war, the vast majority of homes that had never had them before had by now bought cleaners. Sales were getting tougher as markets became more and more saturated. In 1958, 3.2 million vacuum cleaner units were sold in the U.S., only a slight increase of 10,000 over 1957—a mere three-tenths of one percent. There were 35 million homes owning vacuum cleaners, and only 14.3 million homes were without them. Canisters still showed the greatest strength, estimated by some at 40 to 50 percent of total sales, with tanks at 29 percent and uprights at 21 percent. The year was dominated by at least three makers introducing "two-in-one" units, combining the suction of the canister cleaner and the beating action of the upright with broad use of turbine and motor-driven rug attachments.[45]

One rug manufacturer recognized that rug cleaning was an endless chore and decided to provide an alternative solution. Florence Knoll (1917–2005), president of Knoll Associates from 1955 to 1959, had a light-colored rug placed inside the entrance to a mall for several months, after which she had the color of this incredibly dirty rug matched, and produced a line of rugs in this exact color for high traffic areas. Dirt made invisible through design! Many commercial carpet buyers found this solution highly appealing.

By this time, mass merchandizing had made door-to-door selling more or less obsolete for many manufacturers. Many two-wage-earner families just didn't have time to schedule demonstrations, and high-pressure sales tactics turned many off. Most preferred to buy household products as they needed them, and to buy them where they normally did their weekly shopping in malls, when and if they decided to do so. By 1958, at Hoover, the days of door-to-door salesmen were being phased out, and dealer-training meetings were being held around the country. Hoover products were now sold in retail department stores, malls, and specialty shops, similar to many other consumer products, and store employees became thoroughly trained on Hoover product features. Eventually, Hoover would establish a chain of 30,000 reputable dealers from coast to coast.[46]

But for others, the days of door-to-door salesmen were not over. Kirby had 4,000 door-to-door salesmen in 1956. Since the 1920s they had been sold only through home demonstrations, to give an individual a chance to try the vacuum before buying, and Kirby would continue to do so for the rest of the century. Nor was door-to-door selling over at Rexair. In 1959, J.V. (Sandy) Sanders purchased the company, which became a separate company, Rexair, Inc., and introduced its Model D2 in 1960. The only difference from the Model D was the addition of a power outlet, allowing the use of a new power nozzle that was privately labeled from the Eureka Company. Sanders refocused the company on direct marketing via independently owned distributors, a door-to-door system known as The Rainbow Opportunity, which remains to this date. Salesmen scheduled an appointment with a prospect to show the Rainbow vacuum cleaner. The prospect was offered a gift for their time and no purchase was necessary. In the tradition of the 1920s, Rexair taught sales and marketing techniques through corporate distributor developers and training seminars. In its 83-year history, Rexair has manufactured only ten sequential models of the Rainbow.[47]

In response to the fast-growing popularity of vinyl and bare floors, floor polisher sales had increased in 1958 by 21 percent over 1957,[48] enticing many manufacturers to participate in this high-growth market potential. Hoover expanded its floor care line in 1959 with its Model 3500 electric Floor Washer and Dryer (design patent 188,411),[49] the first of its kind in the industry. Project engineer was Don Krammes (d. 1973), who later in the 1960s would replace Pierce Daiger as vice president of engineering. A sticklike design with extended, aircraft-style landing gear-type wheels to guide and support the unit at 45 degrees, the 3500 dispensed water onto vinyl or linoleum floors and permitted stand-up scrubbing with a nylon bristle floor nozzle before vacuuming floors dry, suctioning the dirty water into a flexible, transparent, disposable bag within a transparent, rigid water container that also held clean water. Hoover made a dramatic introduction of the product with sponsorship of a "spectacular" CBS television program welcoming popular entertainer Arthur Godfrey (1903–1983) back to television before an audience of 50 million, after his six-year recovery from hip surgery. The 3500 would be followed in 1960 with Hoover's twin-brush rug and floor shampoo polisher, and in 1961, Hoover would begin to manufacture, bottle, and market its own line of floor wax and rug shampoo for Hoover floor care appliances.

The latest popular vacuum cleaner accessory was the power nozzle. Up to this time, except for Air-Way's 1935 Chief, most upright cleaners had driven the agitator/brush roll with a drive belt run by the same fan motor that generated suction. But canister and tank cleaners had been at a disadvantage without such a brushing action on carpets, until now. In 1959, Electrolux debuted the first canister or tank cleaner to use two motors by adding its Power Nozzle PN-1 to its Automatic F cleaner. One motor created the suction and the other turned the brush roll in the floor nozzle. This was a canister cleaner that could claim effective cleaning on carpets, similar to upright cleaners, as well as above the floor. The PN-1's effectiveness in cleaning earned it a number one rating by *Consumer Reports*. The total sales price was $168 ($99 for the cleaner, $49 for the power nozzle), plus $19 for an optional cord reel.[50]

Competitive motor-driven power nozzles on Whirlpool's Imperial Mark XII; its sister brand, the Lady Kenmore Whispertone; Lewyt's 107; and Sears 7006 also appeared in 1959,[51] as well as Eureka's Vibra-Beat nozzle on its Mobile-Aire cleaner, which had rows of vinyl "beads" which vibrated and made much noise.[52] Many other manufacturers used the privately branded Preco Power Brush, available from Preco Inc., Los Angeles, California, at $19.95, which could be used on any cleaner as an attachment in which normal airflow powered the brush.

In 1960, Electrolux introduced its Model G, the first with a wrap-around bumper, built-in cord rewinder, automatic stop-and-open action, flip-over attachments, and power nozzle; these were the same features as in its Automatic F, but in the Model G, they were built in and not optional. It was available in two colors, turquoise or bronze, and would be sold until 1968.[53]

Due to the growing popularity of canister cleaners, General Electric had suspended production of its upright cleaners and tank cleaners in 1951, when it introduced its Model 815 canister, the first with the GE swivel top, and which sold for $99.95. Succeeding these, from 1952 to 1954, were its C-1, C-2 ($89.95), and C-3 canisters, the latter featuring a detachable dolly on which the cleaner rolled about, for holding accessories. In 1956, GE had introduced its Model C-4 canister with a new, low silhouette, double-action floor tool; in 1957, would begin production on its C-6, the first equipped with casters; and in 1958 on a deluxe canister designated as Model C-7. The series would continue on through C-8, C-9, C-10, and C-11. GE's new canister vacuum launched in 1957 was its turquoise R-1 Roll-Easy, a horizontal cylinder that had on each end 12-inch rubber-tired wheels, slightly larger in diameter than the cylinder. The hose connected to the side of the cylinder, so that it could easily pull the cleaner in any direction. Designed by well-known female industrial designer, Freda Diamond (1905–1998), it presented a dramatic, futuristic appearance. In 1959, GE would restore an upright to its line by launching its Model U-1 with a 4,000-rpm power brush and disposable bags, and in 1961, GE would debut its Model U-4 deluxe upright.[54]

1956 General Electric Model V12C4. Courtesy MIAD.

Lewyt was reaching the end of its trail.

Its last vacuum cleaners were its Model 97 of 1957–1958, and Model 105 in 1959 — both variations of its 1955 Model 77. Many attribute its downfall to its introduction of a power nozzle on its 1958 Lewyt Electronic Model 111, one of the first that did not offer exterior or "pigtail" wires to carry current to the nozzle. The 111 ran 24-volt power stepped down from 120 volts via a secondary winding in the motor, ran current through the coils of the hose, and used the metal wands to carry the current to the nozzle. Over time, the insulation between the primary and secondary windings in the motor deteriorated, and some users were shocked while using the machine. Resulting lawsuits are said to have caused the business to fail, and other manufacturers abandoned this type of connection.[55]

In 1960, Alex Lewyt sold his vacuum cleaner company to Shetland Company Inc. of Lynn, Massachusetts, which changed its name to Shetland-Lewyt. The original company was a small 1949 offshoot of the Signal Manufacturing Company. The Shetland Company, which initially made an all-purpose drill, sander, and polisher, had been headed by Robert Lappin since then. Starting in 1952 the company had produced a twin-brush floor polisher, and in 1956 had introduced the Floorsmith, a combination twin-brush floor polisher and rug shampooer, which became a best seller at $59.95. In 1960, Shetland-Lewyt would introduce a completely redesigned, lightweight "stick vac" cleaner with a revolving brush in its nozzle, a 9½-pounder called the Shetland Upright. The stick cleaner would become an immediate success at $29.95 and would be produced until 1972. In 1961 Lewyt would sell his financial interests to Lappin, who would expand the line of stick cleaners and launch a completely redesigned canister cleaner, which would also carry the Shetland-Lewyt brand.[56]

Again in 1959, Consumer Union tested 46 vacuum cleaners for effective carpet cleaning and published results in its *Consumer Reports* magazine. This was the most comprehensive report ever. Rather than highlighting the best buy or

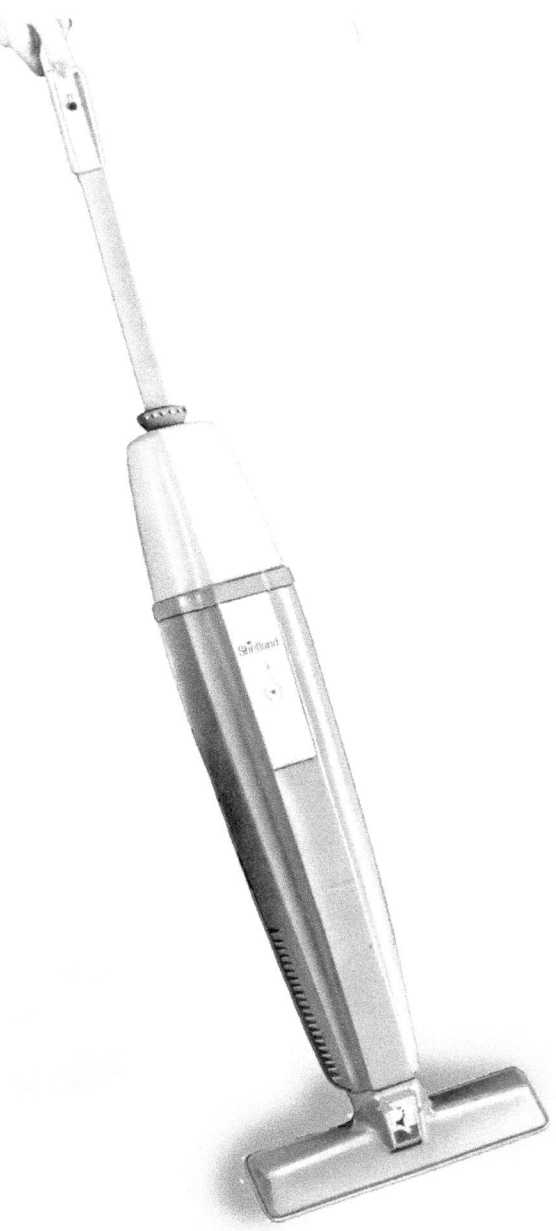

1964 Shetland-Lewyt 9½-pound Upright. IHA/Lifshey.

best cleaners, this report classified models as "Acceptable-Good," "Acceptable-Fair," or "Acceptable-Fair-to-Poor." Certain model cleaners "meriting first consideration" were "check-rated." In tank cleaners, those check-rated as "Acceptable-Good" were: RCA Whirlpool Imperial Mark XII E90 ($129.95-$149.95); Sears Lady Kenmore 7006 ($119.95); Eureka Mobile-Aire 1010 ($89.95); Singer C2 ($89.95); and Electrolux Automatic F ($99.95). The first two had motor-driven revolving brushes in the nozzles. In uprights, "Acceptable" and "check-rated" cleaners included: Hoover Convertible 66 ($109.95) and Hoover Convertible Special 31 ($89.95). The article also reported that the Food and Drug Administration recently accused Health-Mor of making unsubstantiated claims for its Filter Queen, and the company agreed to destroy sales literature that claimed it would "prevent streptococci infections and such diseases as tuberculosis, scarlet fever, diphtheria, bronchopneumonia, smallpox, measles, tetanus, and asthma." This sounded better than health insurance! Detailed descriptions of all 46 cleaners reviewed can be found on the Web.[57]

Stick and hand vacs led the trend into a "second cleaner" market for the vacuum industry. Users liked the idea of a lightweight, easy-to-transport cleaner for quick pick-ups of surface dirt, rather than their heavy, unwieldy, primary cleaner, even though such smaller cleaners had less power and cleaning efficiency. The traditional motivation for frequent and regular deep carpet cleaning, so common before the war, was fast fading. In busy, space-age America, users were more tolerant of concealed dust and too busy with other interests or family commitments to take the time required for deep cleaning. Cleaning priorities had changed. Other young homeowners, shopping for their first cleaner, liked the much lower cost of such lightweight cleaners and found them entirely adequate for their needs.

For example, in 1960, Bissell introduced a stick vac, a lightweight vacuum cleaner that could be handled like a broom, not unlike the rationale it promoted for its carpet sweeper. The stick vac competed directly with similar models built by General Electric and Regina, but filled the need many people had for quick and easy pick-ups.[58]

Hoover first responded to these needs in 1961 with a redesigned stick vac, the lightweight Model 2940 Lark[59] (design patent 194,025).[60] Hoover also decided that in order to increase sales and profits, it needed to convert itself from a domestic business with a few loosely held foreign subsidiaries into a worldwide operating company. Sales in Britain had fallen by one-third and France had suffered significant losses. In January 1961, in order to better control expanding global operations, Hoover president Herbert W. Hoover, Jr., with his executive vice president Oscar Mansager, organized Hoover Worldwide, headquartered in New York, as a joint subsidiary of all Hoover companies in Europe, Latin America, Canada, and England. Oscar retired from the parent company, relocated to New York, and was replaced in North Canton by his brother, Felix Mansager. Over the next few years, Hoover Worldwide would address a number of organizational challenges. As one member said, "We went to work on this company and started to tie the strings together on Madison Avenue."[61] An excess of cash would be used to diversify and expand the line with new, innovative products. New and improved manufacturing plants would be built, and new management personnel would be hired.

Hoover's Model 2100 Portable canister cleaner of 1962, known for obvious reasons as the "suitcase" cleaner, stored all its cleaning tools and hose on the inside of a lid that opened for access, then could be closed while the cleaner was towed around the room on its side by its hose. This concept of on-board tools would not become popular on upright cleaners until the late 1980s. When closed for storage, the Portable could be carried like a suitcase and stored upright and compactly in a closet. It also had a built-in cord reel. The Portable was among the products proudly displayed at an exhibition of American design at the Louvre in Paris in 1963, organized by the Industrial Designers Institute (IDI). IDI was formerly the American Designers Institute (ADI), which had changed its name in 1951 and instituted the first national design awards program. Displayed along with the Portable in Paris was the Studebaker Avanti, designed by Raymond Loewy; IBM's electric typewriter designed by Eliot Noyes (1910–1977); and a Lincoln Continental Mark II,[62] designed by a team headed by our familiar vacuum cleaner designer George W. Walker, by now vice president of styling at Ford. The Portable won IDI's national design award the next year and was soon followed by a variation without tool and hose storage, called the Slim-Line cleaner Model 2000, an even smaller and lighter suitcase.

In 1963 Hoover achieved a major breakthrough in cleaner design with its Model 1100 Dial-a-Matic (design patents 198,435 and 198,436),[63] a combination upright and canister cleaner, or, more correctly, an upright cleaner that also functioned as a canister. It was, in fact, the first "clean-air" upright, achieved by placing the motor after the air had passed through the bag and gone through a triple filtered air stream. A lightweight, multiple high-speed "clean air" fan system, previously used only in canister cleaners, replaced the traditional upright "dirty" or direct air fan, which, because of dirt and small objects passing though it, had to be strong and heavy. The Dial-a-Matic could switch at a touch between traditional carpet cleaning and above floor cleaning, canister style. In essence, it was a canister cleaner mounted on an upright cleaner. Also, the Dial-a-Matic was the first upright with a rigid plastic bag housing. This created an entirely new innovative and unique look for upright cleaners, which traditionally had soft fabric or plastic bags.[64] The Dial-a-Matic was the cleaner used by Samantha (Elizabeth Montgomery) in the TV series *Bewitched*, from 1964 to 1972. It also enabled Hoover to briefly reclaim industry leadership, but since its mechanical innovations were not patentable, competitors such as Eureka and Kenmore soon replicated the Dial-a-Matic's appearance design and mechanical characteristics.

Now, after many years, the traditional functional difference between canister and upright cleaners became blurred. Canisters (and tanks) now had power nozzles with agitators, and uprights had high-speed airflow and hoses for effective above-floor cleaning. Both could claim to do the same job as the other, and consumers would not have to buy

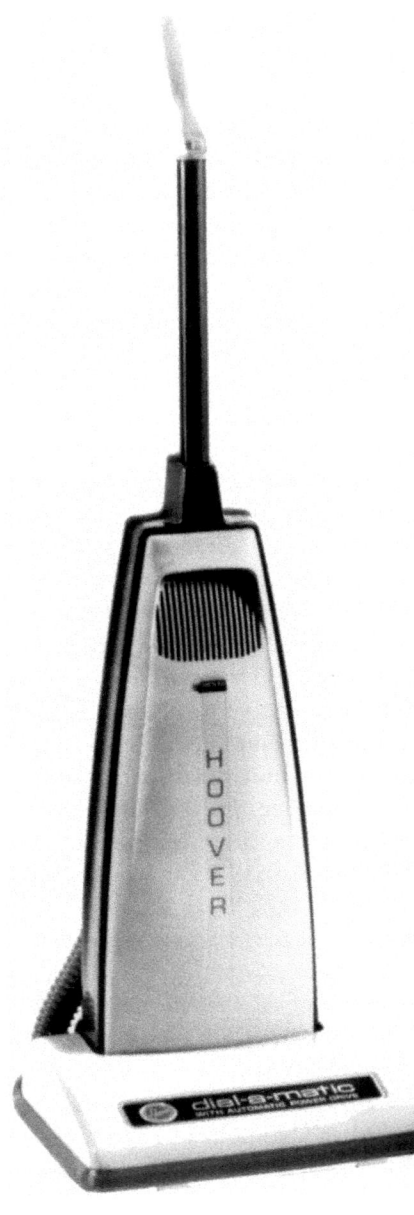

1963 Hoover Dial-a-Matic. HHC.

one of each type to perform the various cleaning tasks needed in the home. In lieu of one of these, they would probably buy a stick or hand-held cleaner for light work.

With domestic profits doubling in 1962 as a result of Hoover Worldwide activities, Hoover stock rose from 25 to 40 and split two-for-one. As a result, the company began diversifying and expanding its product line. It entered the U.S. major appliance market in 1963 with its redesigned compact, spin-drying wash machine, imported from its manufacturing plant in Wales, Great Britain (earlier models since 1953 did not spin-dry). Designed for compact European apartments, which had neither the room nor hot water supply required by larger U.S. types, the machine was 29 inches wide, 16 inches deep, and 31 inches high. The following year, Hoover began manufacturing these machines in Canton, Ohio, in white (Model 0510) and coppertone (Model 0512). In about 1970, color options would include Harvest Gold and Avocado Green.[65]

At the same time, Hoover began expanding its small appliance business over the next few years from electric irons and fry pans to include can openers, hair dryers, hand mixers, coffee pots, toasters, and even carpet sweepers. This effort, which would put Hoover in direct competition with major companies such as Westinghouse, Sunbeam, and Bissell, would be accelerated in 1969 by the acquisition of the National Enameling and Stamping Company (NESCO) of Two-Rivers, Wisconsin,[66] made famous by its portable roaster oven in the early 1930s, and the Knapp-Monarch Company of St. Louis, a major manufacturer of electrical appliances founded in 1925 by Andrew S. Knapp (d. 1961). His son, Robert S. Knapp, would head the new Knapp-Monarch Division of Hoover.

Hoover's extensive expansion of new product lines was far too much for its small industrial design staff. Up until 1955, the Dreyfuss office in New York, with its huge staff, had been able to support any level of development Hoover needed. But now, with the death of Russ Swann, the design staff was down to one. So in 1961, Hoover began to increase internal staff and engaged the industrial design firm of Smith, Scherr, and McDermott (SS&M) of Akron, Ohio, to assist in the appearance design of new products. In 1963, Hoover replaced SS&M with industrial design consultant Robert Hose (1915–1977), who had worked for Henry Dreyfuss Associates (HDA) in New York since 1946 and had become a partner in the firm, but who resigned in 1961 to consult independently. Hose had worked in the early 1950s with Hoover as an account executive of HDA, and was well known by Hoover Engineering vice president Daiger. More importantly, Hose's fees were much more reasonable than Dreyfuss's had been, since he had no staff or overhead, and was much more personable, easy-going, and collaborative with engineers than Dreyfuss had been. He was also a past president of the Society of Industrial Designers, and was well known in the design profession. The small, but growing, in-house Hoover staff did most of the industrial design work, with periodic consultation with Hose.[67]

In August 1963, another new name appeared in the vacuum cleaner market. Whirlpool entered into an agreement with newly established Oreck Corporation, an independent marketing organization, and appointed Oreck as exclusive distributor of vacuum cleaners under the Whirlpool name for a period of five years. David Oreck (b. 1923), Oreck's founder, said he planned to use a "specialty approach." After World War II, in which he had served as pilot, navigator, and bombardier for B-29s in the Army Air Force, Oreck had worked for RCA as a distributor for seventeen years. In 1963 he acquired an abandoned design for an upright vacuum cleaner from Whirlpool and a failing RCA distribution facility in New Orleans. The Whirlpool cleaner, at eight pounds, was a third the weight of other machines and had failed because competitors used this fact to criticize its effectiveness and durability.

Oreck decided to market the cleaner to hotels where light weight was a key factor, and where it met with considerable success over the years, since 50,000 hotels would eventually use them. Early on, hotel employees asked to buy machines for their own use, and Oreck began selling them to the general public. Whirlpool gave Oreck exclusive rights to market the cleaner throughout the U.S., and gave him free reign to redesign it.[68]

Under pressure from competition, some manufacturers were forced to withdraw vacuum cleaners from their line because they were less profitable than other products. Hamilton Beach, a division of the Scovill Manufacturing Company since 1923, still had a 1962 line of vacuums including a Model 35 square canister, a Hatbox canister that stood on end on stairways, and its Port-A-Vac that doubled as a hand vac or stick vac. But in 1965, Hamilton Beach exited the vacuum cleaner business to concentrate on its extensive and highly successful line of small appliances. In 1990, Hamilton Beach would merge with Proctor-Silex, becoming Hamilton Beach/Proctor-Silex Inc., the largest manufacturer of small appliances in the U.S.

Other manufacturers grew their profitability through acquisitions. From 1964 to 1973, Scott & Fetzer Company would diversify its manufacturing range through an acquisition spree, emerging with 31 businesses, with products including chain saws and trailer hitches, although Kirby vacuums were still a major and successful line of its products. The Kirby Company would become a wholly owned subsidiary of Scott & Fetzer during this period, continually adding new innovations to its vacuums. In other 1964 news, Singer introduced a canister cleaner and floor polisher, both designed by Henry Dreyfuss, and Regina debuted a 4½-pound, hand-held vac designed by Samuel D. Han. All three won IDI national design awards that year.

In the world of industrial design, three independent organizations — the American Society of Industrial Design (ASID), the Industrial Designers Institute (IDI), and the Industrial Designers Education Association (IDEA) — merged to become the Industrial Designers Society of America (IDSA) in 1965, with an estimated membership of 650 professionals. IDSA's first president was Henry Dreyfuss. IDI had conducted annual national design awards since 1951 to recognize outstanding designs, a program which ended in 1965. That year, IDSA, in cooperation with the U.S. Information Agency (USIA), sponsored a trade exhibit in London, displaying 300 U.S. products, packages, and corporate identity programs. Also that year, the National Endowment for the Arts (NEA) was created by an act of Congress to support the arts, and it included a design arts program to fund grants in architecture and industrial design.[69]

By 1965, Hoover's global product line had grown to 83 models, made by 22 plants in 12 countries, but profits were less than 2.5 percent of sales, 75 percent of which came from overseas operation.[70] There arose considerable friction between Hoover Worldwide and the North Canton organization. In 1966, as a result of a proxy battle to oust Herbert Hoover, Jr., Felix N. Mansager, the brother of Oscar Mansager, became chief executive and president of the company. Felix had replaced his brother Oscar as executive vice president of Hoover Worldwide in 1963 when Oscar retired. He pushed to increase U.S. domestic sales through product diversification and expansion of manufacturing facilities.[71] New products that year included a new line of 2900 series stick vacs called the Handivac (design patent 207,443) and a redesigned Pixie (Model 2800) hand vac, as well as a new lightweight, 10-inch upright cleaner, Model 1340 (design patent 211,604)[72]; a hand-held upholstery shampooer (design patent 205,584)[73]; and a hand-held shoe polisher. In 1966, Hoover's popular upright Convertible cleaners were upgraded in appearance with new motor hood designs for Convertible Models 1010 and 1060 (design patent 209,189).[74]

Along with its expansion of product lines, Hoover had been increasing its industrial design internal staff since the mid–1960s. By 1970, Arne Diehl would be added to the staff of six. He would head the group in 1972, would transfer to the marketing division in 1984, and in 1998 would be named company director of product planning.

I would be remiss if I did not mention another young industrial designer of note hired in the mid–1960s by Hoover Ltd. in Great Britain. Bill Moggridge (1943–2012) graduated from the Royal College of Art and worked for Hoover Ltd. until he would open his own industrial design office in 1969. In 1979, he would emigrate to the U.S. and establish a design office, ID Two, in Palo Alto, California, with compatriot Mike Nuttal. In 1990, ID Two would merge with Matrix design firm and David Kelley (b. 1951) to form IDEO. By 1993, IDEO would be the largest and most successful design office in the U.S., and by 2004, it would have a staff of 350, an annual revenue of $62 million, and offices in Palo Alto, San Francisco, Chicago, Boston, London, and Munich. By 2009, IDEO would have won 37 national design awards since 2004, more than any other design firm. And in 2011, Bill Moggridge would become director of the Smithsonian Institution's Cooper-Hewitt National Design Museum in New York.[75]

Despite the saturation of the vacuum cleaner market, the industry expanded. By 1966, people had begun to acquire more than just one vacuum cleaner. One was for deep cleaning, and another for quick pick-ups. That year, 15 percent of households owned more than one cleaner, a tripling of the 5 percent that existed only six years earlier. The trend was fueled by the success of handheld and stick vacs. Stick vacs, during that same six-year period, increased in annual unit sales from 100,000 to 800,000, a seven-fold increase. GE's MV-1 handheld vac had launched in 1964, its SV-1 stick vac in 1966, and by 1967 had sold half a million units.[76]

Bright colors and dramatic graphics dominated the 1960s, following the trends of the era. In 1966, Shetland-Lewyt added a new line of canister vacs called the Fashionables, a series of eight cleaners in different colors and bold graphic designs, to its extensive line of stick vacs. By 1967 the company was the largest producer of electric shampooers, and had also become influential in lightweight stick cleaners and in low-end canister cleaners. That year, Smith Corona Marchant (SCM) Corporation bought Shetland-Lewyt and operated it as a division. In 1969, Shetland introduced a stick-type floor washer called the Electra Sponge and soon added a number of table appliances made by Proctor Silex Division of SCM, but bearing the Shetland label.[77]

Wet cleaning had become the new dimension in floor care. The Shop-Vac Corporation of Williamsport, Pennsylvania, introduced its first Shop-Vac in 1967. The canister-style Shop-Vac was the first popular, effective, wet/dry vacuum cleaner. Martin Miller, founder of Craftool, had invented the five-gallon drum cleaner for cleaning in commercial shop environments in the 1950s. It also happened to be the first vacuum to target a primarily male market, and continues to dominate the wet/dry vacuum market to this day. While this was not the first to use the principle of water filtration, such as Kirby in 1906 and Rexair in 1937, Miller is credited with bringing the wet/dry vacuum to the masses.

Bissell was also in the wet cleaning business. In the ten years since the 1957 introduction of its Shampoomaster, sales had foundered because few homes were dirty enough to require regular carpet shampooing. Bissell's revenue had grown five-fold by 1967, but not because of this new product. Surprisingly, the growth was a result of a burst of demand for its carpet sweepers as secondary carpet cleaners, with the same "quick pick-up" purpose as stick vacs and hand vacs.[78] This was further evidence that users were trying to avoid regular deep cleaning and looking for an easier and faster way to pick up prominent surface dirt.

Responding to industry demand for expanded wet floor care, Hoover introduced an entirely new five-product line of floor polishers and rug shampooers in 1967, led by the remarkable Floor-a-Matic rug and floor conditioner Model 3600 (design patents 211,108 and 211,109),[79] which not only permitted the power scrubbing of floors, but also vacuumed up dirty scrub water. It could also shampoo carpets and rugs, damp-mop, apply wax, polish, and buff.[80] By 1968, Hoover earnings were at 7.3 percent of sales.[81]

Electrolux AB of Sweden in 1968 sold its American Electrolux Company subsidiary to Consolidated Foods Corporation, providing Electrolux with funds to design a completely new vacuum, its Model 1205. It was the first with a solid state circuit board, the first with an "electric hose" with direct-connect technology, and the first canister with a telescopic electric wand. The bag chamber was modified for maximum airflow and filtration, and overheating of the motor by clogging of the hose was prevented with a small air vent. The Model 1205 was the most technically advanced cleaner on the market.[82] However, the sale of American Electrolux meant that Swedish Electrolux was no longer allowed to use the name Electrolux in the U.S.[83]

Even the usually conservative vacuum cleaner industry had a sense of humor. In the 1960s, Electrolux successfully marketed its vacuums in the U.K. with the slogan "Nothing sucks like an Electrolux." Americans assumed the slogan was a blunder because of the negative American informal meaning of the word, but this was also well known in the U.K. at the time, and the company hoped the double entendre slogan would gain attention. It did.[84]

Air-Way in 1968 was again manufacturing its own machine when it introduced its first power nozzle in its Sanitizor and initiated the first Tandem Air design: one motor to operate the suction fan of a Lamb motor in the canister and another to operate the suction fan and brush roll in the power nozzle.[85] The nozzle was a modified version of Eureka's 1400-series upright, made by Eureka, which instead of a bag and handle, had a bellows/wand setup so it could be connected to the hose end of the Air-Way and used as a power nozzle.[86] Air-Way would make its own power nozzle in 1972.[87]

Vacuum cleaners became popular television entertainment in 1968 when Stan Kann

1971 Electrolux Model 1205. IHA/Lifshey.

(1924–2008) appeared on the *Johnny Carson Show* to demonstrate some of his vintage vacuum cleaners, after being recommended by Phyllis Diller. His cleaners had been disassembled for shipping, and when he began to assemble them for the show, he was told that such props had to be assembled by union workers, and they were. But as Kann went on camera with his prepared spiel, the first cleaner fell apart, the second lost a wheel, the bag fell off another, and another lost its handle. Through each disaster, Stan, doggedly and deadpan, never broke stride but resolutely continued his prepared demonstrations. For a full 16 minutes, Carson and the audience were in complete hysteria, and pundits said it was one of the wildest and funniest spots the show had ever had.[88]

Kann perfected his accidental comedy routine and would return to the *Johnny Carson Show* 77 times, demonstrating vacuum cleaners and other gadgets. He would become a surefire, last-minute replacement for any guest who cancelled. He would be on the *Mike Douglas Show* 89 times, and in the '70s, '80s and '90s, would appear on the *Merv Griffin Show*, the *Dinah Shore Show, Bill Cosby Show*, and the *Gypsy Rose Lee Show*, among others. He would wear a suit decorated with antique cleaner bags and would inspire dozens of antique vac collectors across the country.

Stan, a dignified and serious professional musician, was the last person one would expect to enter show business. At age 15, he began to study classical organ, majored in classical organ at George Washington University, and was musical director of a Methodist church for 22 years. Four times a day and seven days a week he played the mighty Wurlitzer organ at the Fox Theater in St. Louis, Missouri, the organ he had restored in 1951 and played until 1974. He was also the organ and musical director on the *Charlotte Peters Show*, a noon talk show in St. Louis where Diller discovered him.

But Stan had another lifetime passion. He began collecting vacs when he was eight years old, and found most of his cleaners at Goodwill Industries, the Salvation Army, or from neighbors discarding their used ones. He repaired neighbor's cleaners while in grade school and sold new ones door-to-door when in high school. Eventually, he had a collection of at least 170 antique vacs in his 1914, 28-room Georgian mansion at 29 Washington Terrace in St. Louis. With his photographic memory, he knew each one in mechanical detail. He operated all of them in a regular rotation sequence to clean his home, and could identify them by the sound of their motors. Stan would move to California in 1974 because of his show business career, and return to St. Louis in 1998 to resume playing the organ and perform concerts throughout the country and England. After a tornado would damage his home in 2006, Stan would begin to sell some of his collection on eBay, and a year after his death, in 2009, 70 vacs from his collection would go up for auction on eBay. Bids would be reported at $22,900.[89]

Even though Stan made some commercials for Hoover, saying their latest cleaner "spoiled his act because it always works," Hoover preferred new cleaners rather than antique ones. In 1969, Hoover introduced its upgraded power-drive Dial-a-matic upright cleaner, Model 1170, which was self-propelled, a feature which made back and forth movement of the cleaner effortless with a "triple-action" handgrip. A system of gears, belts, and cables used the power of the cleaner to enable any slight movement of the sliding handle grip forward or backward to automatically engage the self-propelling mechanism. A headlight would be added to the non-power-drive version on Model 1151, and on the power-drive model 1176 in 1970.[90] Hoover introduced its Model 404 Swingette (design patent 219,752)[91] handheld cleaner in 1969 that had the same power as a canister cleaner. About the same size of a large lunchbox, with a carry handle, shoulder strap, and hose, it was easily portable around the house and extremely compact for storage. A redesigned tank cleaner Model 509 for England (design patent 215,489)[92] was also introduced with a square cross section and modern lines.

The 1960s were good for vacuum cleaners, since the market grew from 3.2 million full-sized vacs per year to 7.1 million, more than doubling market size. The growth was attributed to the national economic boom, an increasing focus on department store and mass-merchant retailing, a dramatic upswing in housing as suburbs sprouted across the nation, and the popularity of wall-to-wall carpeting and vinyl floors.[93] During the 1960s, many vacuum cleaner manufacturers replaced traditional stamped steel and cast aluminum parts with those of high-impact plastics such as Lexan polycarbonate or ABS (acrylonitrile-butadiene-styrene). These materials were both tough and lightweight, allowing vacs to become much lighter and easier to handle.

But the 1960s were not good for the nation. They had been filled with national tragedy, including the assassination of two Kennedys, Martin Luther King, and deadly race riots resulting from the civil rights movement. Social change was dramatic as baby boomers became of age, with anti-establishment demonstrations against the Vietnam War resulting in students being shot and killed. Youth rebelled with miniskirts, rock bands, beards, long hair, and a hippie free sex culture that in some instances resulted in murder cults. Women's liberation discarded both bras and traditions of marriage, and "housecleaning" and "housewives" were no longer terms of respect or endearment. Many regarded corporations as greedy destroyers of the environment. The older generations struggled to adapt to these dramatic cultural changes, and in many instances, fought against them. At the Hoover Company, for example, facial hair and casual dress by employees was forbidden, a practice not untypical of many conservative manufacturing organizations. Social consciousness had been raised, but the national psyche had been dealt a serious blow, and both the public and business needed a pat on the back.

CHAPTER 6

Globalization 1970–1990

The year 1970 began the new decade with a triumph of national technology, shortly after U.S. astronauts landed on the moon. There was also much to celebrate in the vacuum cleaner industry. Business had doubled in the past ten years. In the past five years, upright sales increased by 18.6 percent, canisters by 10 percent, stick vacs by almost 50 percent, and polisher-shampooers were enjoying spectacular growth of 15 percent just this year, with major suppliers being Hoover and Regina. In overall floor care products, Hoover had 30 percent of industry volume in units, followed by Sears at 12 percent, Eureka 10 percent, Electrolux 9 percent, GE 8 percent, Kirby 5 percent, and Singer 3 percent. Shetland, Regina, Sunbeam, Royal, Frantz, Bissell, Health-Mor, and Japanese importers Brother, Sanyo, Panasonic, and Hitachi shared the remaining 21 percent. Thirty-eight percent of all households now owned two vacuum cleaners, and 15 percent owned three.[1]

During the 1960s, there had been a renewed public awareness of air and water quality, which continued into the 1970s. Citing rising concerns over environmental protection and preservation on July 9, 1970, President Richard Nixon submitted an executive order to Congress creating the Environmental Protection Agency (EPA) to comprehensively regulate environmental pollutants in the air, water, and land. Public concern included vacuum cleaner owners who increasingly became aware of the quality of air coming from their cleaners, and manufacturers increased their use of HEPA bag and filter materials, as well as other efficiency features and advertisements, including health-related brand names, to promote cleaner and healthier exhaust air.

Some manufacturers had been promoting these qualities since the late 1930s, including Health-Mor. By 1970, Health-Mor Industries (HMI) had developed a loyal customer following. Its Filter Queen, with a multitude of attachments, could function as an air filter, air freshener, paint sprayer, hair dryer, sander, polisher, and upholstery and carpet cleaner. Health-Mor's 161 employees generated about $9 million in annual sales with $1.3 million in income. Health-Mor president John Licht concluded that retail sales cannibalized traditional direct sales, so he developed an alternative model, the Princess, specifically for the retail market. Over the next decade, HMI sales tripled to $30.1 million, with income at $3.3 million.[2]

Kirby had also been doing well by continuously improving and refining its Model C upright cleaner, originally introduced in 1935. Typically, it produced only one model at a time, and following in sequence were: Model 505 (1945–1947); Model 510 (1950–1951); Model 516 (1956–1957); and as air quality became an issue in the 1960s, its Dual Sanitronic 50 (1965–1967). The last of this generation, introduced in 1967 and sold until 1970, was the Dual Sanitronic 80.[3] In 1970, Kirby, in the tradition of Alfred Fuller, used input from customers and distributors to make a wholly new vacuum cleaner, the Kirby Classic, which was an instant success and would be produced through 1973. Its introduction, ironically, nearly coincided with the death of James Kirby in 1971. Success of the Kirby Classic would force the company to expand its manufacturing facilities outside of Cleveland for the first time, and in 1972, Kirby West would begin operations in Andrews, Texas, doubling the company's manufacturing capabilities. The Kirby Classic Omega would be produced from 1973 to 1976. In 1974, Ralph E. Schey would become president of Scott & Fetzer Company, Kirby's parent. He was a venture capitalist who would trim and restructure the company, and within four years would reduce it from 31 to 20 divisions. Kirby, of course, would remain, and from 1976 to 1979, would produce the Kirby Classic III.[4]

In another classic company, John M. Bissell, the founder's grandson, assumed leadership of the company in 1971. Prior to this, Bissell had begun expanding its product line. In 1965, it had acquired the Wood Shovel and Tool Company in Ohio, making garden tools. In 1970, Bissell acquired a Swiss electric shaver company and in 1971 took over a number of printing companies that became Bissell Graphics. John Bissell felt that the company was risking its best business, floor care, and reintroduced a new rug shampooer, the Bissell Electro-Foam Carpet Shampooer, similar to its Shampoomaster of 1957 through 1967, but this time as an electric device. In 1974, Bissell would purchase the Penn Champ Company, manufacturer of aerosol cleaners and fabric shampoo to protect and grow its floor care line. Bissell also would develop another token line of vacuum cleaners, to keep a foothold in the market.[5]

Another company kept its foothold through the courts. Oreck Corporation had been the exclu-

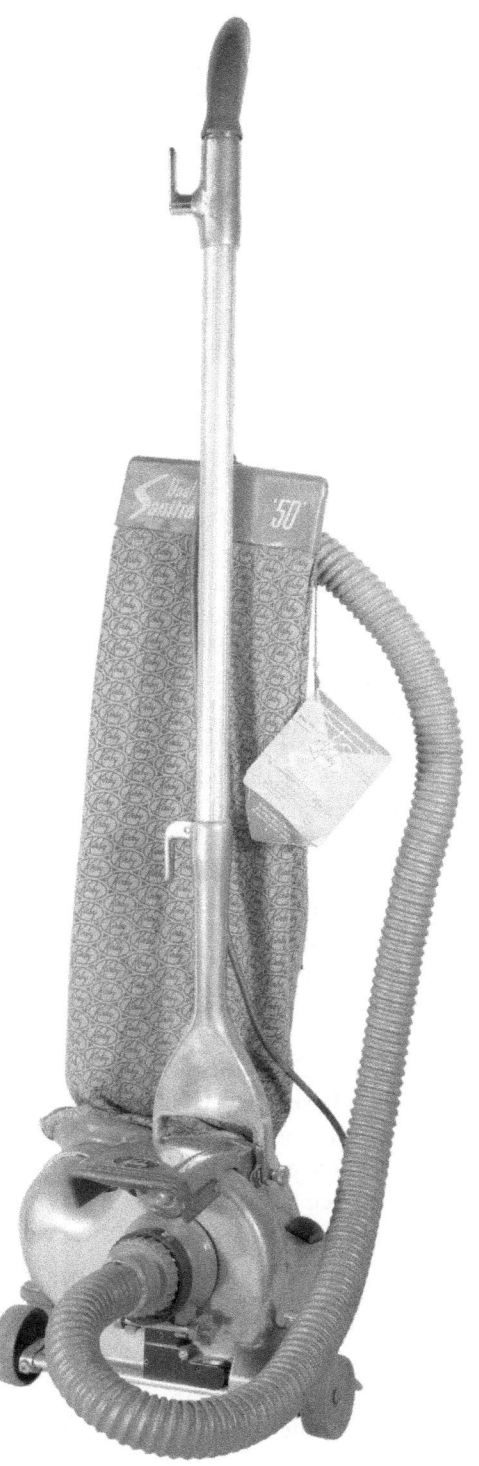

1965 Kirby Dual Sanitronic 50. Courtesy MIAD.

sive distributor of Whirlpool and Sears/Kenmore vacuum cleaners since its mutual agreement of 1963, mentioned earlier. In 1971, the year Oreck sold 78,203 cleaners, mostly by mail order, Whirlpool terminated its relationship with David Oreck, ending his distributorship of Whirlpool and Sears/Kenmore vacuum cleaners. In 1972, Oreck sued both Whirlpool and Sears, claiming "conspiracy in unreasonable restraint of trade" to exclude Oreck from the vacuum cleaner market. Oreck claimed that Sears had urged Whirlpool to cancel the contract, and as we saw in 1957 with the Birtman takeover by Sears, there was good cause for suspecting Sears of manipulation of its vacuum cleaner suppliers. The case was decided in Oreck's favor in a 1976 District court, and although a divided U.S. Court of Appeals in 1977 would remand the case for a new trial, in 1978 it would affirm the original 1976 decision as a violation of antitrust law.[6]

In 1972, Air-Way made a decision to make its own power nozzle, which did not use a separate suction fan. This new metal RugMaster was durable and low to the floor, and it would be purchased and sold by Water-Matic, Fairfax, and Royal as private brand labels.[7] In the 1970s, Air-Way would produce its 12 millionth cleaning appliance and would begin making other related accessories such as a rug and carpet shampoo unit, a moth control device, a sprayer, a vaporizor, and even a hair dryer.[8]

The built-in home vacuum system invented by David Kenney in 1902 was still going strong in 1972, 70 years later. There were at least half a dozen firms producing built-in systems, including Whirlpool, Black & Decker, Beamco, VACUFLO, Nutone, and Cor/Vac. The obvious appeal of these systems, despite their higher initial cost, was three-fold: First, with the power unit in the basement, cleaning could be done with a minimum of disturbing motor noise in the upper floors. Second, users needed only to move the hose and nozzle around to each location, not drag the entire portable power unit and dirt collector. Third, there was superior suction from a larger and more powerful motor in the basement than provided by portable units.

In spite of the steady increases in the market as a whole, not all vacuum manufacturers were thriving. Three important vacuum cleaner manufacturers dropped out of the business in 1972, indicating that competition was as intensive

Hoover Celebrity Model S3001 canister. HHC.

as ever. Parent company Smith-Corona-Marchant announced that Shetland-Lewyt would be discontinued and put up for sale, stating: "A study indicated that the future of the Shetland-Lewyt floor care operation does not appear promising even if we instituted major changes."[9] Westinghouse Electric Corporation, which had only recently had introduced its Converto-Vac II upright, announced that its electric housewares operation, including vacuum cleaners, had lost $7 million in 1971, and Westinghouse was putting it up for sale. Scovill Manufacturing Company bought the business, presumably for its Dominion and Hamilton Beach operations. By 1975, Westinghouse would be bought by White Consolidated Industries, forming the Mansfield Products Company of White Westinghouse.

General Electric's housewares division also announced: "After a thorough review we have concluded that there are opportunities ... which provide a greater return than exists with floor care products."[10] Accordingly, GE sold its operations in 1972 to a group headed by Raymond Finberg, former vice president of Shetland-Lewyt, of Elkins Park, Pennsylvania, who named it the Premier Electric Company, a name recalling and honoring the original company founded by the Frantz brothers in 1910 and sold to GE in 1915.[11]

In another nod to tradition, Hoover introduced its Celebrity Air-ride canister cleaner model S3005 in 1974, which functioned identically to its 1956 Constellation by floating on its own exhaust air, but had a flattened circular flying saucer shape rather than the spherical shape of the Constellation. The Celebrity had a powerful 800-watt motor and a large, 13-quart, disposable dust bag, which wrapped around the motor. Another model in the line, the S3001, had an on-board tool caddy but instead of Air-ride, relied on large side wheels and casters. It was advertised as "the most powerful home vacuum cleaner you can buy."

A more significant corporate change in 1974 occurred when Hoover earnings collapsed from $33 million to $8.7 million, as a result of the overcrowded appliance market. Felix Mansager retired in 1975 and was replaced by Merle R. Rawson, who moved Hoover Worldwide from New York back to North Canton and

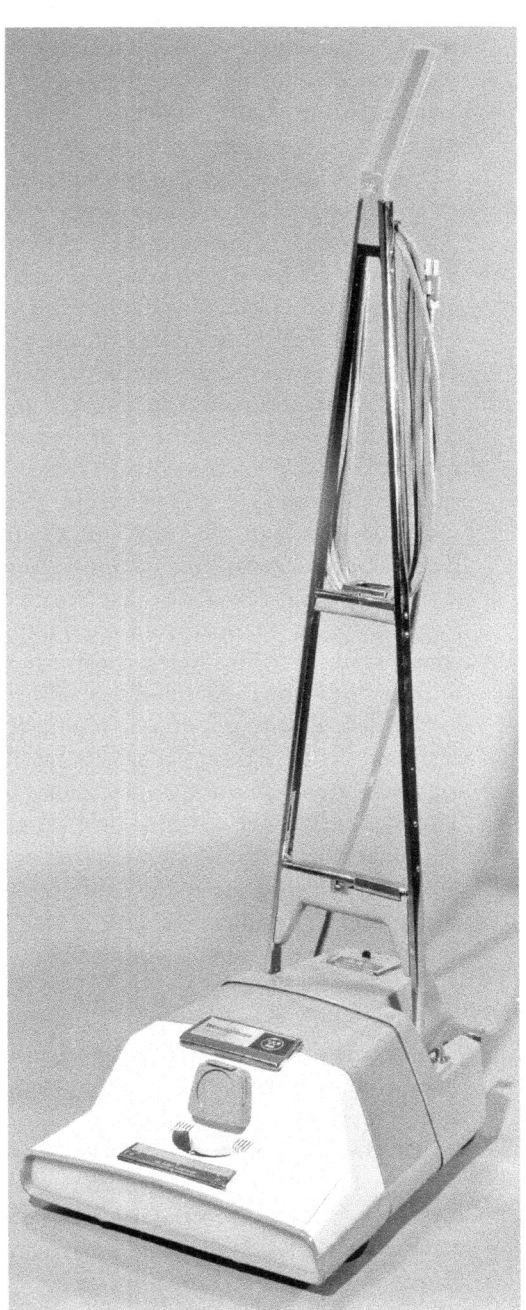

1965 Westinghouse Converto-Vac II. IHA/Lifshey.

in 1977 would sell its small appliance business to concentrate on Hoover's core products in floor care, primarily vacuum cleaners, in a market that had been highly saturated for years.[12]

As mentioned earlier, Eureka-Williams, as a division of National Union Electric Company, had a complete 65-product line of vacuums by 1971. In June 1974, the Swedish company, AB Electrolux, unable to use its name in the U.S. since 1968, and anxious to re-enter the U.S. market, purchased National Union and changed Eureka-Williams's name back to the Eureka Company. Over the next twenty years, Eureka would expand its operations to manufacturing plants in Mexico, Texas, and Illinois, producing vacuums under the Electrolux brand name, as well as its own.[13]

Electrolux had made its Model 1205, described earlier, from 1968 to 1973. Subsequently, its power nozzle was upgraded from its 1959 PN-1 to a new PN-2, which had a brush bar that allowed for adjustment to compensate for bristle wear. Motor life was also doubled. A new power nozzle, PN-4, designed especially for shag carpets, was featured on the Super J model that debuted in 1976. It had the most powerful motor ever used in an Electrolux. In 1978, Electrolux launched its first line of Heavy Duty upright cleaners. The following year, Electrolux's lower-priced Model L, available since 1963 in white, blue, bronze, or gold, would be discontinued, and a special model was introduced that eliminated the "pigtail" connecting cord at the hose to the machine. The Super J would be given a new name and color to celebrate the Moscow Olympic Games of 1980, during which it would be the official vacuum cleaner. Called the Olympia One, the J would be colored in chocolate brown.[14]

The decade of the 1960s had significantly transformed both business and society at large. Business, pressured by competition both domestic and foreign to produce and sell more products at lower cost, had resorted to acquisitions, mergers, and expansion in order to grow profits. Many discontinued unprofitable lines instead of innovating dramatic new concepts in the 1970s. Many consumer products, including vacuum cleaners, had become ordinary household commodities, similar to food, hardware, and clothing. Vacs were no longer perceived as the unique examples of technological invention that they were in the first half of the century. They were available everywhere, at costs ranging from lowest to highest, from dozens of manufacturers, and they all performed satisfactorily for most users. The vacuum cleaner had become a generic household tool, little different from a toaster, a wash machine, a dishwasher, a refrigerator, or a power tool.

Nor had there been any substantial innovative appearance designs to dramatize otherwise generic products since the 1950s. The industrial design profession, which in the past had energized the market with fresh, innovative designs, appeared to have been in the doldrums since 1968. That February, *Fortune* magazine featured an article titled "The Decline of Industrial Designers," which claimed that "radical innovation in design is now the exception rather than the rule," and characterized much current design as "same old, same old." The profession was also criticized throughout the 1970s by its most youthful cohort for selfishly serving the interests of "big business" and for its failure to address the social, ecological, and environmental concerns of the times.

Vacuum cleaner manufacturers were no longer seeking innovative industrial designs, but relying only on their traditional strategy of competing solely on the basis of comparative mechanical performance and special features. Traditional suction, airflow, rug agitators and beaters, power nozzles, and raw horsepower were the major competitive weapons of cleaning effectiveness. Features such as disposable bags, full bag ejectors, headlights, filtration systems, adjustability, flexibility, low profile, color, etc., were among the many conveniences or attractions offered, but these were relatively minor incremental improvements, lacking drama or ingenuity. Most vacuum cleaners looked like and functioned much the same as any other. There were no new earth-shaking technologies or innovative visual designs to lift products above generic adequacy in the saturated market.

Typical of the times was the Hoover Concept I cleaner, introduced in 1978, with outstanding cleaning effectiveness and a power drive feature on some models. It had a feature that brushed along baseboards and reached hard-to-get corners. Retail price was $279.95. Hoover, however, at this time, became subject to serious acquisition threats. In 1979, enticed by Hoover's low stock price of $11, J.B. Fuqua, owner of Atlanta-based Fuqua Industries, targeted the company for a takeover bid, by offering family members $22 per share. In a court battle initiated by Hoover, a federal judge ruled that Fuqua's bid by letter to family members, rather than to stockholders, constituted an illegal tender offer. When the deal fell through, stockholders, who could have doubled their investments, initiated class action suits against Hoover, and to appease them, Hoover promised to look for other buyers, which indeed it would do.[15]

By 1979, changes in society were a generation away from the postwar era of the 1950s and 1960s, but vacuum manufacturers did not seem to notice that their new, younger customers were dramatically different. The new generation of baby boomers, unlike their parents, no longer regarded technology or inventions as special qualities to be amazed at, but expected the latest technology as a matter of course in anything they bought. Women, long considered as housewives, the primary target of vacuum manufacturers since 1908, now regarded the term pejoratively since women's liberation encouraged women to advance in education, professional careers, and personal independence. Many men were now using vacuum cleaners while their wives were busy as the breadwinner of the house. Perhaps most importantly, with busy lifestyles, desire for more leisure or family time, and casual living styles, "deep down" and frequent cleaning was no longer the high priority of many households as it had been in previous generations. Certainly, there were far fewer house-cleaning Saturdays. Catherine Beecher would probably roll over in her grave, were she aware of the modern attitudes and dependence on appliances by housewives, as opposed to the nineteenth century's devotion to women's work ethic and responsibilities to the home.

There were reasons for such changed attitudes. Much of the drudgery of household tasks had been eliminated or reduced. Automatic washers and dryers took most of the labor out of clothes washing and drying. Synthetic textiles eliminated ironing. With family members at work or school most days, homes accumulated less dirt and debris. Preprogrammed ranges cooked meals while homeowners worked. Precooked or frozen dinners required no food preparation at all except for microwaving. Dishwashers took care of the dishes. Housework had become not only easier, but could be accomplished by any family member, not just the "housewife."

The baby boomer generation of homeowners was also better educated, more analytical, and more averse to household drudgery than their parents. While vacuum manufacturers focused their efforts on the mechanical effectiveness and efficiency of the vacuum cleaner itself, many consumers held a broader view of what Ruth Schwartz Cowan, in her 1983 book, *More Work for Mother*, called the entire "work process." This referred not merely to the specific household task itself, but to all the related elements required to perform the task.

Analyzed in this manner, the "work process" of vacuuming includes not just the act of pushing the vacuum cleaner about the room to clean, but all related activities, including: going to the closet (in another part of the house) to get out the cleaner; unwinding the lengthy cord; finding proper attachments and attaching them as needed (for each separate task); finding an electrical outlet to plug in the cord (in each room cleaned); actually running the cleaner (the single task intended); emptying the dirt container or disposing of a full bag; cleaning and replacing the emptied container or installing a new bag; going to the electrical outlet to pull out the cord (in each room cleaned); winding up the electrical cord onto its proper storage hooks; gathering up the hose and tools and replacing them into their proper places; and finally, returning the cleaner to its storage position in the closet (in another part of the house). Housewives of the 1930s may have gone through this entire "work process" routine for any minor isolated surface dirt without even thinking about it, but many baby boomers were unwilling to expend such effort or time except for special occasions or important visitors.

Although many vacuum cleaner manufacturers failed to understand these changes, user need for high performance and power had become significantly less, and desire for convenience and simplicity had increased dramatically, as had tolerance for minor dust and dirt. Vacuum cleaning, for many, was not the serious business it once was, and "deep down cleaning" was no longer the primary objective. These changing social perceptions set the stage for new categories of vacuum cleaners in the 1980s and 1990s by manufacturers that recognized these social trends, as we will see.

It seems appropriate at this point to digress from this reasonably chronological story and to flash back in history, in order to trace the evolution of one of the most successful new vacuum cleaner concepts of the late 1970s. This example will also illustrate the typical, complex, multidisciplinary development process of any new product designed for mass production, involving marketing, industrial design, engineering, finance, tooling, production, distribution, and sales.

Just two years after W.H. Hoover bought Murray Spangler's Electric Suction Sweeper Company and introduced his Model O, two young men, S. Duncan Black (1883–1951) and Alonzo G. Decker (1884–1956), in 1910 founded the Black & Decker Manufacturing Company (B&D) in Baltimore, Maryland. In 1916, they patented and introduced the world's first portable power drill with a pistol-grip and trigger switch that set them on a path that would lead to a huge international power tool business, first in professional tools, and after World War II, in inexpensive consumer tools. By 1970, one could buy a ¼-inch drill for $7.99. Tools were sold at B&D dealerships around the world.

Like the Hoover Company, Black & Decker was a family-run company and from 1964 to 1975, its chief executive officer (C.E.O.) was Alonzo G. Decker, Jr. (1908–2002), who had followed Robert D. Black (1897–1981) in this position. Both were sons of the cofounders. In 1961, the company had pioneered in rechargeable battery technology using new GE nickel-cadmium (Ni-cad) batteries and had introduced the world's first cordless rechargeable drill in its professional line of products.

Alkaline and zinc-carbon battery-operated products, such as toys, flashlights, and clocks, had been in use for many years, but were plagued with the lack of, or diminishing, power and the frustration of frequently having to replace the run-down batteries, which gradually lost power over time. Ni-cad batteries were different. Their terminal voltage changed very little as they discharged, so could remain at full power until needing to be recharged. The improved GE batteries not only allowed recharging, but were powerful enough to operate larger motors for longer times, such as power tools. In 1962, B&D debuted four additional professional cordless drills and a hedge trimmer, and in 1968, B&D built the Apollo Lunar Surface Drill, used on Apollo 15 (1971) and Apollos 16 and 17 (1972), to obtain core samples from the moon.

1974 Black & Decker Mod 4. Author.

By 1971, B&D introduced its first cordless consumer tool, a grass shears, which, within a few years, led to a multiproduct combination package of cordless, hand-held products, including a ¼-inch drill, shrub trimmer, sealed-beam lantern, and a small, hand-held vacuum cleaner. All four were powered by a single, detachable power handle with four Ni-cad batteries, which could be recharged in a separate charger base that converted 120-volt household power to the 12 volts DC needed by the batteries. This four-product package deal, called the Mod 4, was introduced in 1974 at considerably less cost than four individual such tools and was hailed as an innovative product. A grass shears had been added at the last minute, so it actually became a five-product package.

The small hand-held vacuum cleaner, part of the Mod 4 package and called the Spot-vac, was cylindrical and very similar to the 1967 Hoover Pixie or other corded hand vacs, but with no hose and just a small rigid nozzle attachment at the business end of the dirt container. It had a disposable paper bag, just as most such cleaners had, inside the dirt container, or cup. Spot-vac's only distinction from many similar hand vacs on the market was that it was cordless. It was in typical B&D consumer power tool colors of orange and white, as were all the products in Mod 4, because they were considered power tools for male users. The presumed purpose of the Spot-vac was to vacuum up small wood shavings or debris from a basement or garage workbench, where B&D power tools were normally used.

The Spot-vac was not the first B&D vacuum cleaner. The company had been offering five- and eight-gallon commercial vacuums made by Shop-Vac under the B&D brand for some years, presumably to clean basements or garages. B&D also offered a built-in system. It is important to note that none of these B&D cleaners were in the mainstream, competitive vacuum cleaner market, because they were distributed and sold only through B&D-authorized B&D power tool service and sales dealers and hardware stores. Thus, they were relatively unknown to most people in the vacuum cleaner trade.

There were also more recent B&D thoughts of vacuum cleaners. Back in 1971, B&D was experiencing extreme competition from Europe in its line of professional power tools. Production was reduced significantly. So B&D began looking for new products that would utilize its existing power tool manufacturing processes, including electric motors, plastic moldings, die castings, and assembly lines. The vacuum cleaner market fit this strategy perfectly and gradually became a more attractive option. It became a high priority in 1972 for Francis P. Lucier, just elected president and director of the company under Alonzo G. Decker, Jr., who continued as C.E.O., when B&D annual sales stood at $350 million. So in 1972, when a position opened for a new manager of industrial design for consumer products at B&D's engineering department in Towson, Maryland, it made sense for B&D to hire someone who was experienced in vacuum cleaner design, and this is how I arrived at B&D, remaining for the next 14 years.

In 1975, when Francis P. Lucier took over from Alonzo Decker, Jr., as C.E.O. of B&D, it was the first time a family member did not hold that position. As one of his first acts, Lucier requested that B&D's industrial design department develop appearance designs for a line of B&D vacuum cleaners. Preliminary engineering studies were done, and final wood appearance models were designed and built for two different canister designs with power nozzles, as well as for a stick vac. Such a line would have involved costly tooling, an entirely new distribution system for B&D, and going head-to-head against the Hoover Company, Eureka, Electrolux, and other vacuum market leaders. After lengthy deliberations, management decided such a course was too high in both risk and cost, and shelved the project in early 1976. Instead, B&D intensified its search for new products by creating a new, separate organization in its engineering department to identify new product ideas and bringing them to critical reviews by a three-person multidisciplinary new products committee, comprised of a single key representative from marketing, engineering, and industrial design.

By early 1977, even though it was hailed by *Fortune* magazine that year as "one of the 25 best-designed products available in the U.S.," it was clear that the Mod 4 cordless group of products introduced in 1974 was not a great success in the market. B&D postmortem market research revealed the many reasons why. First, there was a business recession during the introduction, and B&D had not effectively promoted the product and its unique new concept to the general public. Many potential buyers were not accustomed to buying multiple products because they traditionally bought only a single product at a time, when they needed it. Other potential buyers, accustomed to existing battery-operated products, such as toys, did not understand that the new batteries were, in fact, rechargeable. They regarded *any* battery-operated product as unreliable and requiring frequent battery replacement. Most users were unfamiliar with the new battery technology, or with the extra step required to recharge batteries. Many forgot to put the power handle into the charger when they finished a task, so when they wanted to use it again, it was dead.

B&D learned several important lessons from this. First, that market research needed to be done *before* investing in a new product, not after. Second, consumers needed to be educated about using new technology. Finally, they learned one more thing—the genesis of B&D's most successful new product in history.

You see, there was one bright positive result in the postmortem research: 92 percent of women users were highly satisfied with the Mod 4 Spot-vac. Many were wives of husbands who had bought the Mod 4 system for their basements or workshops. Wives reported that when they had minor spills of cereal or food on the kitchen floor, or perhaps dead bugs or spiders, they went down in the basement or to the garage to get the Spot-vac, because they found it easier than going to the hall closet for their canister vac, lugging it to the kitchen, unwinding and plugging in the cord, and after a two-second pick-up, unplugging the cord, rewinding it, and lugging the canister back to the closet.

There were, of course, some negative complaints from users. Lack of power was the main complaint. Some thought charging time was too long. Others disliked the messy paper disposable bag, or the clunky nozzle. Some didn't know where to store it, or how to recharge it. Some thought it did not look like a woman's product. In spite of these concerns, 92 percent were quite satisfied.

To the new products committee, this looked like a promising new cordless product for women, and it was initially dubbed the Kitchen Vac. In contrast to almost all other B&D products intended for men, this needed to be one aimed primarily at women, to be used upstairs in the home, not the basement. Based on preliminary research, marketing suggested that 50,000 units could be sold annually, which was about 20 percent of the entire current hand-held vacuum cleaner market of 250,000—a reasonably profitable new product potential. B&D engineering was charged with improving power, and industrial design was charged with redesigning the Spot-vac to address other user problems.

After a few concept sketches and analysis by industrial design, a final concept sketch of a Kitchen Vac, revised significantly to correct user problems, was completed for review in March 1977. A fundamental problem was where to store the cleaner, since most such competitive vacuums were stored in a closet. But no closets contained an electrical outlet; consumers needed to keep the Kitchen Vac on constant charge when not in use, and more importantly, to provide constant power to the separate charger. Therefore, the vac and its charger had to be stored in plain sight, adjacent to an existing electrical outlet. Moreover, it had to be returned to this location after each use to be recharged. So the sketch proposed a charger base mounted permanently on a wall to which the vac could be returned, immediately beside an electrical outlet.

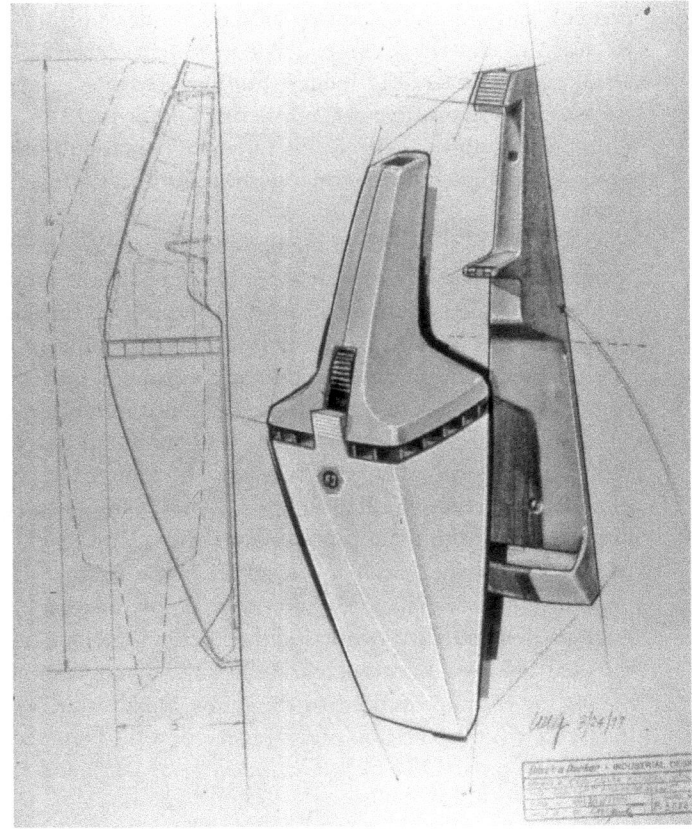

Kitchen Vac concept sketch, 1977. Author.

Since the vac and its charger had to be near an electrical outlet on a wall, and since the Spot-vac was a cylinder about five inches in diameter, the assembly mounted on a wall, with the charger base, would protrude over six inches away from the wall, subjecting the cleaner to being knocked off from its charger base by people passing by. So the sketch changed the cylindrical cross section of the Spot-vac to a rectangle that, with its charger, was flattened against the wall, protruding only four inches out from it. The handle above the motor housing and the nozzle below it were both angled toward the wall, to minimize these protrusions being accidentally hit by passing traffic. This configuration created an appearance significantly different than most other hand-held vacs.

The separate nozzle attachment for the Spot-vac was a problem, both as to where to store it, and regarding concern that any typical attachment tool became unsightly with dirt after use. So the separate nozzle was eliminated. The sketch showed the business end of the dirt container flattened and tapered to form a wedge shape, the narrow end of which itself functioned as the nozzle, and this open nozzle end was concealed in a pocket of the charger base to hide its dirty appearance after use. The wedge shape of the dirt container echoed that of a typical dustpan, to suggest its intended use. A rubber flapper valve allowed dirt to enter the dirt container, but not fall out by gravity.

The new products committee received the concept sketch enthusiastically and authorized development of a wood appearance model to refine it in three dimensions, concurrently collaborating with engineers to determine physical size requirements for batteries, charger, dirt cup and motor. Marketing used the model, finished in September 1977, in focus groups to describe and show the concept to potential users. The physical model was amazingly small and compact, and focus group participants immediately embraced the concept. The response was so positive that marketing's first-year sales projections were doubled to 100,000 units. On January 12, 1978, B&D management authorized a final go-ahead for full development, and an independent, multidisciplinary project team was assembled and went to work.

Engineering improved the batteries, balancing power versus run time to maximize performance, and estimated that three Ni-cad batteries would produce 150 hours, or five years of 15- to 20-second bursts of usage. The project team decided to eliminate the replaceable filter bag, and instead, used a periodically washable and reusable filter. To empty the vac, one just snapped off the dirt cup and emptied it directly into the trash. No replacement bags were needed — just a rinse of the removable filter with water from time to time.

Industrial design reshaped the wall-mounted charger base into a "nest," similar to the 1965 Trimline wall telephone (designed by Henry Dreyfuss Associates for Bell Labs), which matched the contour of the phone to nest into the shape of its wall-mounted base, and invited the user to hang up the phone there after use. This same principle encouraged the user to return the Kitchen Vac to its home charger. The colors were also changed to a popular almond, consumer houseware color, instead of B&D power tool colors, and the graphics deliberately downplayed the B&D brand name; both changes intended to reposition the vac from the power tool market into the housewares market. A company contest soliciting names for the vac resulted in the name Dustbuster, which was tested with the public to insure its communication of aggressive power. The catchy name would be later echoed in a highly popular 1984 film, *Ghostbusters*, which made $238 million, and its sequel, *Ghostbusters II*, in 1989.

Marketing had a much bigger problem. B&D had to somehow transfer its brand image from the basement to the upstairs, and from male buyers to female buyers. This would require massive promotion within the trade and with consumers, as well as require a completely new distribution system to reach department stores and discount houses, in addition to B&D's usual hardware store and service center distribution system. Altogether, it was a huge capital investment with high financial risk.

By July 1978, a run of prototype models had been produced for mechanical testing and for marketing focus groups to demonstrate an actual functional product with potential customers. The prototypes were 14 inches long and weighed 1.4 pounds. In August surveys with consumers, several commented that the Dustbuster didn't *sound* very powerful, which of course, it wasn't. Still, 60 percent said they would definitely or probably buy such a product. This raised marketing's first year sales estimate to 250,000, equal to the entire current annual market for hand-held vacuums. Manufacturing had to scramble to more than double its tooling, and engineering tweaked the fan to make a bit more noise, so it would *sound* more powerful.

Marketing feverishly produced trade literature, packaging, and displays for the unveiling of the Dustbuster at the August 1978 Hardware Show in Chicago. It was a big hit, and was chosen as the "Best New Product in Show" by trade publications. Consumer research tested the Dustbuster against three of the top-selling current appliances at the show and it was preferred by 35 percent, compared to a corn popper (5 percent) and a Mr. Coffee coffeemaker (17 percent), and was exceeded only by the very popular GE Toaster Oven (42 percent). This encouraging result caused marketing to again double its first-year sales estimates to 500,000, and production to again double its tooling capacity.

A final pre–Christmas sales test of an initial 6,400 preproduction Dustbusters was conducted in December 1978, in Kansas City, Missouri, which was a total sellout. An additional 4,000 were also sold, for a total of 10,400. Sales projections and tooling were raised to 750,000. Production sales began in January 1979 and by February, when major account orders were tallied up, it was clear that over a million units would be sold that first year, at $19.95 each — the highest first-year unit sales of any product in B&D history.

Women apparently had voted with their pocketbook for an inexpensive, lightweight, quick and convenient hand vac to pick up surface dirt and spills, even though it was not very powerful. The Dustbuster functioned absolutely no differently than the hand vacs that had preceded it for the previous 60 years, except, of course, that it was cordless, and had a dramatically different configuration because of its technology. It had become the home's electric dustpan, a reasonable alternative to more powerful and expensive vacuum cleaners, which required time to set up, took up valuable storage space, and were heavy to transport and maneuver for small jobs. It became clear that many busy families simply did not have the time or inclination for frequent, thorough, and "deep" vacuuming. There were better uses of valuable time and resources.

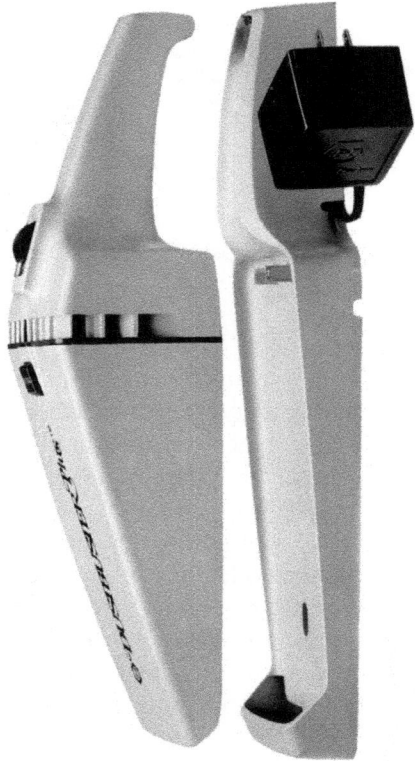

1979 B&D Dustbuster.

United States Patent [19]

Gantz

[11] Des. 257,661
[45] ** Dec. 23, 1980

[54] PORTABLE VACUUM CLEANER
[75] Inventor: Carroll M. Gantz, Baltimore, Md.
[73] Assignee: Black & Decker Inc., Newark, Del.
[**] Term: 14 Years
[21] Appl. No.: 967,361
[22] Filed: Dec. 7, 1978
[51] Int. Cl. .... D7—05
[52] U.S. Cl. D7/164
[58] Field of Search D7/164; D15/32, 55; 15/344

[56] References Cited
U.S. PATENT DOCUMENTS
D. 209,274 11/1967 Richards et al. D7/164
D. 247,341 2/1978 Pelly D7/164
4,011,624 3/1977 Proctt 15/344

Primary Examiner—Catherine E. Kemper
Attorney, Agent, or Firm—Edward D. Murphy

[57] CLAIM

The ornamental design for a portable vacuum cleaner, as shown.

DESCRIPTION

FIG. 1 is a perspective view of a portable vacuum cleaner, showing my new design;
FIG. 2 is a right side elevational view thereof;
FIG. 3 is a left side elevational view thereof;
FIG. 4 is a top plan view thereof;
FIG. 5 is a bottom plan view thereof;
FIG. 6 is a front end elevational view thereof;
FIG. 7 is a rear end elevational view thereof.

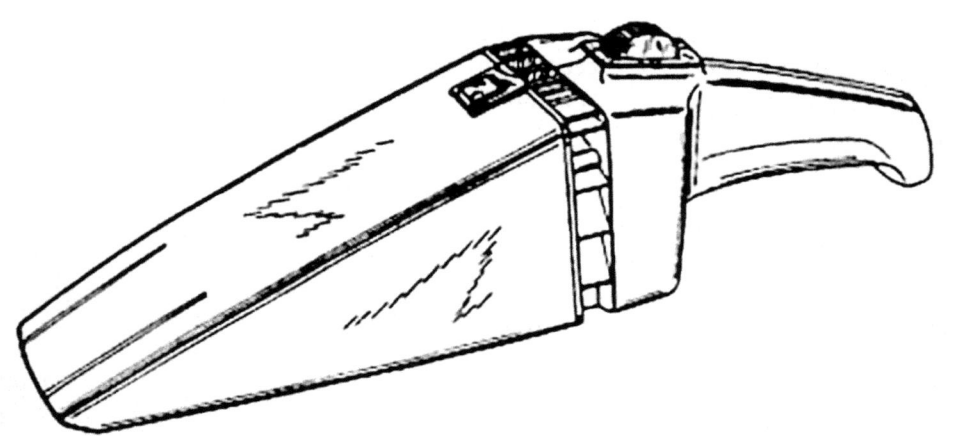

1980 Dustbuster design patent D257,661. U.S. Pat. Off.

Speaking of resources, it is relevant to note that B&D developed the Dustbuster from concept to production in a mere 18 months, illustrating how rapidly mass-produced products were brought to market in 1979 when driven by competition and compressed development processes; a far cry from the four-year development cycle common in the 1950s. Efficiencies in design, specification, tooling, and distribution had nearly tripled.

The impact on the market was instantaneous and pioneered an entirely new category of cordless hand vacs, more than four times the size of the original category of hand-held vacs. Competition rushed in to fill the obvious consumer demand, and within a year the market was flooded with imitations of the Dustbuster.

Fortunately, B&D had applied for a design patent (its first) for the external appearance of the vac and a utility patent for the unique storage and charging system. Both were filed in late 1978, and were issued as utility patent 4,225,814[16] on September 30, 1980, and design patent 257,661 on December 23, 1980. This protection would prove invaluable in a few years, when design patent infringement lawsuits would begin. Design patents are intended to prevent competitors from copying a new appearance "typeform" and capitalizing on its market success, just as utility patents prevent competitors from copying successful mechanical devices or mechanisms.

The Dustbuster illustrates the important principle of industrial design known as a "typeform." The term refers to the general physical form of a particular product, in this case, a hand-held portable vacuum cleaner. For years, the general external form of this type of product had been essentially a cylinder containing a motor, with a carry handle parallel to and beside the cylinder. On the front of the cylinder was a tubular socket into which a cleaning tool or hose was inserted. Over the years, consumers recognized this configuration as the way hand-held portable vacuum cleaners *should* look, and thus is referred to as an established typeform. This, in turn, encouraged manufacturers to carefully and conservatively follow the popular and traditional typeform because that is what people expected.

The Dustbuster dramatically changed this traditional hand-held vacuum cleaner configuration by combining a new wedge-shaped front forming its own nozzle with a rectangular motor cross section and a handle extending rearward. In effect, it set a new typeform which replaced the old, and in the process, communicated visually, and probably unconsciously, to the consumer that there was something new and different about this product. This is exactly what manufacturers hope to accomplish with slogans, advertisements, fancy names, new functional features, and super-large graphics on the front of their products. The Dustbuster demonstrated that changing a product's typeform can be just as effective as words, and it avoids much of the traditional and ubiquitous commercialism and hype that in modern culture has become annoying and off-putting for many. Art museums and the industrial design community itself often regard new typeforms as more creative, innovative, and artistic, which they often are, and recognize these qualities with design awards.

The success of the Dustbuster had a dramatic impact on B&D, encouraging it to diversify beyond power tools into household products. In 1980, B&D established a new household products division in Easton, Maryland, at the site of its outdoor products manufacturing plant, with its own industrial design staff headed by myself. Along with a new line of products, additional versions of the Dustbuster were soon introduced: a 12-volt car vac, similar to the Dustbuster but in a black color; and a larger size intended for use in the shop, the Collector, similar in form but with a power cord, in 1981. In 1982, a B&D Dustbuster Plus came out with five Ni-cad batteries, improving power and run time.

This example of the Dustbuster illustrates that a successful consumer product is not the result of an instantaneous flash of genius that many imagine. Rather, it is a lengthy and complex process of high stakes trial and error, perfecting and refining an initial concept. Even when the result is a market failure, the lessons learned can lead to a more successful result.

This is the same multidisciplinary development process followed by many consumer product manufacturers, including those of floor care, such as Bissell. In 1980, Bissell reintroduced its carpet shampoo concept, a refinement of its 1971 Electro-Foam shampooer, in a simple household extraction device called the Carpet Machine. The following year, Bissell rolled out another wet carpet cleaner called Carpet Magic, its first deep-cleaning carpet machine, but it failed to meet expectations and was phased out.[17]

Despite some successes, the decade of the 1970s was difficult for many U.S. manufacturers. The U.S. was in a recession in 1970, and again from 1973 to 1975. Competition was intense as companies struggled to hold or gain market share with lower and lower costs. Chrysler and American Motors were saved from bankruptcy in 1979 by government bailouts, and Chrysler, Ford, and General Motors posted billion-dollar losses in 1980 and 1984. By

1980, Japan made more cars than the U.S., and in 1982 Honda became the first Japanese manufacturer to produce cars on U.S. soil. Lower labor costs were available in China, Taiwan, or South Korea, and many firms took advantage by subcontracting to these manufacturing locations. An era of corporate mergers and acquisitions began, which affected many brand names. Responses to these challenges from companies included cost cutting, layoffs, sell-offs, and closings.

A typical scenario occurred at Sunbeam. Sunbeam had been producing a complete line of vacs in the 1970s, including two uprights, a heavy-duty canister, five household models, three stick vacs, and two tanks, but trouble was brewing. In 1980, Sunbeam acquired the Oster Manufacturing Company, founded in 1924 by John Oster, Sr. Oster was originally in Racine, Wisconsin, but by 1980, was located in Glendale, near Milwaukee, and headquartered in Chicago. However, in 1981, Allegheny International Inc. (AI) of Pittsburgh would purchase Sunbeam and move its headquarters from Chicago to Florida, where it became the Sunbeam Appliance Division of AI, with sales of well over $1 billion. But by 1988, AI would file for bankruptcy, and a new entity, Sunbeam/Oster Corporation, would emerge. In 1996, "Chain-saw" Albert J. Dunlap, already famous for his cost-cutting reductions in personnel, would try to turn around the failing company, but would be accused of fraudulent accounting, and in 2001 the company would file for Chapter 11 bankruptcy. The following year, it would emerge from bankruptcy as American Household Inc. (AHI), and Sunbeam Products, Inc., would become a subsidiary of AHI. Subsequently, in 2004, AHI would become a subsidiary of the Jarden Corporation.[18]

It is relevant to note that in the acquisition of any company with a strong brand identity, that is, one with a long history of consumer product recognition, the brand name survives, even though another manufacturer in another location might make the product. This happens because the public attaches enormous value to a name that has demonstrated customer satisfaction with performance for many years and believes that this reputation will be continued despite change in ownership or manufacturing location. Without doubt, the acquiring company is anxious to benefit from this residual past reputation and is reluctant to change the brand name. Brand names are worth their weight in gold and in acquisitions.

Also during this period, the computer "information age" matured and changed everything. It reached public awareness in 1977 with the development of personal computers by Apple, followed by IBMs in 1981. "The computer" was *Time* magazine's Man of the Year in 1982. In 1983, the first hand-held cell phones became available. In industry, Computer Numerical Control (CNC) was available in the mid–1970s and radically changed manufacturing. In the machining of production dies, curves became as easy to cut as straight lines, 3-D structures were relatively easy to reproduce, and the machining steps requiring human action were greatly reduced. Computer Aided Design (CAD) systems such as Catia (1981), Auto CAD (1982), and Pro/Engineer (1988) would become available to design products by computer. The days of hand drawing specifications for final products was over, significantly reducing the need for traditional draftsmen and eliminating entire drafting departments. Manufacturers, with large investments in new electronic and computerized equipment, would speed up both design and production processes. But this required costly time, retraining, and capital investment by industry, hurting short-term profits.

During yet another recession from 1979 to 1982, profits were also of concern to the people that sold vacuum cleaners to the public. In 1981, independent vacuum cleaner retailers organized to communicate, to assist each other in areas of common interest, and to hold annual conventions. J.R. "Dick" Beall (1939–1996) and Charles M. Dunham founded the

Vacuum Dealers Trade Association (VDTA), dedicated to supporting independent vacuum cleaner retailers, of which there were many thousands struggling in a tough economy.[19]

Industrial design again became a powerful competitive weapon during this recessionary period. Tom Peters, author of the 1982 best-selling business book *In Search of Excellence* said in an interview with *Newsweek* magazine, "The fire of design is as important to survival as efficient organization." As an example of this, the stagnant U.S. auto industry was revived by the introduction of Ford's 1985 Taurus and Mercury's Sable with an "Aero" softer look, and many products adopted a similar "Euro look" of softer, cleaner design with minimal graphics. In industry, many manufacturers established appearance and graphic standards for consistency, called "Corporate Identification Manuals." With globalization becoming a reality, B&D developed visual standards for a global design look for its multinational product line.

Design was leaving the postwar modern period and entering the so-called postmodern era that was led by Italian Memphis furniture design in 1981, which defied all the established rules of modern "good design." By this time, "good design" of the 1950s through the 1970s had permeated the market with products that, while functional and simple in form, all looked pretty much alike. Designers were seeking innovative new forms that were more expressive emotionally. The ubiquitous and generic black box designs of the 1970s were replaced by designs that were less geometrical and simplistic, and more humanistic, whimsical, and even playful. This trend was labeled postmodern design. For example, Philips's 1986 Roller Radio looked more like a motorcycle than a radio, just to avoid looking like a black box.[20]

Vacuum cleaners entered the world of postmodern installation art when artist Jeff Koons (b. 1955) created a conceptual work entitled *New Hoover Convertibles, New Shelton Wet/Drys 5-Gallon, Double Decker*. The cleaners (two each) were displayed and sealed in an eight-foot-high clear Plexiglas box illuminated from beneath with fluorescent light tubes and would sell at auction in 2008 for $11,801,000. The work was only one of Koons' celebrated series known as *The New*, first exhibited in the window of the New Museum in New York in 1980. The series all featured vacuum cleaners similarly encased in Plexiglas, lit as if on display, and included these titles:

New Hoover Celebrity III's, 1980
New Hoover Deluxe Rug Shampooer, 1979
New Hoover Convertible, 1980
New Hoover Deluxe Shampoo Polishers, 1980
New Hoover Quickbroom, New Hoover Celebrity IV, 1980
New Shop/Vac Wet/Dry Vacuum Cleaner, 1980
New Hoover Celebrity IV, New Hoover Convertible, New Shelton 5 Gallon Wet/Dry, New Shelton 10 Gallon Wet/Dry Double-Decker, 1981–1987
New Shelton Wet/Drys Triple-Decker, 1981
New Shelton Wet/Dry Double-Decker, 1981
New Hoover Convertibles, Green, Red, Brown, New Shelton Wet/Dry 10 Gallon Displaced Double-Decker, 1981–1987
New Hoover 1981–1987
New Hoover Convertibles, Green, Blue, Double-Decker, 1981–1987

Koons' intellectual intention was to invoke the sense of simple aesthetic wonder we experience as children, free and untainted by hierarchies of taste, convention and prejudice we learn as adults. The vacuum cleaners, traditionally used to collect dirt, instead became ironically enshrined in a sealed and spotless container.

Joining the postmodern design trend, a whimsical vacuum cleaner was introduced in England in 1981. It was the Henry, a consumer canister cleaner in multiple bright colors with large cartoon faces, the hose serving as the trunklike nose of a smiley face below its large, impish eyes. Henry had been created by Numatic's owner/founder Chris Duncan in 1968. Numatic International Ltd. was founded in 1969 in the Somerset town of Crewkerne in southwest England when it produced its first commercial vacuum cleaner for cleaning coal, oil, and gas-fired boilers. If vacuum cleaner industry giants considered the Dustbuster as a nonserious, puny, and inadequate entry into the serious vacuum cleaner field, they were probably stunned when they saw Henry. As Electrolux found in the 1960s, mentioned earlier, the British sense of humor is generally sharper and subtler than that of the American.[21]

But this too would change. Monty Python's Flying Circus in the 1970s helped. Back in the U.S., Eureka in 1982 introduced its brightly colored, cute, and somewhat whimsical Mighty Mite, a short, powerful, tank-style cleaner. Eureka industrial designers Samuel E. Hohulin and Kenneth R. Parker designed it with large rear wheels nested in automotive-style fender recesses. In 1983, Eureka debuted a self-propelled upright cleaner to echo the Hoover Dial-a-Matic. In the 1980s, Eureka also introduced a line of highly popular Boss upright cleaners that were aggressively advertised and competitively priced. 1984, Eureka's 75th anniversary year, was said by the company to be its best sales year ever. Sales had increased 211 percent over the previous decade, five times faster than the industry average. But Eureka continued to trail Hoover, its traditional rival.[22]

In 1983, Singer, which made models for Sears, began promoting its own brand, and soon would have five uprights and three canisters in its line. Around that time, Singer also acquired Chicago-based Maxi Vac, Inc., a maker of wet/dry vacuums, boosting its manufacturing and research capabilities. In 1986, Singer's sewing machine business would be

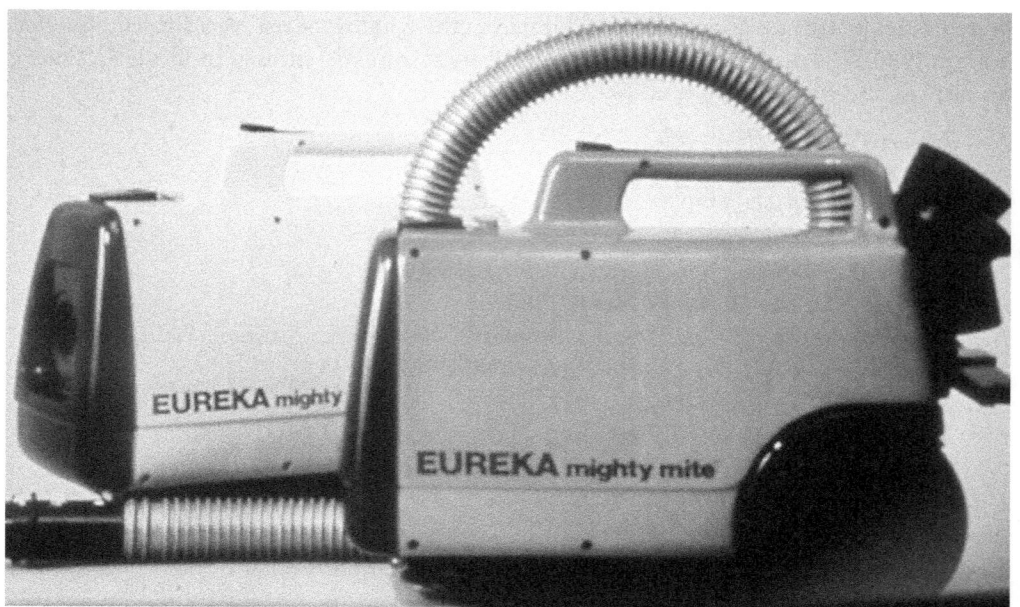

1982 Eureka Mighty Mite.

essentially lost to foreign competition and imports, and it would spin off its sewing machine division as a separate company, SSSM Inc., and abandon all 1,600 company-owned stores and service centers. In 1987, Paul Bilzerian, a corporate raider known as the "greenmailer," would buy Singer for $15 per share. He would sell 8 of Singer's 12 divisions for $2 billion. But in 1989, Bilzerian would be convicted of securities and tax violations. Singer would be renamed the Bicoastal Corporation and would agree to pay $55 million to settle fraud charges.

Meanwhile, Singer Sewing Machine Company (SSMC), the original sewing machine organization, managed to stay in business. In 1989, it would be purchased by Semi-Tech Global and renamed The Singer Company, and in 1991 would be sold to Semi-Tech Corporation. In 1993, Semi-Tech Global would acquire G.M. Pfaff AG, the second largest sewing machine company in the world, and would hire Singer to manage the company and work together to develop new products. In 1997, Singer would acquire Pfaff. But Singer would continue to decline, with a loss of $200 million in 1998, and the elimination of 6,000 manufacturing jobs. In 1999, the company would file for reorganization under Chapter 11 of the U.S. Bankruptcy Code.[23]

Imports were increasingly gaining market share in the U.S. In 1983, the German brand Miele expanded into the U.S., establishing headquarters in Somerset, New Jersey. Meile made all sorts of high-end household appliances, including vacuum cleaners, and had always been a family-owned and run company, founded in 1899 by Carl Miele and Reinhard Zinkann. It exported to many global markets and was represented in 37 countries. This was a period when "global" became a corporate watchword, as companies sought to expand markets and global reach in order to avoid acquisition by bigger fish. Not to do so had become high-risk behavior.

The markets had also been impacted for some time by replacement purchases, that is, older generation consumers replacing cleaners they had been using for 30 to 50 years. These discarded, used, but still operable cleaners became available for little or no cost to many baby boomers, who had an interest in antique cleaners that they recalled from their childhood and began collecting as an inexpensive hobby. To encourage, benefit, and provide communication opportunities to these somewhat unconventional enthusiasts, the Vacuum Cleaner Collectors Club was cofounded by Robert Tabor and John Lucia in 1983,[24] the same year that Brooks Robinson (b. 1937), former third baseman of the Baltimore Orioles, nicknamed "the Human Vacuum Cleaner," was coincidentally inducted into the Baseball Hall of Fame. In 1994, the Vacuum Cleaner Dealers Trade Association (VDTA) and the Sewing Dealers Trade Association (SDTA) established the Vacuum & Sewing Hall of Fame to honor men and women of the past and present that had made significant contributions to the industry.[25]

Electrolux debuted its Silverado model in 1983, changing its 1980 Olympia One color to dark gray, but remaining the same internally. In 1984, Electrolux introduced its Diamond Jubilee model to celebrate the sixtieth anniversary of its 1924 initiation of sales in the U.S. It was in an ivory color with dark gray accents. Ironically, it presaged a decline in quality over the next decade, as Electrolux management decided to use a cheaper motor with a shorter life.[26]

Most global vac manufacturers had been busy developing Dustbuster type hand-held vacs, but not all were cordless. Back in 1979, when the Dustbuster hit the market, Hoover C.E.O. Merle Rawson, responding to a trade press question for a comment, sniffed that Hoover would never make battery-operated products, using the derogative term alluding to toys. He stated that the use of questionable performance cordless units would jeopardize

Hoover's reputation for quality products. Fast forward to 1983, when Hoover introduced its Concept Two Model U3301 upright cleaner, a derivative of its 1979 Concept One, which had replaced the Dial-a-Matic, but which used the traditional soft bag for dirt collection. Front and center on a rigid housing of Concept Two, easily removable for above-floor cleaning, was a Dustbuster-like hand-held cleaner, called the Help-Mate. It was not cordless, but 120 volts and corded, confirming that Merle Rawson meant what he said in 1979: no battery-operated products for Hoover. In 1981, between the two Concept models, Hoover had introduced a line of commercial duty Conquest upright cleaners.[27]

By 1983, dozens of cordless 12-volt products similar to B&D's Dustbuster had appeared around the world. Those offered in Europe and/or the U.S. included: Joiner's Super Cleaner, Joiner's Turbo Cleaner, Philips's Cleanette, Euros's Akkusauger, Siemens's Mini Sauger, AEG's Liliput, Steinel's Vacu-Fix 3000, Bosch's Akku-Sauger, and Royal's Charg-A-Vac, among others. From Asian sources came Minwa's Mini Cleaner, Hiraoka's Super Auto Cleaner, Anex's Car Vac, Good Hope's Auto Cleaner, Forda's Dynamic Car Vacuum, King Nat's car cleaner, Makita's Handy Vac, Sanyo's Let's Clean! and AK Industrial's car vac. There were at the same time rumors of similar products in the works by Norelco, Panasonic, Sony, and Sunbeam.

The legal test for design patent infringement is somewhat subjective, but reasonably rational. In essence, if an ordinary person, in seeing the "imitation" product on display in a neutral color without trade name, recognizes it erroneously as the original product, the "imitator" is considered legally as "confusing" the public as to identity, and the imitation product is considered an infringement. This is why competitors must change the appearance in substantial ways. Most Dustbuster imitators, in varying degrees, avoided design patent infringement by altering the appearance adequately. Still, most copied the Dustbuster's dirt cup tapering to a wedge-shaped nozzle and the handle extending behind the motor.

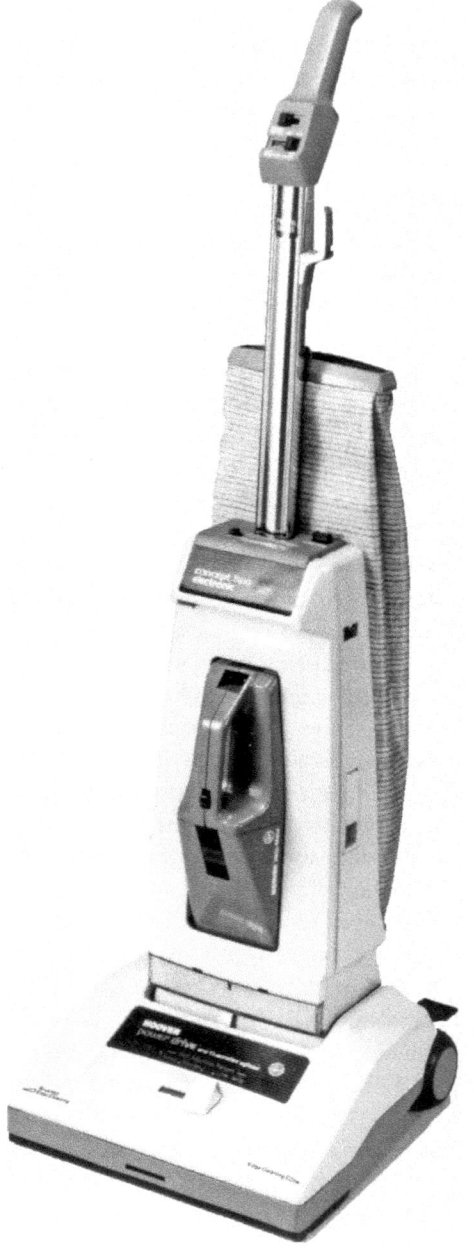

1983 Hoover Concept Two Model U3301. HHC.

Some, like General Electric's Cordless Vac were voluntarily withdrawn from the market after introduction. Others, such as European C.E.P.s Turboduster and Sanyo's Handi Butler were almost identical copies of the Dustbuster. C.E.P.'s Turboduster, in fact, was so identical that its dirt container fit perfectly on a B&D Dustbuster. B&D by 1983 had initiated design patent infringement lawsuits against more than a dozen manufacturers, but in a case that went to court that year against C.E.P., B&D won and C.E.P. was forced to destroy its tooling and cease manufacture.

This legal action by B&D deterred many other competitors. Hong Kong, Taiwan, Korea and Japan were also deterred from entering the market by cheaper, higher quality models made in the U.S.

Meanwhile, serious domestic competition for the Dustbuster was developing. In 1981, a group of investors headed by John Balch had purchased the Royal Appliance Company to implement a fresh, new marketing strategy. This included an aggressive advertising strategy and new distribution into mass retail accounts such as Kmart and Wal-Mart.[28] Leading this growth was the 1984 introduction of an innovative line of corded hand vacuums, based on a redesign of Royal's Princess/Prince hand vac. The new vacuums were called Dirt Devils, made under the Royal name, at least initially. The Dirt Devil had a rotating brush for cleanup on stairs and furniture, plus a hose for hard-to-reach spots. Around the same time, Royal also introduced its Dirt Devil Broom Vac. It was not, however, a stick vac as it claimed, nor was it rechargeable. Basically, it was a Dirt Devil hand-held vacuum with a larger fan, wider brush roller, and a handle that allowed it to be pushed over the floor. Over the next decade or so, it would become the largest selling hand-held vacuum in the U.S., selling more than 25 million units. In 1991, Balch would make Royal a publicly traded company. In 2002, Techtronic Industries (TTI) would acquire Royal Appliances Manufacturing Company for $105.5 million, including the Dirt Devil line.[29] By the end of the century, profits would grow from $5 million in 1981 to $408 million in 2000. The Dirt Devil brand would become so powerful that the Royal name itself would disappear from the product.[30]

Royal Dirt Devil.

The trade press reported that from 1981 to 1984, the hand-held vacuum market had grown 300 percent in both units and dollars, and sold six to seven million units per year, with 90 percent being cordless. It was the hottest category in the business, and grew 157 percent in the last six months of 1985. In a March 1985 article, *Time* magazine called the Dustbuster the "Market Buster," and "the vac that roared." B&D had the lion's share of the market, about 80 percent. In just seven years, from 1978 to 1985, B&D's worldwide household product sales grew from zero to equal that of its established power tool business, and by 1987, B&D worldwide sales would be $1.791 billion! As a result of this demonstration of immediate and long-term market success, in 2009, the Dustbuster would win the Industrial Designers Society of America's coveted Catalyst Award, recognizing design's power to "effect positive change in the world over a period of time."

Late in 1983, B&D announced it had acquired GE's housewares division in Bridgeport, Connecticut, for $300 million, a deal finalized in November 1984. Jack Welsh, GE's C.E.O., did not feel the division fit his vision for GE, and defended his decision with the comment "In the twenty-first century, would you rather be in toasters or CT scanners?" Within a year, all B&D personnel in B&D's household products division were relocated from Easton, Maryland, to Connecticut and merged with GE employees under the name of Black & Decker Household Products Group (HPG) in Bridgeport. By 1986, B&D had converted over 100 former GE brand products into the B&D brand, had initiated a completely new B&D trademark and trade image, and had built a new Household Product Group headquarters building in Shelton, Connecticut.

All this had been accomplished under the leadership of Laurence J. Farley, C.E.O. of B&D since January 1983, when his predecessor, Francis P. Lucier, had stepped down. In 1986, Farley was replaced by Nolan D. Archibald, B&D's Chief Operating Officer (C.O.O.), who Farley had hired in 1985, along with Archibald's own management team. Within six months, Archibald informed the board that he and his team would leave if he was not made chief executive. He was. In 1986, Archibald promptly replaced former B&D management of the household products division in Shelton with former GE management or new hires, transferred industrial design from engineering to marketing, and initiated a massive personnel reduction, including most of the B&D engineers and designers who had developed the Dustbuster. In explanation for his actions, Archibald said that B&D had to become a "marketing-oriented" company. Archibald continued to head B&D until 2009, when he would orchestrate the sale of B&D to the Stanley Works, which would become Stanley Black & Decker.

An organization that would strike terror in many U.S. corporations was founded in 1985, China-based Techtronic Industries (TTI). It was listed on the stock exchange of Hong Kong Limited in 1990 and headquartered in Hong Kong. TTI acquired businesses in power tools, outdoor power equipment, floor care appliances, solar powered lighting, and electronic measuring products, many of them well-known brand names, and established a customer servicing network in North America: TTI Floor Care North America.[31]

Hoover started down the widening acquisition trail in 1985, when it received an unsolicited $40 per share acquisition offer by Chicago Pacific Corporation, a shell corporation created from the remains of the Chicago, Rock Island and Pacific Railroad in 1984, which used the capital from liquidation to acquire nonrail companies, including the Hoover Company and Rowenta A.G. The Hoover board initially rejected the offer, but when the bid was raised to $43 per share as the deadline arrived November 1, the board accepted the offer and sold Hoover for $534.6 million.[32]

The vacuum cleaner market in 1986 was expected to become part of the "sweeping changes" as new lifestyles and newcomers challenged the industry. Consumers now wanted two or three vacs in their home, the extras often being a hand-held or stick cleaner. Sharp Electronics, Whirlpool (with its own brand), Miele Appliances, and Moulinex/Hamilton Beach had entered or were about to enter the American market, and Panasonic and Sanyo were beefing up their lines to strengthen their position here. Eureka and Hoover commanded the total market with 60 to 70 percent, each claiming 32 to 35 percent. Mass merchants, department stores, and catalog showrooms dominated the retail vacuum business. Competitors were expected to move toward higher-priced, fully featured units. The average price of such high-end units was $200, with some, in the 8 percent of the market still using door-to-door sales, reaching $700.[33]

The upright category, with sales of 4.3 million units, was 51 percent of the total market, but Sharp, Whirlpool, Sanyo, Metropolitan Vac, and Bionaire challenged entrenched Hoover and Eureka for share of a saturated market that was expected to grow only at a modest 5 percent pace, mostly replacement sales.[34]

Canister cleaners, with sales of 3.2 million units split evenly between straight suction and those with power nozzles, comprised 34 percent of the market. In 1985, Hoover had introduced its high end Dimension series of canisters, but newcomers Sharp, Whirlpool, Miele, and Moulinex/Hamilton Beach were all lines made overseas where canisters were fully featured, high-end models. Shop-Vac, Sears, and Genie Products dominated the wet-dry segment of the canister market, over half, with sales of 1.7 million units.[35]

Stick vacs, with sales of 1.3 million units, commanded 15 percent market share, but unit sales had grown since 1983 at a high annual rate of 13.5 percent and was expected to increase market share 10 to 15 percent in 1986 alone. A number were cordless, such as Eureka's Quickup, which would have added another 200,000 units. Other contenders challenging Regina's Elecktrikbroom were estimated at 60 percent of the market and included Panasonic, Moulinex/Hamilton Beach, Dazey, and Hoover, which led the pack. Hybrid vacs were also appearing, such as Bissell's successful Three-Way vac, intended for use on stairs and on the second level of homes; Dazey's Vac-Man, which could be operated as a hand-held, stick, or canister vac; and Panasonic's two models that could be operated as sticks or hand-vacs.

Hand-held vacs, with sales of about 7 million units, mostly by B&D, appeared to be the highest category of all, but some said it had peaked. B&D said that sales had leveled off, and expected growth only in the lower end, represented by its Mini-Buster, just introduced. Price wars had reduced the cost of some to $10. Competitors included Cosmo, Douglas, Eureka, Norelco, and Conair. Conair introduced a wet-dry version, and Cosmo had one with a utility light. Douglas made both cordless and corded versions, the later being between 10 and 15 percent of the hand-held market.[36]

We must take a side trip to England at this juncture, to follow the initial development of a vacuum cleaner concept that would transform the market more dramatically than any before. As early as 1979, James Dyson (b. 1947), a 1970 design graduate of the Royal College of Art (and a baby boomer) had begun thinking about vacuum cleaners. He had already invented the Ballbarrow, a modified version of the wheelbarrow using a ball to replace the wheel; the Trolleyball, a boat launcher with ball wheels; and a Wheelboat vehicle with enormous, ball-like tires that could travel 40 miles per hour on land or water. So he was already a successful business entrepreneur, as well as a creative industrial designer.

He got the idea of "cyclonic separation" of dirt from the spray-finishing room's air filter in his Ballbarrow factory, because it was constantly clogging with powder particles. He designed and built an industrial cyclone tower, which removed the powder particles by exerting forces greater than 100,000 times that of gravity.

Cyclonic filtration had for years been used in commercial and fixed central vacuum systems, but the only portable cleaners that used such a system were in the U.S.: the Newcombe Bagless patented in 1922 and introduced in 1928, which in 1936 was refined by Rexair; and the 1939 Filter Queen 200 canister type, made by the P.A. Geier Company, maker of Royal brand cleaners and distributed by Health-Mor, in Cleveland, Ohio. Ed

United States Patent [19]

Dyson

[11] **4,373,228**
[45] **Feb. 15, 1983**

[54] VACUUM CLEANING APPLIANCES

[76] Inventor: James Dyson, Sycamore House, Bathford, Bath, Avon, BA1 7RS, England

[21] Appl. No.: 140,497
[22] Filed: Apr. 15, 1980

[30] Foreign Application Priority Data

Apr. 19, 1979 [GB] United Kingdom 7913690

[51] Int. Cl.³ A47L 9/16
[52] U.S. Cl. .... 15/350; 15/346; 15/352; 15/353; 55/345; 55/429; 55/449; 55/459 R
[58] Field of Search .... 15/331, 335, 345, 346, 15/350, 351, 352, 353; 55/459 R, 459 A, 459 B, 449, 429, 345

[56] References Cited

U.S. PATENT DOCUMENTS

1,220,641	3/1917	Kent	55/459 R X
1,664,092	3/1928	Squires	15/346 X
1,759,947	5/1930	Lee	15/351 X
2,071,975	2/1937	Holm-Hansen et al.	55/459 R X
2,300,266	10/1942	Smellie	15/350 X
2,824,335	2/1958	Moffat	55/459 R
2,867,833	1/1959	Duff	15/350 X
3,308,609	3/1967	McCulloch et al.	55/449 X
3,425,192	2/1969	Davis	55/345
3,484,390	12/1969	Case	15/346
3,520,589	8/1950	White	55/459 R X

FOREIGN PATENT DOCUMENTS

494786 11/1938 United Kingdom 15/335

Primary Examiner—Chris K. Moore
Attorney, Agent, or Firm—Stevens, Davis, Miller & Mosher

[57] **ABSTRACT**

The invention relates to vacuum cleaning appliances.

The appliance of the invention includes a cyclone unit which is operable to extract dust and other dirt from the air flow therethrough and to deposit the extracted dust and other dirt in a chamber outside the cyclone and separate from the air flow through the casing of the appliance. The extracted dirt is removed from the appliance by separation of the cyclone unit from the casing.

The appliance is convertible to act both as an upright type cleaner or a cylinder type cleaner.

8 Claims, 11 Drawing Figures

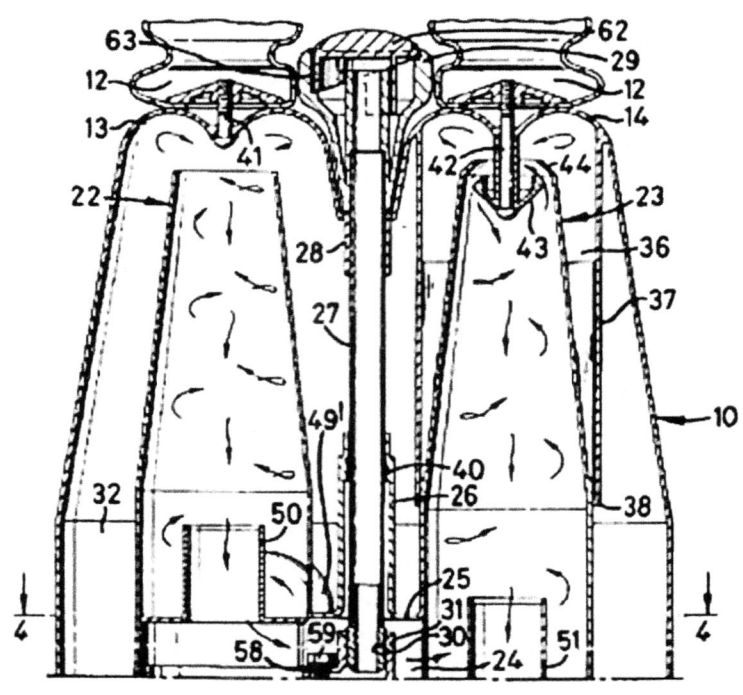

1986 Dyson patent 4,593,429. U.S. Pat. Off.

Yonkers had patented the system in 1937 and sold the patent to Health-Mor. These were both canister-type cleaners, and although they are still in production, their cyclonic patents had expired long ago. Dyson was probably unaware of either, as they were not well known, even in the U.S.

Dyson wondered if the same cyclonic principle of his cyclone tower could be applied to vacuum cleaners. He had been frustrated by his Hoover Junior upright cleaner's diminishing performance, as dust kept clogging its filter bag. Typically, this could reduce suction by 50 percent. On April 14, 1979, he filed an application that was published as English patent 18197, and on April 15, 1980, filed the same application that was granted February 15, 1983, as U.S. Patent 4,373,228.[37]

Over the next five years, Dyson would develop 5,127 prototypes, and by 1983 had developed a pink prototype G-force cleaner, but no manufacturer or distributor in the U.K., including Hoover, would introduce it because it would disturb the valuable market for replacement cleaner bags, worth about $500 million annually. Later, Hoover's vice president for Europe, Mike Rutter, would say on U.K. national TV: "I do regret that Hoover as a company did not take the product technology off Dyson; it would have lain on the shelf and not been used."[38] Unable to find any interest by major vacuum manufacturers, Dyson licensed his pink G-force to Apex, Inc., which launched it in Japan through catalog sales at a price of $1800, where it would win an International Design Fair prize in Japan. Dyson was granted another U.S. patent (4,593,429)[39] for his concept on June 10, 1986.[40] Still, he was only getting started. It would take ten more years before his name would become synonymous with vacuum cleaners.

Back in the U.S., an old name in vacuum history was getting new life. In the years following the war, competitors had out-performed Regina, and its profits had dwindled. Regina had introduced an "air-pulse" nozzle in 1971 that created dirt loosening vibrations with pulsating air, but Regina's problem with profitability continued and had inspired it to broaden its sales base by adding other cleaning products to its line. This strategy was unsuccessful, and in about 1984, an investment group founded in 1920 to acquire the Regina name, General Signal Corporation, now headed by Don Sheelan, purchased majority interest in Regina. Sheelan introduced the Regina Steemer and the Regina Housekeeper, both rug-cleaning machines, a rapidly expanding market, in the hopes of improving the company's bottom line. These moves were quite successful, as Regina advanced to the head of the rug-cleaning market. The Housekeeper was the first modern upright vacuum cleaner to carry on-board tools, which made above-the-floor cleaning handier. Royal would follow Regina with on-board tools on its Dirt Devil upright in 1990, and Hoover would do the same in 1991 with on-board tools on its Elite II upright (see below).

In 1986, a number of vacuum cleaner manufacturers made news. Rexair introduced its Model D4, the first major redesign of the Rainbow since 1955, and is considered by some as the best Rexair ever made.[41] That same year, Electrolux made a large advance into the American market with the acquisition of White Consolidated Industries, the home of such brands as Frigidaire and Westinghouse. Electrolux thus became an international name in major appliances. Concurrently, B&D introduced upgraded Dustbuster and Dustbuster Plus models with added batteries and an increased dirt capacity of 25 percent, but they were still in the external form protected by the original design patent.

In 1988, Hoover introduced its three-model line of Elite upright cleaners, the latest evolution of its best-selling cleaner for the last 30 years, the original 1957 Convertible, which had sold 30 million units over that period. Some Convertibles continued to be sold by Sears for a few years more,⁴² but the new Elite Models U4471, U4473, and U4455 quickly replaced the Convertible as Hoover's top sellers; they were designed with ease of manufacture and cost reduction in mind, since they were produced by automated processes.⁴³ In 1991, Hoover would introduce its Elite II, a line of uprights with on-board tools and a side-mounted hose for above-the-floor cleaning capability.⁴⁴

New vacuum cleaner introductions continued as usual. Douglas launched a five-model line of ReadiVac cordless stick vacs in 1988, as well as a stick vac called the Power Broom 2.⁴⁵ In 1986, Berkshire Hathaway had bought Kirby's parent Scott & Fetzer for $315 million. Berkshire managers made absolutely no changes to the Scott & Fetzer business or management. The next year, a vacuum cleaner named Kirby was one of the cartoon characters in the 1987 animated film *The Brave Little Toaster,* released by Walt Disney Pictures. Kirby in 1989 would rename its top cleaner, the Legend II, which had been produced in 1981 as the Heritage, in the mid–1980s as the Heritage II, and in about 1988 as the Heritage II Legend. The next Kirby introduction in the 1990s would be its Generation G Series, which would feature a "Tech-Drive" variable power assist that would eliminate 90 percent of the effort required to move the unit back and forth.⁴⁶

Acquisitions continued to plague the vacuum cleaner business landscape. Singer had experienced a loss of $20 million in 1987 and was acquired by Paul Bilzerian, who in 1988 sold 8 of Singer's 12 divisions. Ryobi Motor Products Corporation of Pickens, South Carolina, acquired the Motor Products Division of the Diehl Manufacturing Company, a division of Singer that made vacuum cleaners. Philip H. Diehl, a German immigrant who was first employed by the Singer Sewing Machine Company, had founded Diehl & Company in 1887, which in 1909 became Diehl Manufacturing Company, and later, the Diehl Division of Singer, providing motors for sewing machines and vacuum cleaners. In 1988 Ryobi took over as the source of Singer products to Sears Craftsman, contributing to the rise of Sears as a major retail force in electric floor care, and continuing to provide vacuum cleaners under the Singer name.⁴⁷

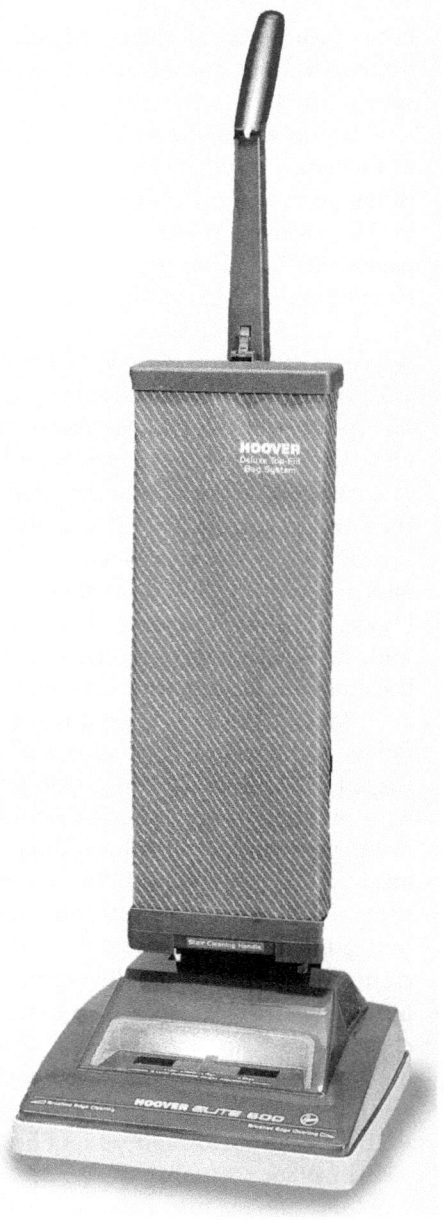

1988 Hoover Elite 600 Model U4473. HHC.

In 1988, Tacony Corporation acquired the Riccar America Company with headquarters in Fullerton, California, and a branch in Atlanta, Georgia, offering a line of vacuum cleaner and sewing machine products through an exclusive dealer network. N.J. "Nick" Tacony (1915–1984) had founded the St. Louis company in 1946 when he began selling and servicing sewing machines in his home. He expanded his business to become a wholesaler of sewing equipment. Nick's son, Ken (b. 1943), joined his father's business in 1970 and became C.E.O. in 1984, overseeing one of the largest American distributors of sewing machines, ceiling fans, vacuum cleaners, and commercial floor care products. In 1989, Tacony introduced the Simplicity line of quality vacuum cleaners.

In January 1989, Chicago Pacific Corporation, the Hoover Company's parent company, was trying to avoid a hostile takeover by an investment group; it was acquired for $1 billion in a friendly buyout by the Maytag Corporation, which sought to evade corporate raiders in the international appliance market. It was a good deal for Maytag, which had no presence in international markets, while Hoover had 13 plants in eight countries that manufactured and distributed washers, dryers, refrigerators, dishwashers, and microwave ovens, as well as vacuum cleaners. Hoover continued to make vacuums in North Canton, Ohio, under the Hoover brand name, but in 1991, would reorganize its European operations into Hoover Europe, and in 1993, would consolidated all vacuum cleaner production in Europe at its facility in Scotland.[48]

During the 1980s, the vacuum cleaner market grew from 7.4 million full-sized units annually to 11 million units. Growth was attributed to a positive economic climate and the transformation of retailing to self-service and mass merchandising. Consumers began to acquire more than one vacuum cleaner per household, a trend exploited by manufacturers and retailers, and electric floor care became a market driven by consumer demand for value.

As the new decade of the 1990s would soon demonstrate with the internet, computerization, and globalization, "the times they were a-changing again," and the vacuum cleaner industry would change with them.

Chapter 7

1990 to the Present

By 1991, there were more than 20 major U.S. electric floor care manufacturers by which 11 million full-sized units were being sold annually. Niche players in the market included Metropolitan Vacuum's upscale canister vacuums that had been around for 50 years, as well as Douglas Quikut, a sister company to Kirby under the Scott & Fetzer Company. Other niche players included Sharp, Oreck, Clarke, Riccar America, and Rainbow.[1]

The most dramatic change in industry was the explosion of computer technology and the so-called Information Age, which surpassed even the Electrical Age of the early 1900s in its impact on our economy, manufacturing, business practices, and social lives. In 1990, only 22 percent of U.S. households had personal computers, but by the end of the decade, the percentage would be 55 percent. This is to say nothing about the impact on manufacturing because of the use of computer aided design (CAD) technology in design and production. Autodesk's Alias industrial design software now enabled industrial designers to digitally manipulate form and color, replacing traditional hand drawn sketches and concept renderings. This process visualized final appearance in three dimensions, allowing designers to revise details and to transfer digital appearance specifications directly to ProEngineer and other engineering software for final production drawings and tooling. Such seamless CAD technology revolutionized the product development process, reducing lead times by many months. The installation of electronics and computers in new products enabled products to be "smart" and accomplish functions automatically and without complex and failure-prone mechanical components.

These developments naturally also affected the vacuum cleaner industry. As an example, Panasonic in 1991 achieved innovations in robotic vacuums and "fuzzy" technology, which sensed conditions such as how high a carpet was, or whether it had actually been cleaned or not. Sanyo-Fisher was developing technology to find dust mites, a common source of allergies, and several of its upscale canisters were said to capture and devitalize these little guys.[2]

James Dyson was still marketing his cyclonic cleaner in Japan with Apex Inc. catalog sales in 1991. Dyson, after being rejected by many major vacuum manufacturers and almost broke, had licensed his patent rights to Amway, which paid him an advance, but Amway changed its mind and wanted the money back. Dyson sued. Dyson then went to a small Canadian company called Iona, which also signed a licensing agreement with Dyson but when Iona saw an Amway dual cyclone called Clear Trak in a Sears catalog, it wanted to negotiate Dyson's royalties downward. He conceded (he had to, he was broke). Amway eventually settled and Iona Appliance's engineering lab began selling Dyson's design for dual cyclonic cleaning technology in vacuum cleaners sold under the trade name of Fantom.[3]

Meanwhile, industrial design was now being promoted directly to the business community. *Business Week* magazine in 1991 began its sponsorship of the annual national Industrial Design Excellence Awards (IDEA), initiated in 1980 by the Industrial Designers Society of America (IDSA), and promoted the award-winning designs each June in a special issue of *Business Week*. These awards demonstrated the power of design by enabling corporations to compete beyond the mechanical and performance qualities of consumer products in the arena of ergonomics, new typeforms, and unique eye-catching appearance. By 2004, this IDEA design awards program would become international, and by 2011, there would be 524 awards out of over 2,000 submissions from 39 countries. A jury of 20 outstanding international practitioners would select the design award winners.

Unfortunately, many vacuum cleaner companies were unaware of the growing power of industrial design, or they failed to use it effectively. They had forgotten the history of the 1930s, when exciting appearance design led many manufacturers out of the Depression, as well as the 1950s, when museums and manufacturers promoted Good Design aggressively to consumers. In too many instances, 1990s' manufacturers introduced new performance features, but continued to use the safe and traditional typeforms that consumers expected, in other words, imitating competition. A typical such manufacturer was the Hoover Company.

In 1992, Hoover introduced its PowerMAX upright cleaner U3745-910, which was self-propelled, as was its predecessor, the 1983 Concept Two. The PowerMAX featured Power Surge, enabling the user to flick a switch to momentarily increase power for tougher cleaning tasks. Cleaning tools were stored inside the front body of the cleaner behind a transparent cover. Two years later, in 1994, this cleaner was replaced with the Power Drive Supreme, Model U6329-930, also with the Hoover self-propelled feature.[4] Both designs were similar in appearance and were well executed, but were highly traditional in the sense that they used a similar rigid form innovated by the Dial-a-Matic in 1963, 30 years before, to house the tools, but still retained the traditional soft bag for dirt collection. In 1994, Hoover also introduced three models of its Steam-Vac, a self-contained, upright carpet cleaner with separate tank for clean and dirty water. Brushes, first stationary, then rotating, were added to later models. Other companies followed into the steam-cleaning category, such as Eureka's Dream Machine.[5]

In 1992, Bissell had decided to diversify because the floor care market had stagnated. To accomplish this, Bissell acquired Chicago-based Maxi Vac, Inc., a maker of wet/dry vacuum cleaners, and began boosting its research capabilities in the deep cleaning market. This paid off in 1992, when Bissell debuted its new carpet shampoo machine, called the Promax, later renamed the Powerlifter because of a copyright battle with Hoover (see PowerMAX, above). In 1993, Bissell introduced another canister cleaner with more attachments and capabilities, called the Big Green Clean Machine. It was promoted with infomercials, a medium not helpful to Bissell's reputation, but highly successful. In October 1993, Bissell introduced a smaller portable version of the device, the Bissell Little Green Clean Machine, first known as the Steam-Mate in the late 1980s. The updated version was designed in house by Bissell's design team led by Steven Umbach. It would become the hottest item in floor care from 1994 to 1995. In 1994, Mark Bissell would replace his father, John Bissell, as president and C.E.O., with John continuing as chairman.[6]

Designs did not always use traditional vacuum cleaner forms, but sometimes injected a sense of humor. In fact, industrial designers in the postmodern era were asking rhetorical questions, such as, "Why should a telephone *not* be shaped like a banana?" This was the revolutionary nature of postmodern design where there were no rules. Eureka in 1993 introduced its hot red Corvette Vac, a hand-held vac to be used in cars and styled to resemble Chevrolet's 1953 Corvette sports car. It was an instant success. Indeed, why should a vacuum cleaner *not* be shaped like a sports car? Eureka also introduced items at the high end to fill all categories of home cleaning. That year, Eureka claimed its highest sales ever, manufactured more than 100 different models of vacs, and held about 20 percent of the $600 million full-sized cleaner market, as compared to Hoover's 35 percent.[7]

1992 Hoover PowerMAX Model U6329-930. HHC.

Sales were not hurt when vacuums once again entered the national entertainment world (remember Stan Kann in the 1960s and 1970s?) In 1993, Robin Williams, in the disguise of *Mrs. Doubtfire*, waltzed around the room with a Hoover vacuum cleaner in one of the film's funniest bits. That same year, the third witch (Kathy Najimy) in the film *Hocus Pocus* rides off on a vacuum cleaner, rather than on a broom. In 1994 12-year-old Fred Stachnik of Milwaukee, Wisconsin, made a big hit on the *Tonight Show* with Jay Leno by demonstrating his collection of 20 or so vacuum cleaners, and also appeared on MTV's *Jon Stewart's Show*, the *Maury Povich Show*, the *Late Show with David Letterman*, *American Journal*, and others over the next year. Fred's incredible knowledge of vacs, intensively developed since childhood by collecting, restoring and fixing vacs, made him a popular TV attraction as "the kid who knows all about vacuum cleaners."

It all started when Fred was 10 and insisted his grandparents take him on a 500-mile road trip to visit the Hoover Company in North Canton, where he received a red carpet treatment and made the local newspapers. Soon, his reputation escalated nationally. After his early TV career, he would be invited to membership in the Vacuum Cleaner Collector's Club, people would send him their old vacs, and he would work in a local vacuum repair shop while in high school. While in college, he would work at Hoover's factory service center near Milwaukee. After graduation from college, in 2007, he would move to Durham, North Carolina, and would work part-time with a local vac store in addition to his full-time professional career.

In 2011, Fred would be working for Wells Fargo in Minneapolis, Minnesota, and his collection would include over 250 vacs.

Meanwhile, in the U.K., James Dyson was finally entering the market big time. In 1993, royalties from the sale of his G-force cleaner in Japan had enabled him to open a research center and factory in Malmesbury, Wiltshire, England, where he began producing his DC01 Dual Cyclone vacuum cleaner under his own name with a strong advertising campaign emphasizing that unlike other vacuum manufacturers, it did not require the continuous purchase of replacement bags. The DC01,

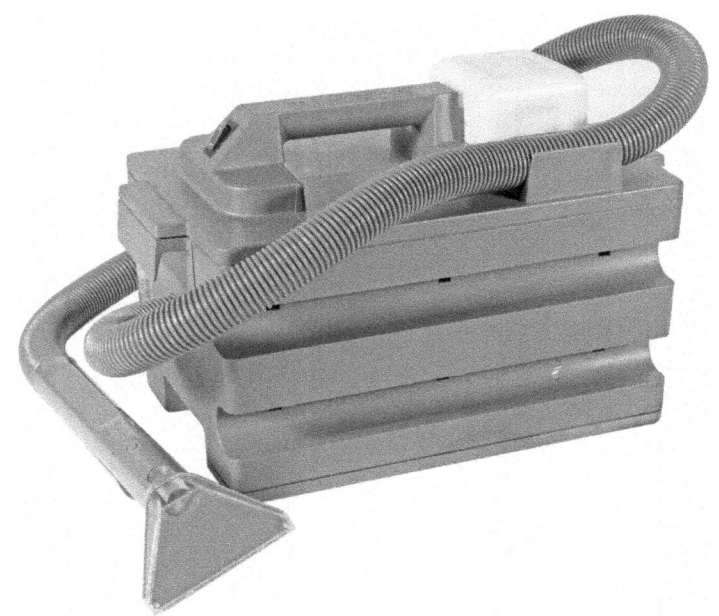

1994 Bissell Little Green portable deep cleaner. MIAD.

priced at £200 (about U.S. $325), twice the price of a conventional vacuum, included a hose to reach the top of the stairs and on-board tools.[8] Soon the Dyson cleaner became the rage of the industry in the U.K., recapturing excitement not seen there since Hubert Cecil Booth's Puffing Billy of nearly a century before. Booth's secret technology was simply "suction" renamed "vacuum." Dyson's not so secret technology was "cyclonic" action in a vacuum requiring no bags, but his real secret was innovative design.

Cyclonic cleaners do not use disposable filtration bags. Instead, the dust is separated into a detachable, cylindrical collection vessel or bin. Air and dust are sucked at high speed into the collection vessel in a direction tangential to the vessel wall, creating a fast-spinning vortex. The dust and dirt particles are moved to the outside wall of the vessel by 150,000 gs of centrifugal force, where they fall due to gravity. In fixed central or ducted vacuum systems, where cyclonic systems are used, the cleaned air may be exhausted outside without further filtration. In portable systems, the cleaned air from the center of the vortex is expelled from the machine only after passing through a number of progressively finer filters at the top of the container. The system does not lose power due to clogging of filter bags, but retains full power until the collection vessel is almost full.

However, the filters must periodically be cleaned or replaced to insure that the machine continues to perform efficiently (the same principle as replacement bags), and the dirt must be emptied by hand, just as before replaceable paper bags were invented, but this did not appear to discourage buyers who were delighted to learn that they no longer needed to buy those expensive, sometimes hard-to-find, replacement bags. Dyson's slogan, "say goodbye to the bag," was more attractive to the public than the claimed suction efficiency of cyclonic action. By 1995, despite costing twice as much as ordinary cleaners, the Dyson DC01 Dual Cyclone technology upright became the fastest-selling vacuum cleaner in the U.K.

That same year, Dyson introduced his DC02 Absolute, the first Dyson dual cyclone canister with both HEPA filtration and a bacteria-killing screen. HEPA, an acronym for High Efficiency Particulate Air, filters are the best-known filters for removing at least 99.97 percent of particulates such as dust, animal dander, smoke, mold and other allergens that are 0.3 microns or larger from the air, thus improving air quality. The Absolute quickly became the U.K.'s second-highest seller.

Next to come out was the DC03, a lightweight, low-profile upright cleaner comparable to a stick vac, but like other Dyson designs, completely different in character from the competition. The year 1995 was the beginning of the Dyson dynasty, which would become another Murray Spangler moment in vacuum history, and which would change the vacuum cleaner world.

Of course, in analyzing the reasons for this success, it is obvious that Dyson cleaners offered a relatively new cyclone technology, combined with a strong user appeal based on eliminating the need for replacement bags. But it is also relevant to consider the degree to which appearance design contributed to its success, particularly because Dyson cleaners are outstanding examples of excellent industrial design. James Dyson, of course, in addition to being a successful entrepreneur and manufacturer, is an academically trained and practicing professional industrial designer.

What constitutes an excellent industrial design? It is actually easier to define what an excellent industrial design is *not*. First, it is not a static definition, because excellence changes with time, peer

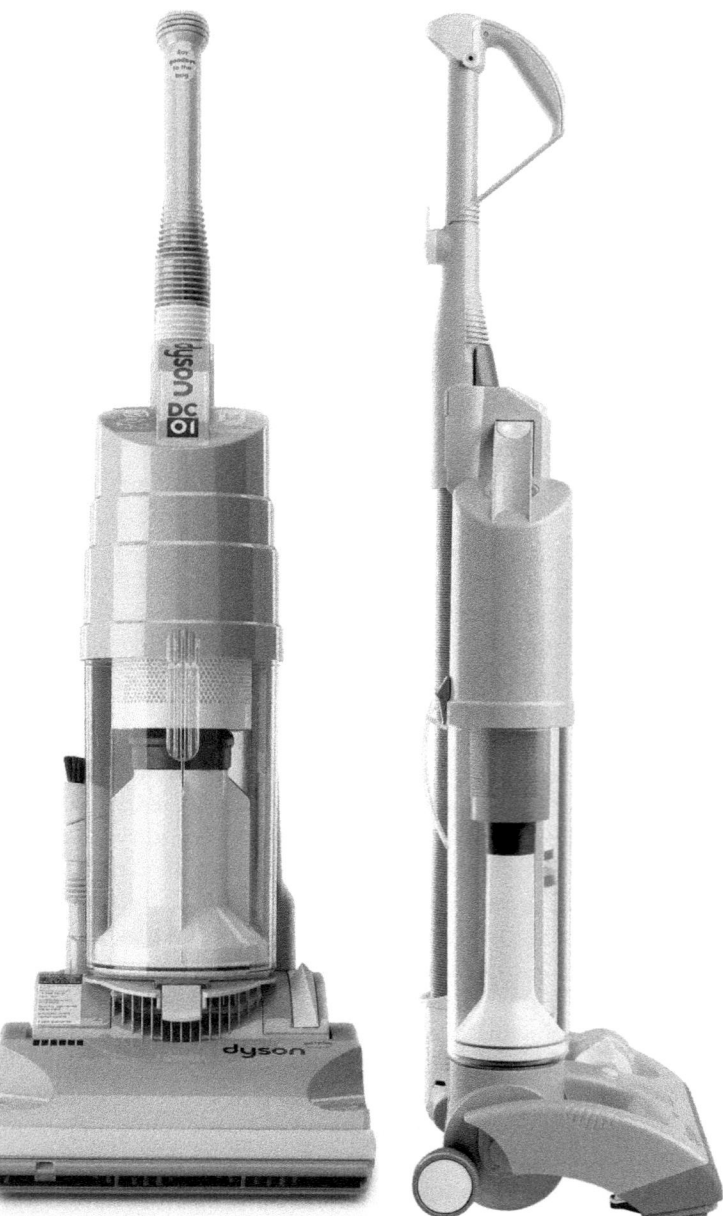

Left: 1993 Dyson DC01. Dyson. *Right:* 1995 Dyson DC03. Dyson.

review, popular acceptance, and the social environment. Excellent designs of the 1930s, or 1950s, or 1970s were no longer considered to be so in the 1990s. Tastes change over time. Second, it is not an imitation of successful competitors, a practice all too common in the industry, whether intentional or not. Finally, it is not possible without professional design expertise. One cannot expect esthetic design judgments to be made by people unfamiliar with abstract visual and cultural principles anymore than one can expect engineering judgments to be made by someone unfamiliar with mechanical principles.

Just as engineers use mechanical principles to achieve a unique and effective functional performance result, industrial designers use abstract visual principles to achieve a unique and effective appearance design result. But whereas functional performance can be measured quantitatively, appearance excellence can only be measured emotionally, similar to art. If it visually and effectively conveys to consumers a sense of uniqueness, functional appropriateness, timeliness to the changing culture, added value, and user friendliness, it probably succeeds. These qualities require perceptive market analysis, innovative thinking, and creative execution of details.

The best definition of excellent industrial design is to look at and analyze the Dyson DC01 and DC02 visually, noticing first that they avoid imitation, and in fact, do the exact opposite. Look at the illustrations to see how the shapes and colors are as different as imaginable from competitive upright or canister cleaners (unfortunately, only color illustrations can show the dramatic yellows, blues and purples). They set new typeforms for cleaners as different as possible from the traditional forms of the previous 30 years and are in no particular style. Notice that they do not rely on names, labels, or graphics to promote their features, performance, or manufacturer. Rather, they rely on their shape and configuration to communicate the unique function of cyclonic action with a bright yellow cylinder, the most important visual detail on the cleaners, within a clear plastic dirt container, or "bin" in which users can actually *see* the cyclonic principle in action. They emphasize mobility with exposed wheels, and identify the important business end with nozzles highlighted in bright yellow accents.

Each element of these designs is configured to express and dramatize its indi-

1995 Dyson DC02. Dyson.

vidual function. Mechanical and visual forms are merged into an integrated, organic expression of elegance. This is obviously the work of a single professional trained in the abstract arts of sculpture and color, fully aware of the changing design trends globally, and with intimate knowledge of the mechanical principles involved. Excellent industrial design is an art form that is inimitable and rare. It is in stark contrast to the traditional addition of artistic cosmetics to the functional mechanical forms developed by engineers, and to the inherent corporate tendency to imitate successful competitors. Other historical cleaner designs that broke as many traditional rules of the trade and innovated new mechanical and visual principles include the Air-Way upright of 1920 and the Hoover Dial-a-Matic of 1963, both described earlier. There are probably others that could be included in this distinguished category.

To express his commitment to principles of historic design in 1996, Dyson would manufacture 20,000 special editions of DC01, DC02, and DC04 (the latter the successor to the DC01) cleaners, inspired by the Dutch De Stijl art/design movement of 1917, which was characterized by color paintings by Dutch artist Piet Mondrian (1872–1944) and furniture by Dutch architect Gerrit Rietveld (1888–1964), both of which dramatically emphasized the integral relationship between color and physical structure. The cleaners were in bright purple, red, and yellow. A promotion, to be sure, but one that focused on the historical principles of design, not on the verbalization of names, features, performance, power, or any of the typical manufacturer promotional language. It promoted industrial design, Dyson's overwhelming advantage over competitors. With the success of his products in global markets, Dyson had become a national celebrity as industrial designer, entrepreneur, and successful business tycoon. In 1997 he published his autobiography, *Against All Odds*, and became a member of the prestigious British Design Council, as well as a trustee of the U.K.'s National Design Museum.[9]

Meanwhile, on the "what's in a name" front, back in the U.S., Royal in 1993 had filed an application to register "The First Name in Floorcare" as a trademark. Hoover began opposition proceedings in 1994, contending that the mark caused confusion with Hoover's slogan, "Number One in Floorcare," used since the mid-1970s. In 1999, the U.S. Patent Office would dismiss Hoover's claim, and an appeals court in 2001 would uphold that decision.[10]

Bissell broadened its line of floor care products in 1996 through the acquisition of the Singer line of upright vacuum cleaners, as well as deep cleaners, from Ryobi Motor Products Corporation, which had been making vacuum cleaners under the Singer name since 1988. This helped Bissell improve its presence in the low-end sector of uprights, as the company's former cleaners were mostly high-end. China's Techtronic Industries (TTI) would acquire Ryobi's U.S. subsidiary from Bissell in 2000, and would continue to manufacture vacuums under the Singer name.

In 1997, Bissell launched its first upright cleaner, the PowerSteamer. Later that year, Bissell built up its line of deep cleaners with the Steam n' Clean model (less than $150); in 1998, the Spot Lifter ($59); and in 1999, the PowerSteamer ProHeat Plus ($299), the first to contain a heating element. Bissell had established itself as a leader in the deep cleaning of carpets by homeowners, as an alternative to commercial carpet cleaning by service trucks. The Hoover Company filed two lawsuits against Bissell in May 1998, alleging patent infringements on certain features of Bissell deep cleaners and upright cleaners.[11] Hoover had introduced three models of its Steam-Vac in 1994, which was the first departure from large, cumbersome carpet cleaning machines that had to be hooked up to a sink. Although many companies use the term "steam" in their products, in actuality it is only hot water.[12] Bissell countersued, but the parties reached a settlement in May 1999, soon after the suits went to trial. The agreement was not disclosed but Hoover stated that the settlement "included an agreement regarding future use of Hoover extractor patents under license." By the late 1990s, deep cleaning machines had replaced carpet sweepers as the core Bissell business. Although the company still held 90 percent of the sweeper segment, that was only 5 percent of its total sales. In 1999, Mark Bissell told Home Furnishing Network's *Weekly Newspaper*,

> Our vision is to continue to be a family-held company ... I have three kids. My brother has three kids. So there are a lot of Bissellettes running around. We hope that someone from the next generation will rise up from the ranks and run the company.[13]

In 1996, the Tacony Corporation of Fenton, Missouri, had purchased Powr-Flite, a vacuum cleaner parts distributor since 1967, and began the manufacture of floor care products in Fort Worth, Texas. In 1997, Tacony brought production of its Simplicity and Riccar vacuum cleaners from Taiwan to the United States. As a result, Tacony was now manufacturing 13 lines of vacuum cleaners at its U.S. facility in St. James, Missouri. This enabled Tacony to proudly place the "Made in USA" label on its products, a distinction fast disappearing on the products of many manufacturers. Tacony would also agree to house an extensive historical collection of 688 vacuum cleaners as a museum, assembled over many years by vacuum cleaner collector Tom Gasko, when Gasko would become a Tacony employee in 2009.[14]

About this time, Sanyo introduced its Transformax 3-in-1 vacuum cleaner Model SC-150, which was designed to function as an upright, canister, and hand-held cleaner by detaching and reattaching components. It was very lightweight, only 7 pounds, but required a power cord.

By 1997, Hoover introduced its Wind-Tunnel Deluxe upright Model U5465-900, in the U.S., emblazoned with promotional

1997 Hoover Wind Tunnel Deluxe Model U5465-900. HHC.

graphics on every available surface. Avoiding the terms "cyclone" or "cyclonic action," because it was not cyclonic but actually a bagged cleaner, Hoover claimed the Wind Tunnel had the ability to "pick up more dirt than any other clean air upright," due to "enhanced and revolutionary design of its agitator cavity." Its embedded DirtFINDER,™ introduced in 1995, communicated to the user with a red light and a green light which indicated the areas of the carpet that were clean, and those that needed more vacuuming. Its features included a six-stage micro-filtration system, as well as covered but visible tool storage, an attached extra-stretch hose, and an additional hose that provided 20 more feet of reach.[15] That same year, Hoover returned to the principle of "clean air" introduced in 1963 by its Dial-a-Matic but which it had abandoned in 1978, when the Dial-a-Matic was discontinued.[16] Hoover would introduce a Self Propelled WindTunnel upright in 1998, the same year it announced the production of its millionth WindTunnel upright cleaner.[17]

The environment, ecology, energy conservation, health, global warming and sustainability had become not only powerful public political policy issues, but were beginning to resonate effectively as sales features for diets, exercise, organic and healthful foods, medications, detergents, and many related products, including vacuum cleaners. Many people were carefully selecting products based on their perception of features or of companies that recognized and addressed these global concerns. No longer were they satisfied with only mechanical and appearance features, but wanted products that expressed their philosophy of life styles and their concerns for the popular political and global issues of the day. They wanted everything to be "green," the term that seemed to encompass all worthy environmental objectives. Vacuum manufacturers began to focus again on their traditional values of health and a cleaner environment for the home. Cyclonic action or not, filters became an important feature to improve and promote.

HEPA filters, developed in the 1950s and described earlier, became popular in household vacs as people became more conscious of pollutants and dust in household air. Hypoallergenic vacuum cleaners would be developed to minimize dust emission. They would consist of three components: the motor, filter, and bag or cup, creating a completely closed system. They would generally use HEPA-type filters to trap a large amount of very small particles that other vacuum cleaners would simply recirculate back into the air in the home. Hypoallergenic bags would come in two pieces, an inner and outer bag. The inner bag would collect dirt and debris, and be thrown away. The outer bag would take the place of a filter, filtering dust and allergens out of the air. Together, the systems not only would remove the dirt, but would clean the air as well.

Many vacuum manufacturers jumped on the ecological bandwagon. Eureka had introduced its Powerline Plus Victory Model 4440 in 1996, and then its Victory EnviroVac in 1997, its first cyclonic cleaner with no bags and a true washable HEPA filter. Eureka claimed it to be a "green" vacuum, meaning environmentally friendly, of course. It was claimed to save 6.25 kilowatt hours of energy, and it would have had the same reduction in carbon dioxide emissions as removing 955 cars from the road, *if only* one-quarter of U.S. homes would switch to it. The cost of the EnviroVac was $72. Rexair in 1998 introduced its Model E series cleaners, with a HEPA filter on the exhaust, in response to a *Consumer Report* complaint that greasy dust passed through the water and was not trapped.[18]

B&D in 1998 sold its household product business to Windemere Durable Holdings, which established Applica Consumer Products, Inc., in Florida. Applica became the exclusive licensee for B&D household products in North, South, and Central America. But B&D retained the one important household product that enabled it to enter the household market

20 years before in 1979: the Dustbuster. By this time, total Dustbuster sales since 1979 were estimated at about 150 million units, bringing in about $6 billion in sales for B&D. In 1999, B&D redesigned its Dustbuster for about the sixth time, this time in a softer form with a closed handle, and with rechargeable batteries totaling 7.2 volts which could be replaced after they lost their ability to recharge efficiently, usually a period of five to seven years.

By the late 1990s, there was a sea change in the global manufacturing market. More and more manufacturers were shutting down their U.S. manufacturing operations and turning to Asian-based third party contract manufacturers. In 1999, Techtronic Industries (TTI) in Hong Kong introduced a new strategy by recreating itself as a brand-name producer. That year, TTI acquired Vax, a leading producer of vacuum cleaners and floor care products for the U.K., New Zealand, and Indian markets. Vax had been founded in 1977 in Worcestershire, England, launching its first product that combined wet and dry vacuuming with a carpet washing function. It was first sold door-to-door, but in the early 1980s distributed through the U.K.'s retail network, as well. TTI would acquire a sequence of name brands, including Ryobi in 2000 and John Deere's Homelite brand in 2001.[19]

According to 1999 industry sales statistics, all U.S. vacuum cleaner sales grew by 4 percent. Although central vacuum systems were only 1 percent of total sales, this particular category increased by 17.6 percent. This was attributed to the ease in which they can be installed (in less than one day) and the benefits of venting microscopic dust outside the living space, keeping the interior free of pollutants. According to Mr. Coghlan of Beam Industries (a central vac manufacturer), 60 percent of U.S. households owned two or more full-sized vacuum cleaners, and 30 percent owned three or more.[20]

By 2000, Dyson's patents for cyclonic action had expired and many manufacturers adopted the technology, but that same year, after extended litigation, Dyson won a major judgment against Hoover U.K. With sales declining, Hoover U.K. had earlier responded competitively to the obvious success of Dyson cleaners by introducing a version of a bagless upright cleaner, its Triple Vortex

1996 Eureka Powerline Plus Victory Model 4440. MIAD.

vacuum cleaner, which it promoted with £5 million. The judgment found Hoover U.K. guilty of infringing Dyson's patent, and required Hoover U.K. to pull the Vortex from the market, as well as pay £5 million to Dyson.[21] At the same time, Iona Engineering's Fantom Technologies, Inc., was still selling Dyson's licensed designs in Canada under the Fantom label.

Hoover U.S. entered the market with its first bagless cleaner in 2000 with the Bagless Upright Vacuum Cleaner models U5280-900, U5288-900, and U5294-900, offering a twin chamber system that held all of the large particles of debris in one empty half of the container; all of the lightweight dust was pushed through a screen and ultimately stopped by a cylindrical HEPA filter. This cleaner was a modified version of the Elite that had been sold in numerous variations and under many different names through the '90s. It was a "direct air" system that blew the dirt into the dirt cup.[22]

Later that year, Hoover's first WindTunnel Bagless upright Model U5750-900 was introduced. It did not have cyclonic action but featured "WindTunnel" technology combined with the "Twin Chamber™ System." WindTunnel technology was in reference to the redesigned agitator cavity, which picked up dirt and prevented the dirt from being redistributed into the carpet as cleaners without WindTunnel technology could often do. It basically picked up more dirt, more quickly, on the first pass, as its clean air system *pulled* the dirt into the Twin Chamber collection tank.[23] It out-cleaned all other bagless uprights on the market, according to tests conducted in accordance with the American Society for Testing and Materials's (ASTM) Standard Test Method F608, the only industry-recognized cleaning effectiveness standard. It also added a powered hand tool that also incorporated WindTunnel technology. The following year, Hoover introduced the Self Propelled WindTunnel Bagless upright with an Embedded DirtFINDER™ feature, its electronic dirt-detection system.[24]

In 2000, Air-Way introduced its all-metal Signature Series Air-Way Centurion 2000, designed by Tom Gasko. It was the first major change to its Sanitizor in about 50 years. The Centurion was in black, chrome, and "purple haze" colors, and featured: a new HEPA bag with 28 layers of cellulose; a Cen-Tec power head with unique bristle rows; a direct connect wand and hose handle; the first Select-A-Flow two-speed motor ever offered by Air-Way; and a Sentry safety light. Eight thousand were sold. Some collectors feel it was the best vacuum ever produced by Air-Way.[25]

Since its first Model C1 in 1946, mentioned earlier, the Interstate Engineering Corporation (I.E.C.) had introduced subsequent models of its Compact canister cleaner, Models C2 through C9, every few years with virtually no change in appearance until about 1982. I.E.C. moved around a few times, then was reorganized and sold off in 2000 to Tri-Star Enterprises, LLC. Tri-Star and Air Storm companies are run by relatives of the original founders of I.E.C., and under those new names they produce vacuum cleaners that closely resemble the original Compact, similar to the 1961 model shown here.[26]

One of the oldest manufacturers of vacuum cleaners came to a regretful end. Following a leveraged buyout and experiencing management changes in the late 1980s, Regina had merged with Electrolux, and Regina's management received an ownership stake in Electrolux, but after a year of sharing research, engineering, and design facilities, and ironing out quality problems, Regina and Electrolux amicably split.[27] An investment group headed by its C.E.O., Don Sheelan, had been managing Regina since about 1984 and had been struggling to increase the company's sagging bottom line. This was critical for Sheelan, since he had mortgaged all his personal assets to purchase rights to the company and needed cash from company profits to pay personal notes coming due. Sheelan had incurred the debt to impress his sister, a very savvy investor in her own right, and with whom he been intensely competitive since birth. Sheelan and his team knew nothing about the industry, but they concluded that the old product line was "over-engineered," and they needed to develop a whole new generation of vacuum cleaners. Needing cash to do so, Sheelan artificially boosted profits for a few quarters by manipulating the cut-off dates for sales and expenses, to drive the stock higher.

Sheelan's scheme worked, and a new generation of vacuums called the Housekeeper was engineered and put on the market with a multimillion dollar ad campaign, and cleaners were flying off the shelf. Immediately, however, there was trouble in the field. Sheelan had replaced the vacuum's metal parts with plastic ones, and they were melting because they couldn't withstand the heat during use. Soon, warehouses were stacked with returns. An annual audit soon revealed the firm's cooking of the books, and Sheelan threw himself to the mercy of the courts. He spent a year in jail and was hit with millions of dollars in civil judgments that he was unlikely to ever pay. Regina folded, and investors and creditors lost $40 million.[28]

Creditors soon sold Regina rights to a sequence of entities, including Venture Management Support (1989); TRC Acquisition (1989); Pass-Port Ltd. (1994); Philips Electronics, North America (1995); and Royal Appliance Manufacturing Company in 2000. Royal repackaged Regina products using the Home Depot brand name.[29] There is still one model left that carries the Regina name: the Regina RG3100, made by Oreck.

In 2001, Eureka debuted its BOSS bagless,

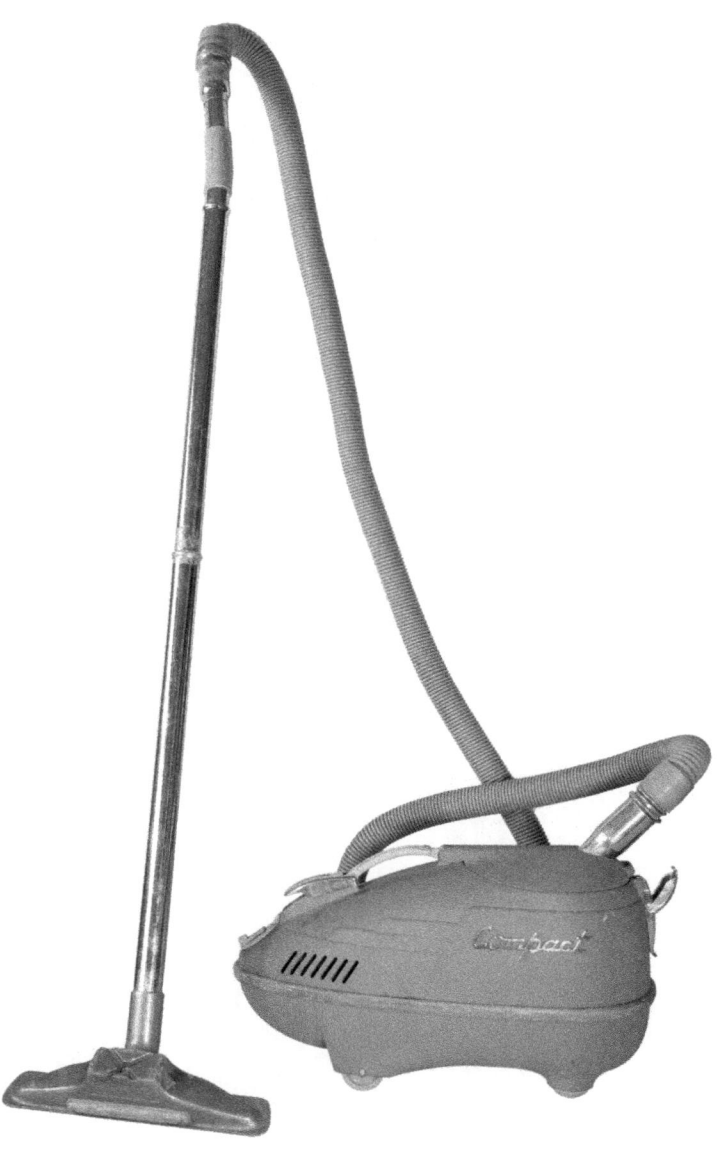

1961 Compact Electra C6. MIAD.

cordless upright cleaner Model 570, which weighed less than eight pounds and let users know when it was time to empty the cup. Sanyo debuted its DirtHunter bagless upright, featuring its DirtCompactor System that compacted the dirt, providing a longer time between disposals. Hoover introduced a SteamVac Widepath with a 14-inch-wide nozzle,[30] and a new product category, Floor MATE, the only product available that could vacuum, scrub (using six rotating brushes), and dry a variety of hard-floor surfaces. Hoover also partnered with Reckitt Benckiser Inc. for cleaning solutions to be used with the Floor MATE. A special formula cleaner for wood flooring was made with Old English and another cleaning solution with Lysol brand disinfectants for other hard surface floors.[31] Electrolux won the Janus design award in France and the Swedish Design Engineering Prize with its Oxygen canister vacuum cleaner, with its ergonomic Backsaver handle that can be swiveled to make it easier to clean overhead areas or low objects without risk of back strain, and its super clean exhaust air (100 particles per liter).[32]

In the period from 1996 to 2004, the futuristic, but prototype, robotic vacuum cleaner originally demonstrated by RCA Whirlpool in 1957 became a commercial reality. A number of companies developed robotic vacuum cleaners, usually with limited or no suction power. These cleaners moved autonomously in "random bounce" mode across a floor, navigating

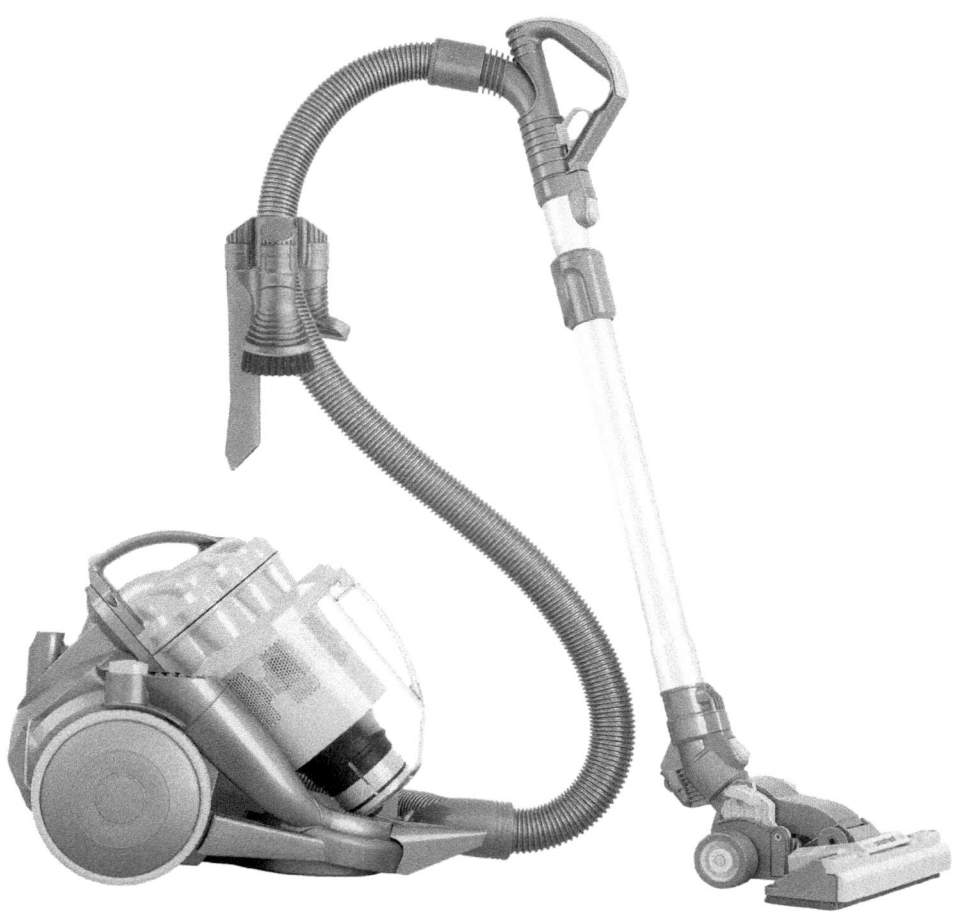

2002 Dyson DC08T. Dyson.

around furniture, collecting surface debris from the floor into a dustbin by using motorized brushes, and returning to a docking station to renew their batteries. In some cases, they emptied their dustbins into the dock, as well. Most were designed for home use, although some were designed for use in offices, hotels, hospitals, etc.

One robotic cleaner, the Trilobite (named after an extinct group of fossils from the Cambrian period that were similarly flat and circular in shape), was developed by Electrolux and introduced in 2001. It was probably the first, since it had been first demonstrated on British television in 1996. It contained a vacuum cleaner and ultrasonic sensors to avoid hitting objects.

Some robots, like the Roomba, were equipped with an impellor motor to create an actual vacuum. Roomba was sold by iRobot and introduced to the market in 2002. It was 13.4 inches in diameter and 3.5 inches high. iRobot was founded in 1991 with the initial name of "IS Robotics" by Helen Geiner (b. 1967), Rodney Brooks (b. 1954), and Colin Angle for the purpose of building robots that do all sorts of things, from climbing walls to squeezing through narrow pipes. In 1993 they got their first big government contract developing the "Ariel Underwater," a crablike robot programmed to detect or place underwater mines. Soon, they changed their company name to iRobot and located in Somerville, Massachusetts. iRobot designed the "Packbot," a 40-pound tanklike robot designed to scout dangerous military territory. They were used at the World Trade Center site in 2001 to search for victims, and later in Afghanistan and Iraq in combat situations.

Other early robotic cleaners included the Robomaxx (no vacuum); the Intellibot; the FloorBot, and Dyson's DC06, which was never released due to its high cost for home use, but which is still under development. By the end of 2003 about a half million robotic vacuum cleaners would be sold worldwide,[33] and by 2011, they could be connected with software to computers or smart phones to increase their capability. A research survey by École Polytechnique Fédérale de Lausanne (EPFL), one of the two Swiss Federal Institutes of Technology located in Lausanne, Switzerland, concluded that after using robotic cleaners for two weeks, only three out of nine researched families liked them. The negatives? They didn't like having to not leave things on the floor, or adjusting furniture to accommodate the robot. We'll just have to wait for a final judgment on their commercial success.[34]

By May 2001, Dyson had 29 percent of the European vacuum cleaner market by volume, and 52 percent by value. It was Europe's top-selling cleaner. That year Dyson developed Root Cyclone technology by replacing one large cyclone with several smaller ones, claiming that this increased suction power by 45 percent. The system first appeared in the 2001 DC07 upright, and subsequently in the DC08 canister cleaner, which, in a variation known as the DC08T, included a telescopic wand.[35] Dyson cleaners were first sold in the U.S in 2002. The many manufacturers that followed Dyson with cyclonic systems included Hoover, Bissell, Eureka/Electrolux, Kenmore, Dirt Devil, Sanyo, and Filter Queen. The cheapest competitive models were no more expensive than conventional cleaners. But most lacked the sophistication and artistic elegance of the Dyson designs in form, color and configuration. Competitors imitated the technical innovations, but not the award-winning industrial design style that had much to do with Dyson's success. In 2004, Dyson's DC07 ($420), Kenmore's Bagfree ($129.99), Bissell's Bagless upright 6595 ($172.82), Eureka's 4885BT Whirlwind True Bagless Cyclonic ($217), and Dirt Devil's Vision Bagless ($89.99) would be tested for cleaning effectiveness by the Wall Street Journal's Catalog Critic. The Dyson was rated "best overall" and the Kenmore, "best value."[36]

Many consumers agreed with the Wall Street Journal's assessment. By 2004, Dyson had captured 21 percent of the American market in dollar sales, a 350 percent jump over 2003, with the sale of 891,000 uprights, the most popular type. That same year, when 18 million units were sold, Hoover fell to 16 percent and Kirby to 14 percent. Dysons retailed for $420 to $620, compared with the $100 or so for the average vac. Hoover's competitive WindTunnel bagless sold for $390.

American Electrolux had continued to deteriorate due to using a cheaper and short-lived motor, which hurt Electrolux's reputation. In North America, the Electrolux name had been long used by vacuum cleaner manufacturer Aerus LLC, originally established to sell Swedish Electrolux products in the U.S. Since 1974, Electrolux-made vacuums had carried the Eureka brand name in the U.S., but in 2000, Aerus transferred trademark rights back to the Electrolux Group. In fact, Aerus would stop using the Electrolux brand name in 2004. While Electrolux continued to make Eureka-branded vacuums, it also began selling Electrolux-branded vacuums. In 2003, the name Electrolux was sold back to Electrolux Sweden, and the American Electrolux USA would change its name to Aerus, a relatively unknown name that would take a while to become readily recognized.[37]

The difficulty of changing to a little-known brand name illustrates why many acquisitions of well-known brands retain their historical names. In the case of American Electrolux, its new name (Aerus) was legally required, but the company continued to use the term "Lux" — the 1912 brand name used by Wenner-Gren — to describe many of its models, and it proudly referred to its history back in 1924 when Electrolux was first imported to the U.S.

In 2003, Kirby was the largest source of revenue for Scott & Fetzer, with about 500,000 sales per year, a third of which were outside the U.S. Scott & Fetzer sold the Kirby cleaners to about 835 factory distributors, who in turn, sold the vacuums door-to-door. Kirby's Ultimate G Diamond Edition concluded Kirby's G Series vacuums in 2005. The series included the G3 (1990–1993), G4 (1993–1996), G5 (1996–1999), G6 (1999–2001), and Ultimate G (2002–2004). In 2006, the Diamond would be replaced by the Sentria, a HEPA-rated vacuum cleaner that used disposable bags, which was a complete home care system that converted into twelve separate units, all powered by the same power plant. It converted into an upright cleaner, canister vacuum, carpet shampoo system, floor buffer, and additional uses.

Unfortunately, Kirby has been subject to relentless criticism by consumer protection agencies, going back to the 1960s and 1970s, for its overly aggressive door-to-door sales, often to people who could ill afford a $1,500 gadget. As of 1999, 15 of 22 state agencies had received more than 600 complaints in just a few years, for violations of state consumer protection laws. Many of these involved older customers who lacked the will to stand up to grueling sales pitches and exorbitant prices. In one example, an elderly couple was unable to remove three Kirby salesmen from their home for over five hours. Between 1998 and 2001, in Alabama alone, more than 100 lawsuits were filed against Kirby, resulting in nearly $2 million in judgments and settlements. In 2001, the West Virginia attorney general obtained more than $26,000 in refunds and credits for dissatisfied Kirby buyers. In 2002, $13,000 was obtained in refunds for 13 senior citizens by an agency. In several states, Kirby was held responsible for rapes committed by salesmen. It is obvious that door-to-door sales, though often effective, can include a serious risk of lawsuits, legal actions, or worse.[38]

Two major industry organizations born at the beginning of the electrical age finally got together. In 2003, the Association of Home Appliance Manufacturers (AHAM), founded in 1915, merged with the Vacuum Cleaner Manufacturers Association (VCMA), founded in 1913. The VCMA was assimilated into AHAM's floor care division. The vacuum cleaner, once one of the first unique household electrical appliances, had finally been subsumed into the infinite landscape of home appliances.[39]

Another historical brand name was also about to be subsumed. Maytag, the parent company of Hoover, lost $9 million in 2004 and became the subject of a takeover battle between a private investment group, a Chinese appliance manufacturer, and the Whirlpool Corporation. In the 1990s, Whirlpool had initiated an aggressive international strategy and became one of the most globally diversified companies in the world. In 1989, it had entered a joint venture with N.V. Philips and marketed its major home appliances in Europe. In 1990, Whirlpool formed a joint venture with Matsushita Electric to operate Whirlpool's 750-employee vacuum cleaner plant in Danville, Kentucky, which made vacs under the Kenmore name for Sears, Roebuck & Company. Joint ventures in China and India opened up the Asian market to Whirlpool. By 1999, it enjoyed revenues of $10.5 billion.[40] On December 22, 2005, Maytag stockholders agreed to sell Maytag to Whirlpool, ending Maytag's 112-year history as an independent company and placing Hoover under the Whirlpool corporate umbrella.[41]

In 2004, Tacony Manufacturing, maker of Simplicity and Riccar cleaners, developed its Tandem Air System, which used two motors: a direct or so-called dirty air motor for floor cleaning, and a clean air motor for advanced filtration above-floor cleaning, a system first used by Air-Way on its Sanitizor in 1968. This created a cleaning combination that was the best of both worlds.[42] The total market size of floor care appliances in 2004 was a unit volume of 8 million, and 58 percent of canister cleaners sold had HEPA filters.[43]

The Eureka Company, purchased in 1974 by AB Electrolux of Sweden, had moved its production from Normal, Illinois, to Juarez, Mexico, in 2000 to benefit from lower labor costs. In 2004 the Eureka Company name was changed to the Electrolux Home Products Division. Eureka brand vacuums continued to be manufactured, but after 2007 they were made by AB Electrolux in Sweden.[44]

Rexair launched its new Rainbow E2 series in 2005, using the world's first "Switched Reluctance Motor" in a vacuum cleaner. It had no carbon brushes, no commutator, no windings on the armature, four field coils and a circuit board that allowed the motor to run at the highest speed of any Rainbow, and which generated a powerful 90 inches of water lift. The motor was claimed to be able to run "forever."[45] Soon after this, Dyson introduced its Digital Motor with similar claims, and Korean LG Corporation, prior to 1995 called GoldStar Company, Ltd., followed suit.

Dyson debuted its DC15 upright, known as The Ball, in 2005, which was available in three variants: All Floors, Allergy, and Animal. The latter recognized the recent and growing numbers of pet owners and their devotion to provide their four-legged wards with the best of care, including healthy food and regular grooming. What better than a vacuum cleaner to help with the latter?

As we know, Dyson had been preoccupied with the concept of a ball since he first developed the Ballbarrow in the 1970s. The DC15 had a large ball instead of wheels, which, in combination with a universal joint on the cleaner head, made it possible to steer the machine to the right or left by twisting the handle. Although the ball was a dramatic innovation that attracted attention to the desirable feature of increased maneuverability, a similar principle had been used for decades on floor nozzle attachments to the end of extension wands for upright and canister cleaners, as well as stick vacs. But never before had it been used on the fixed nozzles of traditional upright cleaners since the 1920 Air-Way, which, although developed to compete with traditional fixed nozzle uprights, was more like a stick vac. Not surprisingly, competitive upright cleaner manufacturers soon would figure out how to achieve similar maneuverability as the DC15, using universal joints and pivoting nozzles, but without using the large, patented Dyson ball.

The DC15 used the same Root 8 Cyclone technology used in the DC14, introduced in 2004. "Root 8" meant that there were eight individual cyclones. This was followed shortly by the introduction of the DC16, Dyson's first hand-held vacuum cleaner, which weighed 3.3 pounds, also using Root Cyclone technology. An Animal version was available with motorized brush-roll for picking up pet hair, as well as a Car and Boat version.[46]

Each Dyson design followed the same excellent industrial design principles established by his DC01. Although the early designs were already quite functional-looking, later designs became progressively more brutally mechanical and functional looking. But they all established new typeforms that were dramatically different in appearance than traditional ones in every category, and they all visually expressed their function effectively through form and color. Truly works of art, they were recognized as such by being included in the permanent collections of many museums, including the London Science Museum, The Museum of Modern Art in New York, the San Francisco Museum of Modern Art, the Metropolitan Museum of Art in New York, and the Centre Georges Pompidou in Paris.

With his name in lights and with speaking engagements around the world, Dyson became the most famous industrial designer on the globe, achieving international visibility not seen since the likes of Raymond Loewy, Henry Dreyfuss, and Walter Dorwin Teague in the 1930s and 1940s. Generously, in 2005, Dyson initiated an annual global student design competition, called the Eye for Why. Organized and run by the James Dyson Foundation, it claimed to "celebrate, encourage and inspire the next generation of design engineers." It was open to graduates or recent graduates in the fields of product design, industrial design, and engineering. This competition, and 20 other regional student competitions, would be rolled into a larger, global competition and would be renamed the James Dyson Award in 2008. National winners were chosen, and the international

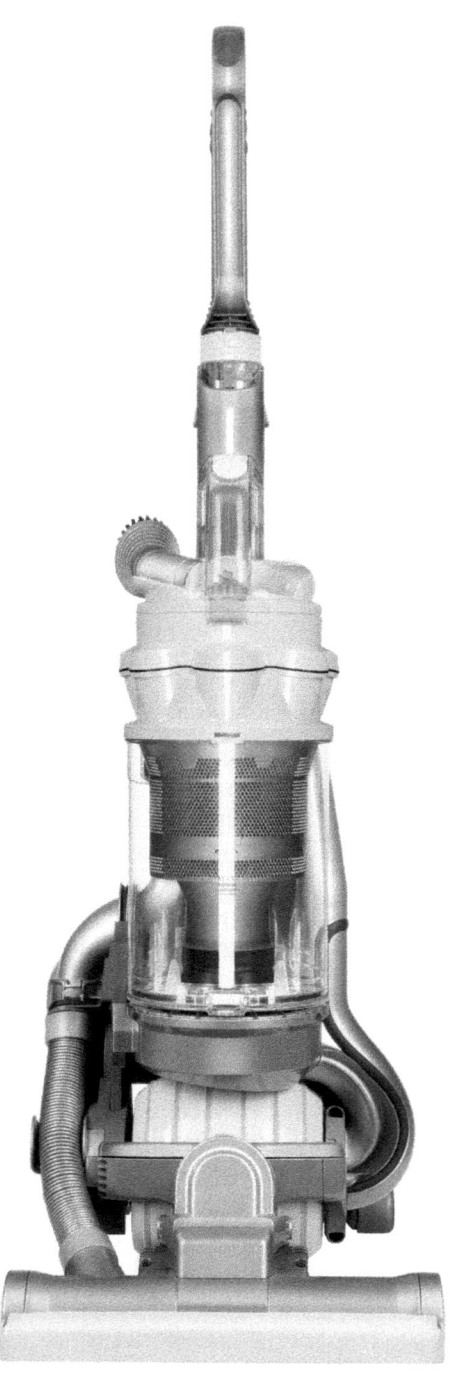

2005 Dyson DC15 The Ball. Dyson.

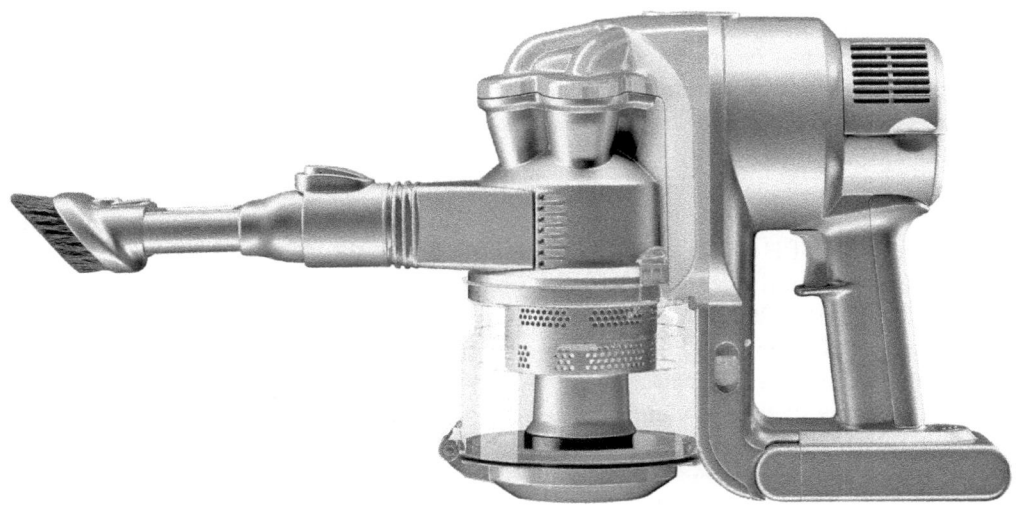

2006 Dyson DC16 cordless hand vac. Dyson.

winner received a £10,000 prize (that's $16,000 U.S.), a trophy, and a visit to a Dyson R&D center. In the meantime, in 2006, Dyson became *Sir* James Dyson when he was knighted by Queen Elizabeth II, and by 2011, he would be a billionaire. It doesn't get any better than that for an industrial designer. Dyson had become the Steve Jobs of the vacuum world![47]

Emer, Inc., founded in Italy in 1955 by Gianni Emiletti, had in the early 1970s expanded its line into floor care, entered the U.S. market in 2005 by forming Emer U.S.A. and introduced a line of floor care products including canisters, power nozzles, drum cleaners, E-ZEE cleaners, and uprights. Its introductory line included a line of special products named Disegno Italiano (Italian Design) with models named Botticelli, Michelangelo, Raffaello, Donatello, and Giotto, to pay homage to the great Italian artists and scientists of history.[48]

What goes around, comes around. In about 2001, a British company had released its Airider, a canister vacuum cleaner that floats on a cushion of air, just like the Hoover Constellation of 1954 — sort of a fiftieth anniversary celebration of a vacuum cleaner hovercraft. Apparently inspired by nostalgia, in early 2006, Maytag Floorcare U.K. (Hoover's ex-parent) reintroduced the classic Constellation, an identical copy of the last model produced in the 1970s, except with a stainless steel finish. It was called the Satellite, and released later that year in the U.S. in pearl white and stainless steel as the Constellation.[49]

Ironically, bad news arrived to Hoover employees in North Canton in 2006 when Hong Kong's Techtronic Industries (TTI) bought the Hoover floor care business and brand from Whirlpool Corporation for $107 million. Whirlpool in the early 2000s had encountered problems with its European operations, and in 2001 suffered major recalls because of fires caused by dishwashers and microwave ovens, resulting in a 2002 loss of $394 million. But Whirlpool worked aggressively to develop successful new products and to eliminate some of its non-major appliance businesses, advancing Whirlpool's goal of divesting itself of "non-core assets."[50]

The deal with Hoover closed in 2007. TTI assumed control of Hoover factories in North Canton, Ohio; El Paso, Texas; and Juarez, Mexico, among others.[51] Fortunately for Hoover retirees, Whirlpool assumed pension and retirement benefit plan liabilities. TTI said it was "looking forward to welcoming Hoover and all of its employees." On September 27, 2007, the Hoover North Canton manufacturing plant was shut down for the first time in 99 years, and Hoover products would henceforth be manufactured in China and Mexico. The number of Hoover employees in the North Canton area had been declining from a peak of 1,750 in 2004, so only 250 remained by 2007.[52] Nevertheless, in 2008, the Hoover Company celebrated the hundredth anniversary of Murray Spangler's and Hoover's Fabulous Dustpan, and the Vacuum Cleaner Collector's Club held its 2008 convention in North Canton in honor of Hoover's one hundredth birthday.

Hoover was not the only well-known brand name acquired by TTI. Its brands included Dirt Devil (Royal) in 2002; Vax floor care appliances (U.K.); Milwaukee Tools, AEG, Ryobi power tools; and Ryobi and Homelite outdoor products. TTI Floor Care North America Division headquarters is located in Glenwillow, Ohio. Despite the traumatic changes, Hoover began an aggressive campaign to develop new products, and its operations were combined with other operations owned by TTI and moved to Glenwillow, Ohio. From there, Hoover created a new line of vacuums, such as its Platinum series, and its newly designed cyclonic vacuums.[53] Hoover's 1994 SteamVac was still a top seller.[54]

The environmental, alternative energy, health, and sustainability movements, building since the 1970s, were beginning to dominate the advertising, promotion, and features of many consumer products, including vacuum cleaners, as manufacturers realized the sales appeal of these positive aspirations. A good example of the trend was Bissell, which in 2006 created its company-wide ForEverGreen team to integrate sustainability into all aspects of its business, which started with the earliest product developments and extended into its architecture, increased use of recyclable materials, rainwater collection, and reduction of energy and water consumption. Bissell partnered with the EPA's Design for the Environment program, which enabled companies to compare and improve performance and environmental costs and risks.[55]

Rowenta, acquired by France's Groupe SEB in 1988, introduced its RO8049 Silence Force Cyclonic canister in 2007, which combined 2200 watt power with a noise level of 69 decibels, innovating a new push for quieter vacuum cleaners, which normally range from 70 to above 90 decibels. Most standard cleaners are in the 72 to 77 range. Highest in noise were the cyclonic cleaners, which set air violently in motion. Although the noise level has absolutely nothing to do with cleaning performance, many consumers erroneously assume that the more noise, the more power, and this may have contributed to the public perception that cyclonic cleaners are more powerful.[56]

Bad news also hit Air-Way as the rising cost of aluminum and steel made its cleaner noncompetitive in cost, and production was ceased in 2008. The disposable bags it invented in 1920, and which were adopted by the entire industry, were in 2008 still available for Air-Way canister cleaners from Air-Tec, a business started by a former Air-Way door-to-door salesman. Air-Tec's Black Knight bags used synthetic allergen material, which fit Air-Way

2007 Rowenta Silence Force Cyclonic canister.

Sanitizors perfectly.⁵⁷ Replaceable bags these days generally can cost between 50 cents to $1.50 each, depending on quantities, a bit more expensive than Air-Way's original "12 for a dollar" in 1920. Obviously, the cost of paper has gone up, but the Gillette "razor blade principle" lives on.

Samsung Electronics introduced its new GTO Hauzen tank cleaner in 2008, along with several robotic cleaners, the VC-RE70V/VC-RE70, and VC-B835R. The former has an integrated camera that keeps track of the route cleaned, thereby avoiding unnecessary repetitive cleaning.

In 2009, Dirt Devil (made by Royal of TTI) engaged world-famous industrial designer Karim Rashid, who has many of his designs in the permanent collections of 14 museums worldwide, to design a series of elegant, sculptural, cordless vacs. Looking not at all like vacuum cleaners, but more like decorative accessories, they were designed for display in plain sight like Dustbusters, rather than stored in a closet. They included the Kwik Cordless Hand Vacuum, the Kurv cordless handheld vac, the Kruz cordless stick vacuum cleaner, the Brum cordless sweeper vac, and the Kone cordless handheld vac.

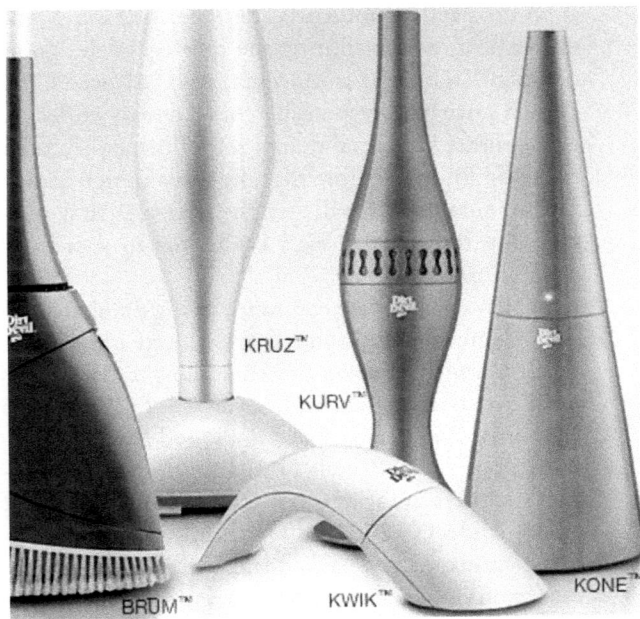

2009 Royal Dirt Devil designer vacs.

The vacuum cleaner industry, like all industries, responds to any new consumer need. A good example is the growing demand for pet care products over the last decade, and the need for vacuum cleaners to remove pet hair, which is particularly difficult to remove from carpets and furniture. *Consumer Reports* (CR) traditionally conducted tests to comparatively measure the efficiency of vacuum cleaners in meeting established standards. These tests normally measure carpet and bare floor cleaning ability, tool airflow, noise, emissions, and handling. In 2010, CR added an additional test to these standard ones — the "pet-hair" test. This test took long fur from the Maine Coon cat, placed five-gram quantities of it in multiple places on a medium pile carpet, then dragged a 19-pound roller over them to push the fur into the carpet to simulate foot traffic, and after 14 passes with a vacuum, weighed the hair left in the carpet, as well as that removed by the vacuum cleaner. Based on the results, a number of competitive vacs passed the test, including Kenmore's Intuition 31100 upright, Kenmore's Progressive 27514 canister, and Panasonic's MC-CG902, all among CR's "best buys."⁵⁸ Now, many vacuum manufacturers promote the pet-hair cleaning performance of their products.

In 2011, Keith McLoughlin, the first non–Swedish chief executive, became president and C.E.O. of Electrolux, and in August, Electrolux Small Appliances NA, a unit of Electrolux Group, moved its North American headquarters from Bloomington, Illinois, to Charlotte, North Carolina. For the first time in the company's North American history, all business units would operate under a single roof. It markets vacuum cleaners and other floor care appliances under the Electrolux, Eureka, Sanitaire and Beam Industries (central vac systems) brand names, and planned to introduce small appliances for home use in fall, 2011. This same year, Electrolux acquired 52 percent interest in the Olympic Group, Egypt's leading appliance manufacturer, and a controlling stake in the Chilean appliance company, Compañia Techno Industrial S.A. (CTI), both part of its strategy to grow in emerging markets. At least one other vac competitor recognized that Dyson's success in the market is the result of excellence in industrial design. In January, 2012, Electrolux CEO McLoughlin appointed Italian industrial designer Stefano Marzano to the newly created chief design officer post, demonstrating the priority of design as a core function of the company. One of design's rock stars is Marzano, who had been C.E.O. and chief creative director at Philips Design since 1991, responsible for all design work at Philips. From 1989, he was vice president of corporate industrial design for Whirlpool International, a joint venture between Whirlpool and Philips. Before that, he was a design leader at Philips since 1973. Hopefully, other vac manufacturers will get the message that outstanding industrial design is the key to successful competition.[59]

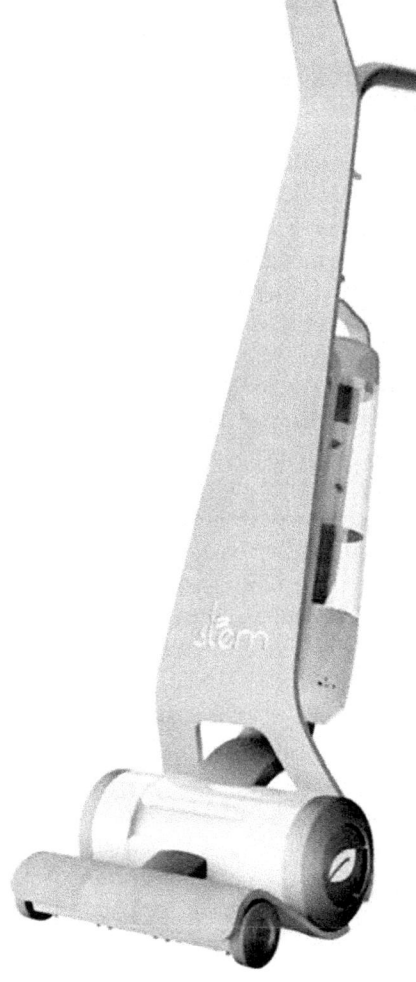

New concepts are constantly being developed. In 2011, Cambridge Consultants announced a new "eco-vacuum" concept known as Stem, calling it the "world's most eco-friendly vacuum cleaner." In 1960, two Cambridge graduates, Tim Eiloart and David Southward, founded a British firm, Cambridge Consultants, in the Cambridge Fen, Britain's equivalent of Silicon Valley. In 2002, Altran, Europe's largest technology consultancy, backed the firm. The Stem vacuum would save energy by automatically varying its power usage, depending on the job being done (for example, using either floor nozzle or hose attachment operation). It also would reduce power usage when no actual cleaning is being done, such as pausing to move furniture. The concept, appearing to feature a cyclonic action, would be similar in size and weight to standard models, would meet all regulatory requirements, and would come at a price point comparable to current premium models.[60] But of course, as often in vacuum land, the concept is not new. In the mid–1980s, the Hoover Dimension 1000 used the same technology to adapt the amount of suction to the task being performed, by monitoring the amount of airflow.[61]

Vacuum manufacturers continue to advertise the ability of vacuum cleaners to remove dust and germs from the air, but sometimes, exaggerated health claims can get them in trouble. A recent federal consumer fraud suit fined Oreck for "false, deceptive, and inaccurate representations" when Oreck claimed

2011 Stem concept.

its products "used scientifically proven technology to eliminate common viruses, germs, and allergens" and "can prevent colds, diarrhea, stomach upsets, asthma and allergies." Several class action suits have been filed by state residents against Oreck for similar "flu fighting" claims.[62]

But from the beginning, the vacuum cleaner industry has been primarily about cleanliness, and finally, there is a Museum of Clean, devoted to our efforts over the centuries to clean up after ourselves. In November 2011, The Don Aslett Museum of Clean opened in Pocatello, Idaho. Aslett, called by some the King of Clean, is owner of Varsity Facility Services, a nationwide facility services provider with 5,000 employees coast to coast. The 75,000-square-foot museum includes historical cleaning devices, including a 1902 horse-drawn vacuum and 250 pre-electric vacuum cleaners.

Each year, of course, brings a new round of vacuum cleaner introductions. The global economy continues to change within the recession that started in 2008 and continued into 2011. These days, it is difficult to identify the locations where vacuum cleaners, or any consumer products, are manufactured, because manufacturers relocate to wherever in the world they can produce most economically to compete, based on labor costs, taxes, and incentives offered by host countries or states.

The choices of vacuum cleaners on the market today are only a bit shy of infinite. Listed below are many of the surviving brand names at the end of the first decade of the twenty-first century. Were you to have gone shopping for a vac in 2011, you would have found the following selections, and these were only some of the many available, at a range of prices depending on whether you bought them online or at dozens of retail outlets. They include a variety of features, power levels, shapes, sizes, and designs, depending on your cleaning needs or desires. This snapshot list is intended only to illustrate the vast landscape of brands, models, and prices, and not to suggest any particular recommendations for purchase.

The latest model **Aerus** Lux (formerly American Electrolux) upright cleaner, was the Guardian with the strongest motor it ever made, and the easiest to use. It had a built-in HEPA filter and sold for $1,200 to $1,300. Additional models included the Lux commercial upright model with a 15-inch cleaning path; the Lux Fresh Era lightweight (less than ten pounds) upright cleaner with AllerGuard filtration; and several models of the classic Electrolux-style tank cleaners, including the Aerus Lux 9000 Guardian Ultra for $1,495; the Legacy; and the Classic.

Air Storm sold its HEPA canister vacuum cleaners with a powerhead brush for $1,100.

Avanti brand offered an eight-pound HMUL-100 HomeMaker Lightweight Vacuum upright for $99.

Bissell's line included the Big Green Complete Deep Cleaner/Vacuum canister for $269; its Healthy Home Vacuum with HEPA filtration upright for $199; its PowerClean Multi Cyclonic Bagless Vacuum with TurboBrush tool, for $199; its Pro Heat Deep Cleaner 25A3 for $129; its Pet Hair Eraser Cyclonic Canister Vacuum for $129.99; its PowerEdge hard floor vacuum; and its upright 82H1 CleanView Helix Bagless Vacuum for $67, among many others.

Black & Decker offered dozens of the latest descendants of the Dustbuster, from the cordless and bagless PHV1810 Pivoting Hand Vac for $53, to the Dustbuster CHV4800 for $16. Included were its canister vac VN1400P for $189.99, its Flex cordless canister FHV1200 for $57, and its stick steam cleaner SM1610 for $64.88, as well as cordless stick cleaners and power mops. B&D also offered a line of nine shop vacs under the DeWalt brand name from $99.99 to $579.99.

Bosch offered a BSG 81380 UC Premium Electro Duo XXL HC canister for $597.95, and a BSG 81370 UC Premium Electro Duo HC canister for $478.95.

Carpet Pro offered a line of uprights including its Heavy Duty Household Upright Vacuum for $160.

Cirrus offered its CR99 Bagged Upright Vacuum for $549, and its Residential Upright Vacuum Cleaner Model CR69 for $359.

Clarke offered a line of commercial upright vacuums including its Single Motor 03017 for $818, and its Carpetmaster 212 for $386.

Daewoo Group made a line of canister cleaners, including its Model RC350B 1500 watt for £29.99, and its Model RC300R at 1300 watts.

Dirt Devil (Royal-made by TTI) models included the M110002 Reaction upright for $177.95, the M140005 Vision Cyclonic upright for $89.95, the M082700 Vision Bagless canister for $135.95, the M084100 Power Stick vac for $ $39.95, the M083400X Swift Stick vac for $39.95, the M0100 Classic Hand Vac for $47.95, the M0914 Extreme Power hand vac for $44.95, the BD10000SLV Kwik Cordless Hand Vacuum for $39.95, the 0216SPC Kurv cordless handheld vac for $57, the Kruz 0313SLV cordless floor vacuum cleaner, and the Kone O213PNK (pink) and 0213BLU (blue) cordless handheld vacs.

DirtTamer made a line of hand-held vacuums including its Filterstream Ultima for $129.50, its Supreme V2400 for $55, and its V2510 Ultima for $65.

Douglas offered a line of corded portable vacs under the Readivac brand name, including stick vacs Power Broom 36310, and the Room Zoom 36829, as well as hand-held vacs Tidy Turbo 36010, 110 Volt Hand Vac 36300, Shoulder Strap 36500, and 12 Volt Auto Vac.

Dyson models included the DC35 Multi Floor Cordless Vacuum Cleaner with digital motor, Root Cyclone technology, and a 22.2 volt lithium-ion battery for $285; the Dyson DC23 bagless Turbinehead Canister Vacuum with Level 3 Root technology, HEPA filter, and the air-driven Turbinehead brush bar for $286; and the Dyson DC41 Ball, just released in December 2011.

Recent **Electrolux/Eureka** products included the 1998 Eureka Enviro Steamer, which cleans floors and carpets with a steam mop that heats water in under three minutes and costs $229; the Electrolux Oxygen canister cleaner at $322; the Electrolux EL8602 Nimble Upright Vacuum bagless with swivel head, $250; the Electrolux 12-pound EL6984 Harmony Ultra Quiet Canister with HEPA filter, $192; the Electrolux EL 1030A Ergorapido Ion Bagless Cordless Stick and Hand Vac, $150; the Electrolux 1022 Ergorapido Ultra stick cleaner; the Electrolux

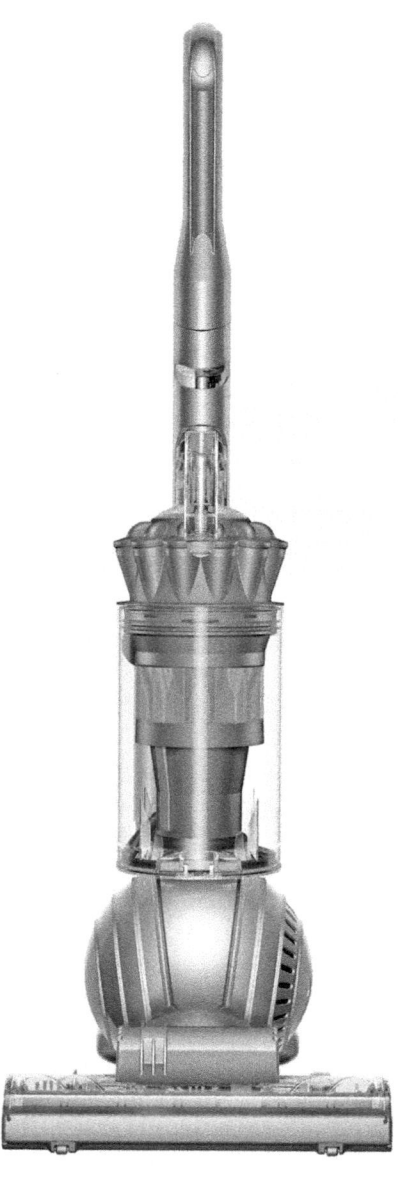

2011 Dyson DC41 Ball cleaner. Dyson.

955A Complete Clean Bagless Canister Vacuum Cleaner, with HEPA filter, $142; the Electrolux UltraSilencer Deluxe canister; the Eureka Boss SmartVac 12A bagged upright Model 4870PZ, with a high filtration bag that claimed to capture dust better than bagless dust cups, $135; the Eureka Model 431F Optima Bagless Upright, weighing 12 pounds, at a cost of $54; the Eureka Model 900A ReadyForce bagged canister cleaner with allergen filtration; and the Electrolux Eureka Model 71A Bagless Hand Vacuum Cleaner with motor-driven brush-roll, on-board hose and tools, and 20-foot cord, which weighed 5.5 pounds and costs $30.

Emer Inc.'s line of vacuums included its Emer KPA01 USA bagless canister for $152, its Emer 908000 Hercules Upright for $159, the Emer Hercules XT Vertical upright for $581.77, and its Emer BP01 Backpack Vacuum for $145.

Euro Pro made a 14-pound bagless Shark Navigator Upright Vacuum Cleaner SAK1001 for $106, an EP601S Bagless Stick Shark for $49.95, and a Shark Cordless Pet Perfect Hand Vac for $39.99.

Fairfax offered its Touch of Class System FFR-2000-BLK for $2,394.

Filter Queen offered its current model, Filter Queen 360 for $844; its previous model, Filter Queen Limited Edition for $649; and its Filter Queen Triple Crown for $589. All were identical except for color and price range.

Haan Corporation made a line of steam mops and multipurpose steamers, including its Multi Model S170 at $139.95, its Complete Model MS30 for $179.95, its Total Model HD60 at $149.95, its Slim and Light Model S135 at $119.95, its Classic Plus Model FS20 at $99.95, and its Agile model S140 at $119.95.

Hoover products (Made by TTI) included the Hoover S3670 Anniversary WindTunnel Canister Vacuum, with HEPA filter, $299.99; the Hoover Model FH50220 MaxExtract 60 Pressure Pro Floor Care upright deep cleaner for carpets and rugs, $229.99; Model UH70120 upright HVR WindTunnel T-Series Rewind bagless with a HEPA filter, $129.99; and Model U2440900W Ultra Lightweight Upright, less than 12 pounds, $60.

The most popular **Kenmore** cleaner was the Kenmore Canister Vacuum Cleaner 2029212 with HEPA filter at $150, and the second most popular, a similar canister (2029319) with a Pet Handi Mate attachment for $200. Others included the Progressive Canister Vacuum Cleaner Model 2171 for $400, the Magic Blue Canister for $100, the 26082 Canister for $60, the Kenmore Intuition 31810 upright bagged silver for $280, the Intuition 31811 upright bagless silver for $280, and the 39000 Quickclean Bagless upright for $50.

Kenwood had a complete line of vacs, including the HV170 and HV190 Handvacs; and a line of canisters including the VC1802, VC1807, VC2200, VC5200, VC6000, VC6200, VC6300, and the VC8800.

Kirby (made by Scott & Fetzer) cleaners available included the Diamond Edition for $849.99, the Ultimate G at $699.99, and the G5 for $499.

Koblenz offered a U-610-N Endurance All Metal Vacuum Cleaner upright for $314.95, a U-310-ZN Endurance All Metal Vacuum Cleaner upright for $238.95, the Model U-110-ZN upright for $205, and the PV3000BR All Purpose Power Vac canister for $87.95.

LG Corporation (formerly Goldstar) had a full line of cleaners, including the LG LUV400T Kompressor Total Care Upright Vacuum Cleaner, bagless with HEPA filter for $424; the LG LCV900B Canister Petcare Plus Vacuum with Kompressor Technology, bagless with HEPA filter at $308; the LG LUV250C Kompressor Compact Pet Care Upright Vacuum Cleaner, bagless with HEPA filter at $249; the LG LUV200R Red Upright Vacuum Cleaner with Dust Kompressor for $227; and the LG LUV300B Upright Vacuum Cleaner

Riviera Blue for $192. Kompressor technology compressed the dust in the bin so it did not have to be emptied as frequently.

Miele offered its S5980 Capricorn canister at $1,199; its Pisces canister at $699; its S5 series Callisto canister vac, S5281IC, at $679; its S4780 Orion canister at $649; its S4580 Luna canister at $569; its Neptune canister S412 at $399; its most popular, the Neptune Polaris w/STB 205-3 Turbo Brush and HEPA system at $399; its S7210 Twist upright in Royal Blue at $429; its S7260 Cat & Dog vacuum; and its S163 upright Universal Mini in black, for $289.

Morphy Richards also offered a complete line of vacs, including its 71080 Vorticity Bagless Cylinder for £199.99, its 73272 Bagless Family and Pets 2100 watt Cylinder for £154.99, and its 73337 Clean Up upright for £129.99.

Nilfisk, founded in 1906, made a variety of commercial duty vacuum cleaners available globally, including the 118 EXP explosion-proof canister and the S3 Industrial vacuum.

Numatic International Limited made **Henry** brand smiley face canister cleaners including Henry Numatic HVR200-22 in yellow and blue for $237, and a green Henry Numatic HHR200-2 Henry Hound Bagged Cylinder with power floor nozzle for $400.

Oreck offered a full line, including its Professional Series Pilot upright for $499.95, its Professional Series Gold upright for $399.99, its Silver Series upright for $249.95, its XL Shield Power Scrubber for $399.99, the Ironman Handheld vac for $299.99, its Edge Cordless Handheld Vacuum for $399.99, and its Deluxe Handheld Vac for $149.99.

Panasonic offered a full line, including its MC-UL675 Bagless Upright for $188.95, its MC-UG693 Upright Vacuum with HEPA Filter System for $178.95, its MC-V5003 upright vacuum for $148.95, its MC-CG885 canister vacuum for $213.95, and its MC-3920 canister for $78.95.

Perfect offered several commercial cleaners: a canister (tank) Model C101 for $899 and an upright Model P101 for $339.95.

Powr-Flite offered a line of wet/dry vacs, including the FM90P Profi 4.5 Gallon Solution Tank-Wet/Dry Vacuum.

ProTeam offered its ProTeam 15 Vacuum-ProCare upright for $399.

Pullman-Holt offered its UV5 Upright Commercial Vacuum' for $398.25 and its Wet Dry Vac 2 Horsepower 10 gallon canister for $228.95.

Rainbow (made by Rexair) offered its newest E2 Blue Rainbow 2 Speed Vacuum canister for $1,080, and its Rainbow SE PN2 Vacuum Cleaner for $229, among many other variations.

Riccar vacuums (made by Tacony) included upright series Vibrance, Brilliance, Radiance, and 8000 Premier; a Lightweight (8 pound) line of uprights including the RSL1A, RSL3, RSL4, and RSL5; a line of canister cleaners, including the Immaculate Premier, the Impeccable Premier, the Pristine, the Charisma Mid Size, the Starbright, the Pizazz Compact, the 1500 Mid Size, the Moonlight Compact, and the Sunburst Sub Compact; and a line of handheld vacs, including the Gem for $38.95, the SupraQuik for $139; and stick vacs, including the OmniClean for $169.95.

Group SEB included **Rowenta**, which marketed Silent Force canisters with decibel level of 69, and Silence Force Cyclonic, along with Silence Force Eco-intelligence that consumed 35 percent less energy than its conventional model and was up to 80 percent recyclable. It also marketed its RO 3841 Compact Power vacuum with a large-capacity bag for $217.49, its RH 8553 Accu-operated stick vac for $193.70, and its Air Force cordless stick vac.

Under the **Royal** (made by TTI) name were the Model 2028 All-Metal Upright for $424.95, the Royal Classic M4000 Upright, and the Royal 1038 Commercial Upright Cleaner.

Royal Philips made five bagless cleaners including the Marathon Bagless cylinder cleaner with a 2000 watt motor; and four bag cleaners, including the FC8917 HomeHero bagged EnergyCare ParquetCare with Tri-active nozzle and the FC8620/01 Expression bagged cylinder cleaner.

Rubbermaid offered a FG9VPH120000 Power Height Upright Vacuum Cleaner for $303.

Samsung offered its Vacuum Cleaner with Variable Speed Control canister for $379, its 5115G HEPA Compact Straight Air Canister Vacuum Cleaner for $220, its Mid-Size Canister Vacuum Cleaner with Metal Telescoping canister for $169, and its Vacuum Air Canister 5100C for $160.

Electrolux also made a line of vacs under the trade name of **Sanitaire.** Offerings included its upright Model SC888K for $189, its Electrolux Sanitaire upright for $103, its S3681D Mighty Mite Canister for $100, and its SC6070 Portable Spot Cleaner for $353.

Sanyo offered its SC-TA3000 Revo Upright Bagless cleaner with twin agitators for $41.

Sebo offered a line of commercial cleaners: the 370 Electronic Upright for $819, the Sebo Automatic X4 X49501AM for $799, Sebo 9825AM Felix stick vac for $489, and others.

Sharp made an EC-TU5306 Vacuum-Twin Energy Upright for $199.

Shop-Vac offered a broad line of wet/dry canister cleaners, including the 405-00-10 Industrial for $146.98, the 9633400 Wet/Dry Vac with Blower for $100, the Ultra Pump 10 Gallon for $81.48, and the 10 Gallon Quiet Plus Wet/Dry for $79.

Silver King offered its Blue Max Air 2000 Complete canister for $3,275.

Simplicity vacuums (made by Tacony) included a line of six upright cleaners: Synergy with Tandem Air Technology (two motors, a bypass for tool suction, and direct air for floor suction) and HEPA filters, Synchrony, Symmetry, 7 Series, Freedom, and Commercial. Four lines of canister cleaners included the Premier Power Team line, including the Gusto and Moxie; the Mid Size line, including the Verve, Cinch, Jessie, and S24; the Compact line, including the Jack and Snap; and the Sub-compact line, the Jill. Additional models, no longer in production but possibly available, include canisters S38, S36, S20, and S18; and uprights Symmetry Premium and 6 series.

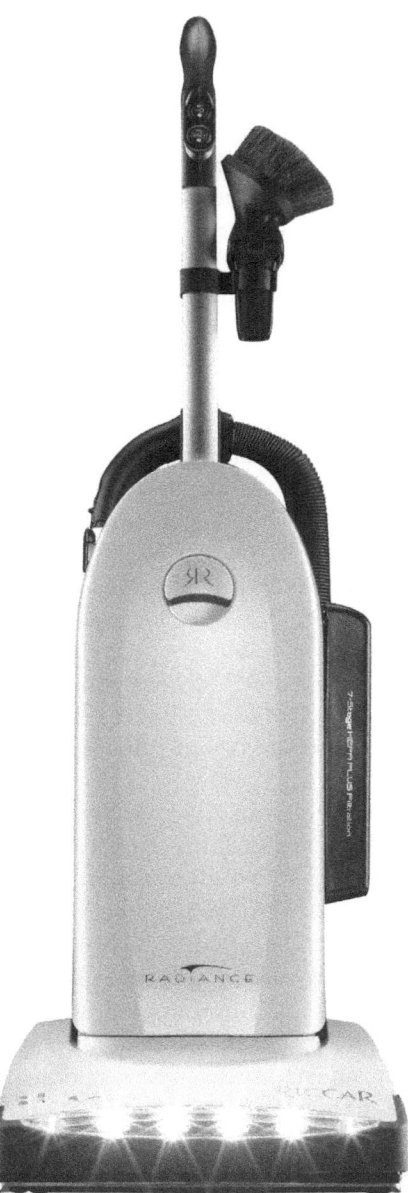

2011 Riccar Radiance. Tacony.

2011 Simplicity Gusto. Tacony.

Singer products (Made by TTI Ryobi) included its AT7739 Lazer Storm Bagless Upright Cyclonic, which worked as three different machines: a handheld vacuum, a canister vacuum, and an upright cleaner. Price was approximately $393.

Tri-star offered its canister Models EXL, MG1, MG2, and A101 in prices ranging from $689 to $1,149.

Vanisher options included its DMV 14 Commercial upright for $199, and a line of 15 gallon commercial wet/dry canisters priced from $559 to $1799.

Vax brand (made by TTI) offered Vax Mach Air U91-MA-B Multicyclonic Bagless Upright Vacuum Cleaner for £110, Vax Commercial VCU-02 Lightweight Upright Vacuum for £49, Vax U90-MA-R Mach Air Reach Multicyclonic Upright for £220, Vax C90-P2-B Power 2 Cylinder Vacuum Cleaner for £64.99, and Vax H90-GA-B Gator Handheld Vacuum for £42.99.

CENTRAL VAC SYSTEMS: Brand names included: **Brute, Filtex, Nutone, Electrolux, Broan, Beam Industries, Dirt Devil, Air Vac, Vacuflo** by HORSEPOWER Products, **Tacony,** and **Sandia.**

ROBOTS: Robotic cleaner options included the **Neato** Robotics XV-11 Robotic All-Floor Vacuum Cleaner for $355; the **iRobot** Roomba 532 Pet Series Vacuum Cleaner Robot, Recharging Base for $319; the Automatic Hard Floor Cleaner by **Evolution Robotics** for $180; the **iTouchless** AV002A Robotic Intelligent Vacuum Cleaner Pro for $125; the **Metapo** CleanMate QQ-1 robotic cleaner for $99.95; **Samsung** SR8855 and SR8845 NaviBots; **Electrolux** Trilobite; **Karcher** RC 3000 Robo Cleaner; eVac Vacuum Cleaning Robot; **Floorbotics** Vacuum Robot; **Zucchetti** Vacuum Robot; and the P3 **International** Robotic Vacuum Cleaner for $65.

Vacuum cleaners listed above number over 250 consumer choices, probably only a fraction of those actually available. If you are shopping for a vac, many of these vacuum cleaner models are reviewed, rated, and priced by category and brand names on the following websites: http://www.appliance-reviews-ratings.com/vacuum-cleaner-reviews.html; http://www.vacuum-cleanersreviews.org/; http://www.vacuumwizard.com/; http://www.vacuum-cleanerszone.com/. The eternal buyer's dilemma, according to *Consumer Reports*, is that vacuum users want vacuums to deep-clean carpets, but also want them to be lightweight and easy to maneuver, and also not too costly. This combination pits power against size, weight, and cost; a difficult combination in a single product, but easily solved by owning multiple cleaners.

Today, the U.S. annual market for household vacuum cleaners is worth about $3.5 billion, while globally the market is expected to reach $13.2 billion by 2015. The United States, Europe, and Asia-Pacific represent the largest regional markets with about 85 percent share of the total sales volume. The Asia-Pacific market is the largest as well as the fastest growing market, with annual sales of about 11 million units. The U.S. maintains a trade deficit in vacuums, as in many other manufacturing industries, with $1.8 billion in imports and exports of only $305 million. The recession that began in 2008 no doubt affected many manufacturers negatively, and that includes the vacuum cleaner industry.

Upright vacuums are most popular in the U.K. and North America, while in Asia and Europe, canisters (also called "cylinder" or "tank") cleaners are preferred. Bagless vacuums and HEPA filters are experiencing continuous growth, mostly in North American households. Increasing consumer demand for more than one vacuum, decreasing household sizes, the high-income baby boomer population, and the under-35-year-old segment are all factors driving the market.

Recent innovations include lightweight vacuums enabled by smaller and more efficient motors, and new polymers that allow manufacturers to reduce costs and weight while improving strength and quality. Increasing health awareness by consumers is driving the demand for efficient, eco-friendly, ergonomic, and low noise-producing vacuum models.

Current developments in the industry include a new cleaning technology known as Air Recycling Technology. Conceived by a British inventor, it uses a pressurized air stream to collect dust from the carpet instead of a vacuum. Working prototypes have been shown to be more energy efficient than vacuums, but the technology is not currently used in any production cleaner. This system uses the same principle that preceded actual vacuum cleaners, typified by John S. Thurman's pneumatic carpet renovator of 1899, which used a blast of compressed air to blow dirt into a receptacle. Perhaps this new technology will prove that Thurman was on the right track, but simply failed to perfect the idea to the point of practicality.

History repeats itself, as we have seen so many times in any technology, including that of vacuum cleaners. The Vacuum Cleaners Collectors Club, comprised of those intimately familiar with its history, concluded in 1997 that there were only four great inventions in vacuum cleaners:

(1) Hoover's Model O of 1908, with its motor-driven, revolving brush roll, was the first to get the carpet clean, and is still the basis for all rug-cleaning devices, but its cloth bag had to be cleaned after each use, and if it was not emptied, it restricted suction.
(2) Air-Way's replaceable bags of 1920 solved that problem, and are still being used by many vacuum manufacturers, despite the many bagless cleaners now on the market, cyclonic or otherwise. Independent companies make replacement bags for hundreds of vacuums no longer made, because of the long life of original products. Air-Way invented a primitive dirt finder, a hollow upright handle that allowed cleaning above the floor, and the power nozzle in 1933.
(3) Electrolux eliminated the fan contacting the dirt, and invented the cord winder and full bag indicator.

(4) Rexair's bagless cleaner of 1930, using centrifugal force, and its unique water filter that never reduced suction, was ignored by the industry in general for generations, but reappeared dramatically near the end if the century as the Dyson cyclonic action, and is now a popular system offered by most major manufacturers.[63]

However well deserved, these are purely mechanical inventions, and there are, or should be, esthetic criteria applied, as well. Personally, I would have no problem adding Dyson to this list as the creator of the most consistent marriage of engineering and industrial design ever seen in vacuum cleaner design. The influence of Dyson design throughout the industry is undeniable.

Although these are some of the great milestones in vacuum cleaner history, we need to look at the other side of the coin. The user of vacuum cleaners today still has to deal manually with the physical disposal of dirt, whether it is accomplished by throwing away a full paper bag and replacing it with a new one, emptying a container full of dirt into the trash, cleaning or replacing filters, or cleaning a dirty water container used as a filter. Users still have to wrestle with power cords or replace rechargeable batteries when exhausted. They still have to manually store, lift, maneuver, and handle vacuum cleaners in the same way our ancestors wielded brooms, sweepers, and manual suction cleaners. Vacuum cleaners still require physical work, unlike a clothes dryer, toaster oven, or computer, where a push of a button is all that is needed to complete a task. As Ruth Schwartz Cowan suggested in the title of her book, many modern appliances create "more work for mother." Perhaps we can look forward to some sort of robotic vacuum cleaner systems in the future that automatically clean each room in a house with no human effort, but I will not hold my breath. What will not change is the basic human instinct to keep households as clean and healthy as possible, and the vacuum cleaner is the best solution to that desire that humanity has devised to this date.

As we look back over 142 years of vacuum cleaner development, both technologically and sociologically, some will always want to know: "Who invented the vacuum cleaner?" It is human nature to seek the historical source of any familiar artifact or activity, and if there exists no obvious and satisfying single answer, we tend to invent a myth that despite the facts, makes perfect sense.

For example, baseball historians know that our national pastime evolved over centuries from the old British lower class cousin of cricket, a stick and ball game variously called "rounders," "feeders," or "base ball," and which was imported to America by early colonists, but most people would rather believe the American creation myth, put forth by A.G. Spalding of the sporting goods company in 1907, that Abner Doubleday, who fired the first Union shot in defense of Fort Sumter and who later became a distinguished Civil War Major General, had invented the game in Cooperstown, New York in 1839. The myth is quite detailed, describing how Doubleday scratched the shape of a baseball diamond with a stick in the dirt and prescribed all the rules for the game. What could be more specific? More dramatic? More inventive? More American? As a result, rural Cooperstown today is the home of the Baseball Hall of Fame, attracting thousands of visitors to the shrine. But it is based on a myth.

So it is with the inventor of the vacuum cleaner, or, I should say, the inventors, for there are many credited. Although carpet sweepers, the predecessors of vacuum cleaners, were invented in England, they hardly qualify under our technical definition of vacuum cleaner, so are usually ignored in vacuum cleaner history. Some historians claim the inventor was Daniel Hess, who proposed a carpet sweeper with a bellows to create suction in 1860, but his concept is not known to have ever been produced.

Others claim Ives McGaffey as the inventor, because in 1869 he produced a marketable lightweight portable device that produced a weak vacuum (then called suction) with a hand-cranked fan. John Thurman often gets credit for the first (gasoline) powered cleaner in 1899, but it was not very portable and used compressed air, not vacuum, to move dirt. Corrine Dufour gets credit by some because her cleaning device in 1899 was the first to be electrically powered, but still blew air with a fan, rather than produce a vacuum.

David Kenney was probably the most influential because his patents controlled the market for years, but his invention in 1901 was a stationary vacuum system, not a portable one for home use. Many credit Sir Hubert Booth, whose device in 1901 was a commercial system like Kenney's, only portable, but still was not a consumer product. He claimed to have been the first to call such a device a vacuum cleaner, which is the title we now regard as definitive of the category. But Franz Burger, several years before Booth, patented what he described as a cleaner that operated with a vacuum.

Skinner and Chapman really produced the first portable electric vacuum cleaner for home use in 1905, but it was quite heavy and unwieldy. James Kirby (in 1906) and Murray Spangler (in 1907) developed and sold portable electric vacuum cleaners almost concurrently with other manufacturers, and both deserve much credit. But Spangler had the good fortune of being supported by the Hoover Company, which made and sold the first commercially successful portable electric vacuum cleaner in 1908 and went on to become the most dominant manufacturer of vacuums in the world for well over a century. So Spangler usually receives top billing as "the inventor," deservedly so.

The truth of the matter is that it is impossible to credit the invention of the vacuum cleaner to any one individual. Each of many inventors took the technology available at the time and tweaked it, revised it, or recombined elements to make it better in some way, and to make it more like the vacuum cleaners with which we are most familiar. This is the way invention works, and why there is always controversy about who invented what, and when.

Unfortunately, to most people, the term "invention," or "inventor" always seems to imply the act of a single individual who discovers an entirely new device with a flash of brilliance. This perception is reinforced by the patent system, which grants legal protection to individuals (legally called inventors) for any "new" technical specification or arrangement. Specific patents, in fact, protected all the inventors listed above. Each of these inventions were unique and different from each other in some specific way, yet none embodied all the essential characteristics we now recognize as a modern vacuum cleaner.

Most inventors would not even consider themselves as such, but simply as someone who improves or modifies existing devices or principles. There are a number of features, such as increased power, durability, cleaning effectiveness, and appearance, that depend less on individual invention than on gradual evolution in materials, motors, user habits, social attitudes, or technology developed by creative multidisciplinary individuals or teams. A more precise descriptive term for a vacuum cleaner inventor of today would probably be "the design and development team."

It is much more difficult today for independent Murray Spanglers of the world to enter the competitive vacuum cleaner arena successfully, but as we have seen, James Dyson, through personal invention, diligence, and capital, was able to do so very recently. The vacuum cleaner industry is but a microcosm of global manufacturing today, servicing the consumer needs of millions. It is clear that it will continue to develop new technology, new applications, and new sales features. Hopefully, there will also continue to be Booths, Kenneys, Kirbys, Spanglers, Replogles, Newcombes, Yonkers, and Dysons who shake up the industry and add a new twist to the rich history of vacuum cleaners.

We began this book by suggesting that unwanted and unhealthy dust was the primary reason vacuum cleaners were invented. Dust was, and is still, the enemy. It's a dirty job, but someone has to do it! As we have described over the years, vacuum cleaners have reduced dust and cleaned household air with ever more effective filters and technology. The federal government joined the effort in 1963 with the Clean Air Act, which was significantly amended in 1970, 1977, and 1990, to protect the general public from airborne contaminants that are known to be hazardous to health. Dust, incidentally, in scientific jargon, is known as "particulate matter."

It is only fitting to close our story with the latest battle in our eternal war against dust. In September 2011, the Environmental Protection Agency was considering tightening air quality standards by reducing by half the acceptable amount of particulate matter from 150 micrograms per cubic meter to about 75 micrograms per cubic meter, claiming that "excessive dust particles could be harmful if not deadly for people, causing cardiovascular or respiratory problems." Although farmers complained that enforcement of such regulations would require costly and time-consuming watering of gravel roads or tilled soil, or cause cessation of work with equipment on windy days, we can rest assured that our watchful government will implement and enforce whatever regulations are necessary to keep us healthy and safe from harm.

We should all do our part. Perhaps the vacuum cleaner industry can develop outdoor vacuum cleaners for farmers and suburbanites, since there is a lot more dust outside the home than inside. With the potential combination of climate change and freakish weather conditions, we could soon be experiencing "global dusting" with unimaginable consequences.

Postscript

Those of you who have more than a passing interest in vacuum cleaners may be interested in the following information related to the subject of this book.

If you are a collector of vacs, or just have an interest in vacuum history, you may want to visit the website of the Vacuum Cleaner Collectors Club at http://vacuumland.org/. A number of individual collectors have websites with photos of their collections, and others have demonstration videos of antique cleaners in action.

There are several outstanding museums that are worth visiting to see and learn more about actual vacuum cleaners, many of which are described in this book:

Hoover Historical Center/Walsh University, 1875 East Maple Street, North Canton, OH. Website: http://www.walsh.edu/hoover-historical-center. Phone: 330-499-0287 or 330-244-4667.

Vacuum Cleaner Museum, Tacony Corporation, Route 66, St James, MO. Website: http://www.vacuummuseum.com/index.asp. Phone toll free 1-866-444-9004.

The Don Aslett Museum of Clean, 711 South 2nd Avenue, Pocatello, ID, near Idaho State University; grand opening, April 2012. Contact Don Aslett, 208-236-6906, http://www.museumofclean.com/wp/.

Stark's Vacuum Museum, 107 NE Grand Avenue, Portland, OR. Website: http://starks.com/about_us/vacuum. Phone toll free 1-800-230-4101.

If you are one of the 18,000 independent retailers, manufacturers, or wholesalers of vacuum cleaners or sewing machines throughout North America, you may want to consider membership in the Vacuum Dealers Trade Association (VDTA) and/or the Sewing Dealers Trade Association (SDTA). These associations publish four magazines, each dedicated to a specialty interest. Visit their website at: http://www.vdta.com/index.html.

Notes

Chapter 1

1. Alphin, p. 8.
2. Cowan, p. 17.
3. Kratch, pp. 41–42.
4. Earl Lifshey, *The Housewares Story*, p. 288.
5. Baynes, Vol. V, pp. 127–131.
6. Ibid.
7. Derry and Williams, p. 580.
8. Anderson, *The Most Splendid Carpet*, chapter 4.
9. Derry and Williams, p. 580.
10. Ibid., p. 581.
11. Israel, p. 2.
12. Pulos, p. 70.
13. Cowan, pp. 64–68.
14. Wikipedia, "Catherine Beecher," http://en.wikipedia.org/wiki/Catharine_Beecher (accessed Aug. 15, 2011).
15. Beecher, *Treatise on Domestic Economy*, chapter XXIX, pp. 302–306.
16. Wikipedia, "Catherine Beecher," http://en.wikipedia.org/wiki/Catharine_Beecher (accessed Aug. 15, 2011).
17. Beecher and Stowe, *American Woman's Home*, introduction.
18. Ibid., chapter XXXVI.
19. Ibid., chapter V.
20. Ibid., chapter XXV.
21. Pulos, p. 134.
22. Beecher and Stowe, *American Woman's Home*, "Appeal" at end of book, after chapter XXXVII.
23. Fenster, pp. 12–24.
24. Thomas Robb, "Black History Month Myth Twenty-six and Twenty-seven," Feb. 21, 2011, http://tarobb.blogspot.com/2011/02/black-history-month-myth-twenty-six-and.html (accessed Aug. 15, 2011).
25. Wikipedia, "Linoleum," http://en.wikipedia.org/wiki/Linoleum (accessed Aug. 15, 2011).
26. Baynes, Vol. IV, p. 404.
27. Ibid.
28. Ibid.
29. Benjamin Franklin, as quoted by Lemay and Zall, editors, in *Benjamin Franklin's Autobiography*, p. 106.
30. Baynes, Vol. IV, p. 404.
31. Giedion, as referenced in *The Vacuum Cleaner*, Vacuum Cleaner Manufacturers Association, 1945, p. 2.
32. Ibid.
33. U.S. Patent Office, No. 3124.
34. Taggart, p. 344.
35. Ibid.
36. Ibid.
37. Answers.com, "Carpet Sweeper," http://www.answers.com/topic/carpet-sweeper (accessed Aug. 15, 2011).
38. Beecher and Stowe, *American Woman's Home*, Chapter XXX.
39. *Executive Documents*, The Senate of the United States, Second Session, 1858–1859, Vol. 986, p. 443.
40. Taggart, p. 345.
41. *Executive Documents*, The Senate of the United States, Second Session, 1858–1859, Vol. 986, p. 440.
42. Taggart, p. 345.
43. Giedion, as quoted in *The Vacuum Cleaner*, Vacuum Cleaners Manufacturers Association, 1945, p. 1.
44. Harris, *Ohio Cultivator*, Volume XIV, 1858, p. 350.
45. The Inline Planet, "The Illustrated History of Inline Skate Design at the U.S. Patent Office, Part 1," http://www.inlineplanet.com/History/shalerparlorskate.html (accessed Aug. 15, 2011).
46. *Executive Documents*, The Senate of the United States, Second Session, 1858–1859, Vol. 986, p. 443.
47. Ibid., p. 444.
48. Ibid.
49. *How the Vacuum Cleaner Began*, the Hoover Company, 1967, p. 3.
50. Taggart, p. 345.
51. Ibid.
52. Giedion, as quoted in *The Vacuum Cleaner*, Vacuum Cleaners Manufacturers Association, 1945, p. 1.
53. Jailer-Chamberlain, *Floor Care Professional*, 1997, p. 33.
54. Ibid., p. 35.

Notes — Chapter 2

55. *Laundry Management, A Handbook for Use in Private and Public Laundries*, Vol. 4 (London, 1902), chapter 23; and "Carpet Beating," as referenced in *The Vacuum Cleaner*, Vacuum Cleaner Manufacturers Association, 1945, p. 4.
56. President Ulysses S. Grant, as quoted in Leslie, *Illustrated Historical Register of the Centennial Exposition*, 1876; and quoted in Pulos, p. 142.
57. Taggart, p. 345.
58. Jailer-Chamberlain, *Floor Care Professional*, December 1997, pp. 33–34.
59. Taggart, p. 346.
60. Ibid.
61. Ibid.
62. Ibid., p. 347.
63. Ibid., p. 348.
64. Ibid.
65. Ibid., p. 350.
66. Ewbank, "About Us," http://www.ewbank.co.uk/about-us (accessed April 15, 2012) and "Help Desk," http://www.ewbank.co.uk/help-desk (accessed April 15, 2012).

Chapter 2

1. "From Sweeping to Suction," *Home Furnishings Daily*, Nov. 25, 1991, p. 72.
2. U.S. Patent 29,077, U.S. Patent Office.
3. *Chronology of Sweeper Patents*, Hoover Historical Center, p. 1.
4. U.S. Patent 91,145, U.S. Patent Office.
5. "From Sweeping to Suction," *Home Furnishings Daily*, Nov. 25, 1991, p. 72.
6. Tabor, *Floor Care Professional*, December 1997, p. 38.
7. Bunch, p. 1871.
8. Tabor, *Floor Care Professional*, December 1997, p. 38.
9. L.K. Acheson, *History of Carpet Cleaning Development*, p. 1.
10. Ibid., p. 2.
11. Ibid.
12. Ibid.
13. Smith, p. 18.
14. Basalla, p. 41.
15. Cheney, p. 24.
16. Ibid., pp. 38–39.
17. Ibid., pp. 44–45.
18. Ibid., p. 45.
19. Ibid., pp. 70–75.
20. Ibid., pp. 48–49.
21. Derry-Williams, pp. 635–636.
22. About.com, "History of Electric Vehicles," http://inventors.about.com/od/estartinventions/a/History-Of-Electric-Vehicles.htm.
23. Hughes, p. 243.
24. Bunch and Hellemans, p. 305.
25. *Chronology of Sweeper Patents*, Hoover Historical Center, p. 3.
26. *Evolution of Vacuum Cleaners*, General Electric, 1971, p. 3.
27. Wikipedia, "Vacuum cleaner," http://en.wikipedia.org/wiki/Vacuum_cleaner (accessed Aug. 15, 2011).
28. VACUUM CLEANER CO. v. I NNOVATION ELECTRIC CO., Inc. (Circuit Court of Appeals, Second Circuit, December 20, 1916), http://ftp.resource.org/courts.gov/c/F1/0239.f1.pdf (accessed Aug. 15, 2011).
29. *Chronology of Sweeper Patents*, Hoover Historical Center, p. 3.
30. Blackburn (Part 3), p. 44.
31. *Evolution of Vacuum Cleaners*, General Electric, 1971, p. 4.
32. *Chronology of Sweeper Patents*, Hoover Historical Center, p. 4.
33. *Evolution of Vacuum Cleaners*, General Electric, 1971, p. 4.
34. *The Vacuum Cleaner*, Vacuum Cleaner Manufacturers Association, 1945, p. 1.
35. U.S. Patent 1,057,347, U.S. Patent Office.
36. Hubert Cecil Booth, *The Origin of the Vacuum Cleaner*, Newcomen Society Transactions, V. 15, p. 93, London, 1936, as quoted in *The Vacuum Cleaner*, p. 6, Vacuum Cleaner Manufacturers Association, 1945.
37. Hubert Cecil Booth, "The Origin of the Vacuum Cleaner," *Newcomen Society Transactions* 15 (London, 1936), p. 93, as quoted in *Chronology of Sweeper Patents*, Hoover Historical Center, p. 4.
38. Ibid.
39. *The Vacuum Cleaner*, Vacuum Cleaner Manufacturers Association, 1945, p. 5.
40. Wohleber article, "The Vacuum Cleaner," *Invention and Technology* (Spring 2006), p. 5.
41. Van Dulkin, p. 34.
42. *Chronology of Sweeper Patents*, Hoover Historical Center, p. 5.
43. *Installed Cleaning Systems Using Suction Only*, Hoover Historical Center.
44. *Evolution of Vacuum Cleaners*, General Electric, 1971, p. 5.
45. *The Vacuum Cleaner*, Vacuum Cleaner Manufacturers Association, 1945, p. 7.
46. *Evolution of Vacuum Cleaners*, General Electric, 1971, p. 5.
47. *Chronology of Sweeper Patents*, Hoover Historical Center, pp. 5–6.
48. Ibid., p. 6.
49. Ibid., pp. 6–7.
50. Ibid., p. 6.
51. Ibid.
52. Wikipedia, "Vacuum cleaner," http://en.wikipedia.org/wiki/Vacuum_cleaner (accessed Aug. 15, 2011).
53. VacHunter, "History of Vacuum Cleaners," http://www.vachunter.com/index.html (accessed Aug. 15, 2011).
54. Ibid.
55. Ibid.
56. Ibid.
57. *Chronology of Sweeper Patents*, Hoover Historical Center, p. 10.
58. VacHunter, "History of Vacuum Cleaners," http://www.vachunter.com/index.html (accessed Aug. 15, 2011).
59. Ibid.
60. Ibid.
61. *Manual Cleaners*, Hoover Historical Center.

Chapter 3

1. HubPages, "Mom's old Iron — Reflection of Technology," http://rodmaru.hubpages.com/hub/Moms-old-Iron-Signpost-of-Technology (accessed Aug. 15, 2011).
2. Matranga, *America at Home: A Celebration of Twentieth-Century Housewares*, p. 21.
3. Lotz, "Richard K. Kirby, The Man Behind Less Work," *Machine Design* (July 1948), p. 108.
4. Blackburn (Part 2), p. 46.
5. Lotz, "Richard K. Kirby, The Man Behind Less Work," *Machine Design* (July 1948), p. 109.
6. Blackburn (Part 2), p. 46.
7. Frank Hoover, *Fabulous Dustpan*, p. 57.
8. *Chronology of Sweeper Patents,* Hoover Historical Center, p. 7.
9. U.S. Patent 889,823, U.S. Patent Office.
10. *Chronology of Sweeper Patents,* Hoover Historical Center, p. 8.
11. Frank Hoover, *Fabulous Dustpan*, pp. 61–62.
12. *Chronology of Sweeper Patents,* Hoover Historical Center, p. 8.
13. Ibid.
14. U.S. Patent 889,823, U.S. Patent Office.
15. Frank Hoover, *Fabulous Dustpan*, p. 63.
16. Ibid., pp. 96–97.
17. Blackburn (Part 1), p. 42.
18. *Chronology of Hoover Vacuums 1908–1940,* Hoover Historical Center, p. 1.
19. *Chronology of Sweeper Patents,* Hoover Historical Center, p. 9.
20. Frank Hoover, *Fabulous Dustpan*, p. 97.
21. Ibid., p. 104.
22. W.H. Hoover, as quoted in 'From Sweeping to Suction," *Home Furnishings Daily,* November 25, 1991, p. 77.
23. Frank Hoover, *Fabulous Dustpan*, pp. 104–105.
24. *Chronology of Sweeper Patents,* Hoover Historical Center, pp. 9–10.
25. *Chronology of Hoover Vacuums 1908–1940,* Hoover Historical Center, pp. 1–2.
26. Ibid., p. 1.
27. *Saturday Evening Post,* October 9, 1909.
28. *Cleaner Sales by Years, 1908–1939,* Hoover Historical Center.
29. *Chronology of Hoover Vacuums 1908–1940,* Hoover Historical Center, p. 2.
30. *The Vacuum Cleaner,* Vacuum Cleaner Manufacturers Association, 1945, p. 10.
31. *House Beautiful,* May, 1909, Lifshey Collection.
32. *Ladies Home Journal,* June, 1909, courtesy Vicki Matranga.
33. *The Vacuum Cleaner,* Vacuum Cleaner Manufacturers Association, p. 10.
34. Blackburn (Part 3), pp. 45–46.
35. *Chronology of Sweeper Patents,* Hoover Historical Center, p. 9.
36. U.S Patent 1,116,850, U.S. Patent Office.
37. Blackburn (Part 3), p. 46.
38. *Vacuum Companies Incorporation Details,* from the 25th Anniversary, Vacuum Cleaners Manufacturers Association, October 21–22, 1938, Hoover Historical Center.
39. *Chronology of Sweeper Patents,* Hoover Historical Center, p. 10.
40. Ibid.
41. Blackburn (Part 4), p. 51.
42. *Chronology of Sweeper Patents,* Hoover Historical Center, p. 10.
43. Lotz, "Richard K. Kirby, The Man Behind Less Work," *Machine Design* (July 1948), p. 109.
44. Vintage Machinery, "Domestic Electric Co., Patents," http://vintagemachinery.org/mfgindex/detail.aspx?id=2775&tab=7; and Blackburn (Part 2), p. 46.
45. Gasko consultation.
46. *Chronology of Sweeper Patents,* Hoover Historical Center, p. 10.
47. Blackburn (Part 5), p. 53.
48. U.S. Patent 1,151,731, U.S. Patent Office.
49. Jailer-Chamberlain article, "This is the Way we Cleaned our Floors: A History of the Vacuum Cleaner," p. 34.
50. *Chronology of Sweeper Patents,* Hoover Historical Center, p. 2.
51. *Chronology of Sweeper Patents,* Hoover Historical Center, p. 11.
52. Ibid.
53. *Report of the Federal Trade Commission on the House Furnishings Industry,* Volume III, Kitchen Furnishings and Domestic Appliances, October 6, 1924.
54. Blackburn (Part 5), p. 52.
55. *Cleaner Sales by Years, 1908–1939,* Hoover Historical Center.
56. *Chronology of Hoover Vacuums 1908–1940,* Hoover Historical Center, p. 3.
57. "From Sweeping to Suction," *Home Furnishings Daily,* Nov. 25, 1991, p. 78.
58. "How the Vacuum Cleaner Began," *Electrical Merchandising* (January 1957), p. 232.
59. *Evolution of Vacuum Cleaners,* General Electric, 1971, p. 9.
60. "How the Vacuum Cleaner Began," *Electrical Merchandising* (January 1957), p. 228.
61. Vintage Machinery, "Domestic Electric Co., History," http://vintagemachinery.org/mfgindex/detail.aspx?id=2775&tab=0.
62. Gartman, p. 40.
63. Gantz, *The Industrialization of Design,* pp. 106–107.
64. Christine Frederick, *The New Housekeeping: Efficiency Studies in Home Management,* p. 81.
65. Hopper article, *House Furnishings Review,* June 1914.
66. Edward B. Moore, May 12, 1913, *New York Times,* as quoted by Pulos, p. 260.
67. Theodore Roosevelt, "A Layman's Views of an Art Exhibition," *Outlook* 103 (March 29, 1913), pp. 718–720.
68. *Chronology of Hoover Vacuums 1908–1940,* Hoover Historical Center, p. 11.
69. Christine Frederick, *Household Engineering,* p. 157.
70. Ibid., pp. 158–161.
71. Ibid., p. 163.
72. Ibid., pp. 156–160.
73. *The Vacuum Cleaner,* Vacuum Cleaner Manufacturers Association, 1945, p. 9.

74. Frank Hoover, *Fabulous Dustpan*, pp. 78–91.
75. Whitehorne, "To Harness the Whole Industry for You," *Electrical Merchandising*, Vol. 17, No. 1 (January 1917), pp. 4–6.
76. *Electrical Merchandising*, November 17, 1917, p. 277.
77. *Electrical Merchandising*, Vol. 18, No. 4 (April 1918).
78. *Chronology of Sweeper Patents*, Hoover Historical Center, p. 12.
79. "From Sweeping to Suction," *Home Furnishings Daily*, Nov. 25, 1991, pp. 78–79.
80. Electrolux, http://group.electrolux.com/en/founding-an-international-company-666/ (accessed Aug. 15, 2011).
81. *Chronology of Sweeper Patents*, Hoover Historical Center, p. 12.

Chapter 4

1. *Report of the Federal Trade Commission on the House Furnishings Industry*, 1924, p. 14.
2. Burke, *The Pinball Effect*, p. 229.
3. U.S. Patent 1,533,271, U.S. Patent Office.
4. Gasko, "The Air-Way Sanitary System: Looking Back at Models from 1920–1941, Part 1," *Floor Care Professional*, November 2009.
5. Blackburn (Part 2), p. 46.
6. *Evolution of Vacuum Cleaners*, General Electric, 1971, p. 10.
7. U.S. Patent Office.
8. Earl Lifshey, *The Housewares Story*, pp. 302–303.
9. International Historic Films, "Look to Germany," http://www.ihffilm.com/b023.html (accessed Oct. 13, 2011).
10. Vacuumland, "Vacuum Cleaner Collectors' Club," http://www.vacuumland.org/cgi-bin/00ShowCollectionGETD.cgi?photoshow=NL-05-00-023.jpg&dir=/NEWSLETTERS/Spring%202000 (accessed Aug. 15, 2011).
11. Frank Hoover, *Fabulous Dustpan*, p. 143.
12. *Chronology of Hoover Vacuums 1908–1940*, pp. 4–5.
13. Blackburn (Part 5), p. 46.
14. *Report of the Federal Trade Commission on the House Furnishings Industry*, Volume III, Kitchen Furnishings and Domestic Appliances, October 6, 1924, pp. 14–15.
15. Electrolux, http://group.electrolux.com/en/founding-an-international-company-666/ (accessed Aug. 15, 2011).
16. Tom Gasko, email message to author, April 5, 2011.
17. Cowan, pp. 181–189.
18. Gasko, "The American Electrolux, Part 1," *Floor Care Professional*, February 2010.
19. *Vacuum Companies Incorporation Details*, from the 25th Anniversary, Vacuum Cleaners Manufacturers Association, October 21–22, 1938, Hoover Historical Center.
20. Lotz, "Richard K. Kirby, The Man Behind Less Work," *Machine Design* (July 1948), pp. 109–110.
21. Ibid., pp. 110–112.
22. Gorman, p. 115.
23. *Cleaner Sales by Years, 1908–1939*, Hoover Historical Center.
24. Gasko, "The Greater Hoover — Positive Agitation, Part 3," *Floor Care Professional*, September 2010.
25. Ibid.
26. Blackburn (Part 1), p. 45.
27. Gasko, "The Greater Hoover — Positive Agitation, Part 3," *Floor Care Professional*, September 2010.
28. Ann Haines, *Parade of Progress—A Review of the Hoover Company's Innovative Vacuum Cleaner Technology*.
29. Mildred Jailer-Chamberlain, "This is the Way We Cleaned Our Floors: A History of the Vacuum Cleaner," *Floor Care Professional*, December 1997, p. 34.
30. C.E. Alberts, "Where Vacuum Cleaners Were 15 Years Ago," *Electrical Merchandising*, June 1931.
31. "Encyclopedia of Vacuum Cleaners," *Electrical Record*, Vol. XLI, No. 4 (April 1927).
32. Funding Universe, "Company Histories and Profiles: Bissell Inc.," http://www.fundinguniverse.com/company-histories/BISSELL-Inc-Company-History.html (accessed Aug. 15, 2011).
33. Harold Van Doren, *Industrial Design, a Practical Guide*, p. 3.
34. Gantz, *The Industrialization of Design*, pp. 143–144.
35. Pulos, *American Design Ethic*, pp. 330–332.
36. Ibid.
37. Gantz, *The Industrialization of Design*, pp. 145–146.
38. Tom Gasko, personal consultation.
39. Blackburn (Part 5), p. 54.
40. Tom Gasko, "The Air-Way Sanitary System: Looking Back at Models from 1920–1941, Part 1," *Floor Care Professional*, November 2009.
41. Hoover Mothimizer instruction booklet, 1929, Vicki Matranga files.
42. *Chronology of Hoover Vacuums 1908–1940*, Hoover Historical Center, p. 6.
43. Blackburn (Part 2), p. 47.
44. Tom Gasko, "A Closer Look at Rexair, Great Invention in Vacuum Cleaner History," *Floor Care Professional*, January 2010.
45. E.L. Hinchliff, "Always Important," *Electrical Merchandising*, June 1930, pp. 56–57.
46. Walter McClure, "Whether in Park Avenue or the Bronx, They Buy Cleaners," *Electrical Merchandising*, June 1931, pp. 74–76.
47. *Electrical Merchandising*, May 1931, p. 72.
48. Tom Gasko, personal consultation.
49. Frank Hoover, *Fabulous Dustpan*, p. 182.
50. *Chronology of Hoover Vacuums 1908–1940*, p. 7.
51. Tom Gasko, personal consultation.
52. Ibid.
53. Tom Gasko, "The Air-Way Sanitary System: Looking Back at Models from 1920–1941, Part 1," *Floor Care Professional*, November 2009.
54. Norman Bel Geddes, *Horizons in Industrial Design*, p. 3.
55. "Both Fish and Fowl," *Fortune*, February 1934, pp. 40–43, 88–90, 94, 97–98.
56. Lesko, "Industrial Design at Carnegie Institute of Technology, 1934–1967," *Journal of Design History*

Vol. 10, No. 3 (1997), the Design History Society.
 57. Tom Gasko, personal consultation.
 58. Ibid.
 59. Clarence Frantz of VCMA to Earl Lifshey, September 24, 1971.
 60. Tom Gasko, personal consultation.
 61. U.S. Patent Office.
 62. *Cleaner Sales by Years, 1908–1939*, Hoover Historical Center.
 63. "Both Fish and Fowl," *Fortune*, February 1934, pp. 40–43, 88–90, 94, 97–98.
 64. Ann Haines, "Henry Dreyfuss Designs for Hoover."
 65. *Chronology of Hoover Vacuums 1908–1940*, Hoover Historical Center, p. 9.
 66. *Cleaner Sales by Years, 1908–1939*, Hoover Historical Center.
 67. *History and Background Information on Eureka Williams Company*, Eureka Williams Company, Bloomington, Illinois, 1971.
 68. Belle Kogan, interview with Mary Babbitt, "As a Woman Sees Design," *Modern Plastics* 13, No. 4 (December 1935), pp. 16–17, 49, 51, quoted by Gorman in *The Industrial Design Reader*, pp. 137–139.
 69. Tom Gasko, personal consultation.
 70. Lowell Thomas, *The Man who Revolutionized the American Home*. Reprinted by Scott & Fetzer ca. 1960s.
 71. *Electrical Merchandising*, August 1932, p. 46.
 72. U.S. Patent Office.
 73. Tom Gasko, "A Closer Look at Rexair, Great Invention in Vacuum Cleaner History," *Floor Care Professional*, January 2010.
 74. U.S. Patent Office.
 75. Ibid.
 76. Jitterbuzz, "Are Vacuum Cleaners Collectible?" http://www.jitterbuzz.com/indvac.html#wags (accessed Aug. 15, 2011).
 77. Baker, *Great Inventions, Good Intentions*, p. 77.
 78. Tom Gasko, personal consultation.
 79. U.S. Patent Office.
 80. Blackburn (Part 4), p. 46.
 81. Tom Gasko, "The American Electrolux, Part 1," *Floor Care Professional*, February 2010.
 82. Blackburn (Part 5), p. 53.
 83. Tom Gasko, personal consultation.
 84. Ibid.
 85. *Chronology of Hoover Vacuums 1908–1940*, pp. 10–11.
 86. *Cleaner Sales by Years, 1908–1939*, Hoover Historical Center.
 87. U.S. Patent D111,333, U.S. Patent Office.
 88. Tom Gasko, "The Air-Way Sanitary System: Looking Back at Models from 1920–1941, Part 1," *Floor Care Professional*, November 2009.
 89. Fred Stachnik, personal consultation.
 90. Tom Gasko, "The Air-Way Sanitary System: Looking Back at Models from 1920–1941, Part 1," *Floor Care Professional*, November 2009.
 91. Gasko, personal consultation.
 92. U.S. Patent 2,198,568, U.S. Patent Office.
 93. "How the Vacuum Cleaner Began," *Electrical Merchandising*, January 1957, p. 232.
 94. *Chronology of Hoover Vacuums 1908–1940*, p. 12.
 95. Blackburn (Part 4), p. 51.
 96. "From Sweeping to Suction," *Home Furnishings Daily*, Nov. 25, 1991, p. 79.

Chapter 5

 1. *The Exciting World of Hoover*, promotional brochure, the Hoover Company, ca. 1966.
 2. "From Sweeping to Suction," *Home Furnishings Daily*, Nov. 25, 1991, p. 79.
 3. Gantz, *The Industrialization of Design*, p. 183.
 4. Ibid., p. 193.
 5. *Evolution of Vacuum Cleaners*, General Electric, 1971, pp. 11–12.
 6. "From Sweeping to Suction," *Home Furnishings Daily*, Nov. 25, 1991, p. 79.
 7. Funding Universe, "Company Histories and Profiles: The Eureka Company," http://www.fundinguniverse.com/company-histories/The-Eureka-Company-Company-History.html (accessed Aug. 15, 2011).
 8. Gasko, "Selling Out the Hoover Salesmen, 1946–1958, Part 4," *Floor Care Professional*, October 2010.
 9. Charles Richard Lester, "The Compact Vacuum Cleaner," http://www.1377731.com/compact/.
 10. Alex M. Lewyt, *Developing a Successful Merchandising Program*, American Management Association, 1950.
 11. *From Brooms to Vacuum Cleaners*, The Lewyt Corporation, ca. 1971.
 12. Health-Mor Industries (HMI), http://www.fundinguniverse.com/company-histories/HMI-Industries-Inc-Company-History.html (accessed Aug. 15, 2011).
 13. *Chicago Tribune*, January 8, 1998.
 14. Fred Stachnik, personal consultation.
 15. Victoria Matranga, *Biographies of the Designers*, Toledo Museum of Art. *The Alliance of Art and Industry—Toledo Designs for a Modern America*, pp. 197–198.
 16. Fred Stachnik, personal consultation.
 17. Gasko, "Selling Out the Hoover Salesmen, 1946–1958, Part 4," *Floor Care Professional*, October 2010.
 18. "Vacuum Cleaners," *Consumer Reports*, July 1951, pp. 293–298, courtesy Vacuum Cleaners Collectors Association, http://www.vacuumland.org/cgi-bin/00ShowCollectionGETD.cgi?photoshow=CR-7-51-002.jpg&dir=/CONSUMER/CU%201951/ (accessed Aug. 15, 2011).
 19. Fred Stachnik, personal consultation.
 20. Tambini, *The Look of the Century*, p. 131.
 21. Gasko, "*Electrolux, 1956–2003, Part 3*," *Floor Care Professional*, April 2010.
 22. Fred Stachnik, personal consultation.
 23. Funding Universe, "Company Histories and Profiles: Bissell Inc.," http://www.fundinguniverse.com/company-histories/BISSELL-Inc-Company-History.html (accessed Aug. 15, 2011).
 24. "Vacuum Cleaners," *Consumer Reports*, 1954, pp. 1–8, courtesy Vacuum Cleaner Collectors Club, http://www.vacuumland.org/cgi-bin/00ShowCollectionGETD.cgi?photoshow=01@1954%20Consumer%20Reports%20Vacuum%20ReportPage%201.jpg&dir=/CONSUMER/CU1954/ (accessed Sept. 13, 2011).
 25. Design Patent No. 175,210, U.S. Patent Office.
 26. Fred Stachnik, personal consultation.
 27. Patent 2,781,103, U.S. Patent Office.

28. Royal, http://www.royalvacuums.com/about.aspx (accessed Aug. 15, 2011).
29. *Evolution of Vacuum Cleaners*, General Electric, 1971, p. 10.
30. Gasko, "A Closer Look at Rexair, Great Invention in Vacuum Cleaner History," *Floor Care Professional*, January 2010.
31. Gasko, personal consultation.
32. Object Mix, http://objectmix.com/cobol/329769-ot-humor-worst-predictions-all-time.html (accessed Aug. 15, 2011).
33. "Vacuum Cleaning and Floor Polisher Market Studies," *Electrical Merchandising*, January 1956.
34. *Evolution of Vacuum Cleaners*, General Electric, 1971, p. 19.
35. Gantz, *The Industrialization of Design*, p. 208.
36. Henry Dreyfuss, *Designing for People*, title page.
37. Gantz, *The Industrialization of Design*, p. 205.
38. *The Engineering and Manufacturing Facilities of The Hoover Company*, promotional brochure, the Hoover Company, 1956.
39. Gasko, "The Modern Electrolux, 1956–2003, Part 3," *Floor Care Professional*, April 2010.
40. Clarence Frantz of VCMA to Earl Lifshey, September 24, 1971.
41. Ann Haines, *Parade of Progress*.
42. Tom Gasko, personal consultation.
43. Gasko, "The Continuation of Air-Way, Innovative Changes through the Years, Part 2," *Floor Care Professional*, December 2009.
44. Earl Lifshey, *The Housewares Story*, p. 298.
45. "Vacuum Cleaners and Floor Polishers — Market Studies," *Electrical Merchandising*, January 1959.
46. Tom Gasko, "Selling Out the Hoover Salesmen, 1946–1958, Part 4," *Floor Care Professional*, October 2010.
47. Rexair, http://en.wikipedia.org/wiki/Rexair (accessed Jan. 29, 2012).
48. "Vacuum Cleaners and Floor Polishers — Market Studies," *Electrical Merchandising*, January 1959.
49. U.S. Patent Office.
50. Gasko, "The Modern Electrolux, 1956–2003, Part 3," *Floor Care Professional*, April 2010.
51. Tom Gasko, personal consultation.
52. Fred Stachnik, personal consultation.
53. Gasko, "The Modern Electrolux, 1956–2003, Part 3," *Floor Care Professional*, April 2010.
54. *Evolution of Vacuum Cleaners*, General Electric, 1971, pp. 11–14.
55. Fred Stachnik, personal consultation.
56. *From Brooms to Vacuum Cleaners*, The Lewyt Corporation, ca. 1971, pp. 4–5.
57. "Vacuum Cleaners," *Consumer Reports*, 1959, pp. 1–12, courtesy Vacuum Cleaner Collectors Club, http://www.vacuumland.org/cgi-bin/00ShowCollectionGETD.cgi?dir=/CONSUMER/CU1959/.
58. Funding Universe, "Company Histories and Profiles: Bissell Inc.," http://www.fundinguniverse.com/company-histories/BISSELL-Inc-Company-History.html.
59. Fred Stachnik, personal consultation.
60. U.S. Patent Office.
61. Siekman, "Hoover's Well-Vacuumed World," *Fortune*, June 1964, pp. 143–213.
62. Gantz, *The Industrialization of Design*, p. 227.
63. U.S. Patent Office.
64. Ann Haines, *Parade of Progress*.
65. Fred Stachnik, personal consultation.
66. Ibid.
67. "The Consumer's Best Friend," *Hoover Worldwide*, Spring 1967, pp. 12–15.
68. Wikipedia, http://wikipedia.org/wiki/David_Oreck (accessed Aug. 15, 2011).
69. Gantz, *The Industrialization of Design*, pp. 230–231.
70. *Hoover News*, Volume 44, No. 3 (March 12, 1965), p. 1.
71. Siekman, "Hoover's Well-Vacuumed World," *Fortune*, June 1964, pp. 143–213.
72. U.S. Patent Office.
73. Ibid.
74. Ibid.
75. Gantz, *The Industrialization of Design*, pp. 289–290.
76. *Evolution of Vacuum Cleaners*, General Electric, 1971, p. 16.
77. *From Brooms to Vacuum Cleaners*, The Lewyt Corporation, ca. 1971, p. 5.
78. Funding Universe, "Company Histories and Profiles: Bissell Inc.," http://www.fundinguniverse.com/company-histories/BISSELL-Inc-Company-History.html (accessed Aug. 15, 2011).
79. U.S. Patent Office.
80. Advertisement, Gantz files.
81. *The Hoover Company Consolidated Annual Report, 1968*.
82. Gasko, "The Modern Electrolux, 1956–2003, Part 3."
83. Funding Universe, "Company Histories and Profiles: The Eureka Company," http://www.fundinguniverse.com/company-histories/The-Eureka-Company-Company-History.html (accessed Aug. 30, 2011).
84. Wikipedia, http://en.wikipedia.org/wiki/Electrolux (accessed Aug. 30, 2011).
85. Gasko, "The Continuation of Air-Way, Innovative Changes through the Years, Part 2," *Floor Care Professional*, December 2009.
86. Fred Stachnik, personal consultation.
87. Gasko, "The Continuation of Air-Way, Innovative Changes through the Years, Part 2," *Floor Care Professional*, December 2009.
88. R.H. Sutherland, "Music to a Collector's Ear," *Hoover Worldwide* No. 15 (Winter 1969).
89. Charles Richard Lester, "Stan Kann," http://www.1377731.com/stan/ (accessed Nov. 15, 2011).
90. Fred Stachnik, personal consultation.
91. U.S. Patent Office.
92. Ibid.
93. "From Sweeping to Suction," *Home Furnishings Daily*, Nov. 25, 1991, p. 83.

Chapter 6

1. Hoffman, "Electric Floor Care Products Making a Clean Sweep," *Home Furnishings Daily*, May 5, 1971.
2. Funding Universe, "Health-Mor Industries (HMI)," http://www.fundinguniverse.com/company-

histories/HMI-Industries-Inc-Company-History.html (accessed Aug. 15, 2011).

3. Fred Stachnik, personal consultation.

4. Wikipedia, http://en.wikipedia.org/wiki/Kirby_Company (accessed Aug. 3, 2011).

5. Funding Universe, "Company Histories and Profiles: Bissell Inc.," http://www.fundinguniverse.com/company-histories/BISSELL-Inc-Company-History.html (accessed Aug. 15, 2011).

6. Oreck/Whirlpool lawsuit, http://webcache.googleusercontent.com/search?q=cache:LlqdQ5lqsOkJ:ftp.resource.org/courts.gov/c/F2/579/579.F2d.126.767631.1173.html+RCA+Whirlpool+robotic+vacuum+cleaner&cd=4&hl=en&ct=clnk&gl=us&source=www.google.com (accessed Aug. 30, 2011).

7. Gasko, "The Continuation of Air-Way, Innovative Changes through the Years, Part 2," *Floor Care Professional*, December 2009.

8. "A History of the Vacuum Cleaner," *Floor Care Professional*, magazine of the Vacuum Dealers Trade Association (VDTA), December 1997, pp. 33–40.

9. Earl Lifshey, *The Housewares Story*, p. 305.

10. Ibid.

11. Ibid.

12. Funding Universe, http://www.fundinguniverse.com/company-histories/The-Hoover-Company-Company-History.html (accessed Aug. 30, 2011).

13. Funding Universe, http://www.fundinguniverse.com/company-histories/The-Eureka-Company-Company-History.html (accessed Aug. 30, 2011).

14. Gasko, "The Modern Electrolux, 1956–2003, Part 3," *Floor Care Professional*, April 2010.

15. Funding Universe, http://www.fundinguniverse.com/company-histories/The-Hoover-Company-Company-History.html (accessed Aug. 30, 2011).

16. U.S. Patent Office.

17. "From Sweeping to Suction," *Home Furnishings Daily*, Nov. 25, 1991, p. 82.

18. Auburn University, "Sunbeam Corporation," http://www.auburn.edu/~stanwsd/sunbeam.html (accessed Aug. 30, 2011); and Funding Universe, http://www.fundinguniverse.com/company-histories/SunbeamOster-Co-Inc-Company-History.html (accessed Aug. 30, 2011).

19. Vacuum Dealers Trade Association, http://www.vdta.com/HOF/Beall.html (accessed Aug. 30, 2011).

20. Gantz, *The Industrialization of Design*, p. 249.

21. Wikipedia, http://en.wikipedia.org/wiki/Numatic_International_Limited (accessed Aug. 30, 2011).

22. Funding Universe, http://www.fundinguniverse.com/company-histories/The-Eureka-Company-Company-History.html (accessed Aug. 30, 2011).

23. Funding Universe, http://www.fundinguniverse.com/company-histories/The-Singer-Company-NV-Company-History.html (accessed Aug. 30, 2011).

24. Robert Tabor, "The Whirlwind, the Oldest Vacuum Cleaner in the World," in "A History of the Vacuum Cleaner," *Floor Care Professional*, magazine of the Vacuum Dealers Trade Association (VDTA), December 1997, p. 38.

25. Vacuum Dealers Trade Association, http://www.vdta.com/hof-list.html (accessed Aug. 30, 2011).

26. Gasko, "The Modern Electrolux, 1956–2003, Part 3," *Floor Care Professional*, April 2010.

27. Ann Haines, *Parade of Progress—A Review of the Hoover Company's Innovative Vacuum Cleaner Technology*.

28. "A History of the Vacuum Cleaner," *Floor Care Professional*, magazine of the Vacuum Dealers Trade Association (VDTA), December 1997, p. 38.

29. Answers, "Techtronic Industries Company Ltd.," http://www.answers.com/topic/techtronic-industries-company-ltd (accessed Jan. 30, 2012).

30. Funding Universe, http://www.fundinguniverse.com/company-histories/Royal-Appliance-Manufacturing-Company-Company-History.html (accessed Aug. 30, 2011).

31. TTI Floor Care North America, http://www.ttifloorcare.com/company.aspx (accessed Aug. 30, 2011).

32. Funding Universe, http://www.fundinguniverse.com/company-histories/The-Hoover-Company-Company-History.html (accessed Aug. 30, 2011).

33. "Changes Sweeping," *Home Furnishings Daily*, Jan. 1, 1986, pp. 74–77.

34. Ibid., p. 74.

35. Ibid.

36. Ibid., p. 76.

37. U.S. Patent Office.

38. The Great Idea Finder, http://www.ideafinder.com/history/inventions/dysonvac.htm (accessed Aug. 30, 2011).

39. U.S. Patent Office.

40. The Great Idea Finder, http://www.ideafinder.com/history/inventors/dyson.htm (accessed Aug. 30, 2011).

41. Gasko, "A Closer Look at Rexair, Great Invention in Vacuum Cleaner History," *Floor Care Professional*, January 2010.

42. "From Sweeping to Suction," *Home Furnishings Daily*, Nov. 25, 1991, p. 83.

43. Ann Haines, *Parade of Progress—A Review of the Hoover Company's Innovative Vacuum Cleaner Technology*.

44. "From Sweeping to Suction," *Home Furnishings Daily*, Nov. 25, 1991, p. 83.

45. *Home Furnishings Daily*, December 7, 1987.

46. Wikipedia, http://en.wikipedia.org/wiki/Kirby_Company (accessed Aug. 30, 2011).

47. Vintage Machinery, http://vintagemachinery.org/mfgindex/detail.aspx?id=1954 (accessed Sept. 16, 2011).

48. Funding Universe, http://www.fundinguniverse.com/company-histories/The-Hoover-Company-Company-History.html (accessed Aug. 30, 2011).

Chapter 7

1. "From Sweeping to Suction," *Home Furnishings Daily*, Nov. 25, 1991, p. 83.

2. Ibid.

3. Abby's Guide, http://www.abbysguide.com/vacuum/discussions/41102-159-1.html (accessed Aug. 30, 2011).

4. Ann Haines, *Parade of Progress—A Review of the Hoover Company's Innovative Vacuum Cleaner Technology*.

5. Fred Stachnik, personal consultation.

Notes — Chapter 7

6. Funding Universe, "Bissell Inc.," http://www.fundinguniverse.com/company-histories/BISSELL-Inc-Company-History.html (accessed Aug. 15, 2011).
7. Funding Universe, http://www.fundinguniverse.com/company-histories/The-Eureka-Company-Company-History.html (accessed Aug. 30, 2011).
8. Wikipedia, http://en.wikipedia.org/wiki/Vacuum_cleaner (accessed Aug. 30, 2011).
9. Dyson, http://www.dyson.com/about/story/default.asp?searchType=story&story=risk (accessed Aug. 30, 2011).
10. Georgetown Law Library, http://www.ll.georgetown.edu/federal/judicial/fed/opinions/00opinions/00-1219.html (accessed Aug. 30, 2011).
11. Funding Universe, http://www.fundinguniverse.com/company-histories/BISSELL-Inc-Company-History.html (accessed Aug. 30, 2011).
12. Fred Stachnik, personal consultation.
13. Website: http://www.fundinguniverse.com/company-histories/BISSELL-Inc-Company-History.html (accessed Aug. 30, 2011).
14. Wikipedia, http://en.wikipedia.org/wiki/Tacony_Corporation (accessed Aug. 30, 2011).
15. Ann Haines, *Parade of Progress—A Review of the Hoover Company's Innovative Vacuum Cleaner Technology*.
16. Fred Stachnik, personal consultation.
17. Ann Haines, *Parade of Progress—A Review of the Hoover Company's Innovative Vacuum Cleaner Technology*.
18. Gasko, "A Closer Look at Rexair, Great Invention in Vacuum Cleaner History," *Floor Care Professional*, January 2010.
19. Answers, http://www.answers.com/topic/techtronic-industries-company-ltd.
20. Linda Abu-Shalback, "A Flurry in Floor Care," *Appliance Magazine*, February 2001, pp. 37–39.
21. Robert Uhlig, "Hoover ban after Dyson patent case," *The Telegraph*, http://www.telegraph.co.uk/science/science-news/4755064/Hoover-ban-after-Dyson-patent-case.html (accessed Jan. 18, 2012).
22. Fred Stachnik, personal consultation.
23. Ibid.
24. Ann Haines, *Parade of Progress—A Review of the Hoover Company's Innovative Vacuum Cleaner Technology*.
25. Gasko, "The Continuation of Air-Way, Innovative Changes through the Years, Part 2," *Floor Care Professional*, December 2009.
26. Charles Richard Lester, "The Compact Vacuum Cleaner," http://www.1377731.com/compact/.
27. "From Sweeping to Suction," *Home Furnishings Daily*, Nov. 25, 1991, p. 83.
28. Joseph T. Wells, "Timing is of the Essence," *Journal of Accountancy*, May 2001, http://www.journalofaccountancy.com/Issues/2001/May/TimingIsOfTheEssence (accessed Aug. 30, 2011).
29. Ibid.
30. Linda Abu-Shalback, "A Flurry in Floor Care," *Appliance Magazine*, February 2001, p. 39.
31. Ann Haines, *Parade of Progress—A Review of the Hoover Company's Innovative Vacuum Cleaner Technology*.
32. "Electrolux Overview," *Appliance Magazine*, February 2001, pp. E-4 to E-54.
33. Wikipedia, http://en.wikipedia.org/wiki/Vacuum_cleaner (accessed Aug. 30, 2011).
34. Emmanuel Barraud, "Domestic Robots: Harmony on the Homefront?" Phys.org, Dec. 21, 2011, http://phys.org/news/2011-12-domestic-robots-harmony-homefront.html (accessed Jan. 31, 2012).
35. Dyson, http://www.dyson.com/about/story/default.asp?searchType=story&story=risk (accessed Aug. 30, 2011).
36. *Wall Street Journal*, October 22, 2004, p. W11C, Matranga files.
37. Gasko, "The Modern Electrolux, 1956–2003, Part 3," *Floor Care Professional*, April 2010; and Electrolux, http://group.electrolux.com/en/founding-an-international-company-666/ (accessed Aug. 15, 2011).
38. Wikipedia, http://en.wikipedia.org/wiki/Kirby_Company (accessed Aug. 30, 2011).
39. Association of Home Appliance Manufacturers, http://www.aham.org/ht/d/sp/i/454/pid/454 (accessed Nov. 4, 2011).
40. Funding Universe, http://www.fundinguniverse.com/company-histories/Whirlpool-Corporation-Company-History.html (accessed Sept. 9, 2011).
41. Wikipedia, http://en.wikipedia.org/wiki/Whirlpool_Corporation (accessed Aug. 30, 2011).
42. Tacony Manufacturing, http://www.madeintheusavacuum.com/vacuum-evolution/ (accessed Aug. 30, 2011).
43. International Housewares Association, *MarketWatch*, Summer 2004, p. 5.
44. Funding Universe, http://www.fundinguniverse.com/company-histories/The-Eureka-Company-Company-History.html (accessed Aug. 30, 2011).
45. Gasko, "A Closer Look at Rexair, Great Invention in Vacuum Cleaner History," *Floor Care Professional*, January 2010.
46. Dyson, http://www.dyson.com/about/story/default.asp?searchType=story&story=risk (accessed Aug. 30, 2011).
47. Ibid.
48. Vacuum Dealers Trade Association, http://www.vdta.com/AssociateMembers/emer.html (accessed Sept. 1, 2011).
49. Fred Stachnik, personal consultation.
50. Funding Universe, http://www.fundinguniverse.com/company-histories/Whirlpool-Corporation-Company-History.html (accessed Sept. 9, 2011).
51. Vacuum Dealers Trade Association, http://www.vdta.com/Magazines/JAN07/fc-hoover-tti.html (accessed Sept. 1, 2011).
52. Ann Haines, personal consultation.
53. Gasko, "Hoover 1958–2010 The End of an Era and a New Beginning, Part 5," *Floor Care Professional*, November 2010.
54. Ann Haines, personal consultation.
55. Matranga, "Housewares History: Bissell, A History of Inventive Problemsolving," International Housewares Association blog, August 1, 2011, http://www.housewares.org/blog/index.php/category/housewares-history/ (accessed Sept. 1, 2011).
56. Digital Versus, http://www.digitalversus.com/rowenta-ro8049-p1039_8429_457.html (accessed Sept. 1, 2011).
57. Gasko, "The Continuation of Air-Way, Inno-

vative Changes through the Years, Part 2," *Floor Care Professional*, December 2009.

58. Consumer Reports, http://news.consumerreports.org/home/2010/10/new-vacuum-cleaner-ratings-tougher-pet-hair-test.html (accessed Dec. 15, 2011).

59. HomeWorld Business, http://www.homeworldbusiness.com/links/news/news.php?ID=18760 (accessed Jan. 10, 2012).

60. Ubergizmo, http://www.ubergizmo.com/2011/08/cambridge-consultants-stem-vacuum-cleaner/ (accessed Sept. 1, 2011).

61. Fred Stachnik, personal consultation.

62. Lawyers and Settlement.com, http://www.lawyersandsettlements.com/blog/ (accessed Dec. 15, 2011).

63. Gasko, "Collecting Dust, A History of the Vacuum Cleaner," *Floor Care Professional*, December 1997, pp. 39–40.

Bibliography

Books

Alphin, Elaine Marie. *Vacuum Cleaners*. Minneapolis: Carolrhoda Books, c/o The Lerner Publishing Group Inc., 1997.

Anderson, Susan H. *The Most Splendid Carpet*. Philadelphia: U.S. Department of the Interior, National Park Service, 1978. http://www.nps.gov/history/history/online_books/inde/anderson/chap4.htm (accessed Aug.15, 2011).

Baker, Eric. *Great Inventions, Good Intentions*. San Francisco: Chronicle Books, 1990.

Basalla, George. *The Evolution of Technology*. Cambridge, MA: Cambridge University Press, 1988.

Baynes, Thomas Spencer. "Brushes and Brooms." *Encyclopaedia Britannica, Dictionary of Arts, Sciences, and General Literature*, Volume IV, 9th ed. New York: Henry G. Allen, 1888, pp. 403–404. Google eBook, http://books.google.com/books?id=waAMAAAAYAAJ&pg=PA403&lpg=PA403&dq=early+brushes&source=bl&ots=9oJ33mftmH&sig=Z4pgYp4wvpAKkMkBrjciU4oBpvc&hl=en&ei=dTd1TbCPMO90QHUtbX9Cw&sa=X&oi=book_result&ct=result&resnum=9&ved=0CDkQ6AEwCDgy#v=onepage&q=early%20brushes&f=false (accessed Feb. 23, 2011).

_____. "Carpet." *Encyclopaedia Britannica, Dictionary of Arts, Sciences, and General Literature*, Volume V, 9th Edition. New York: Henry G. Allen, 1888, pp. 127–131. Google eBook, http://books.google.com/books?id=HKgMAAAAYAAJ&printsec=frontcover#v=onepage&q&f=false (accessed Aug. 15, 2011).

Beecher, Catharine E. *A Treatise on Domestic Economy, for the Use of Young Ladies at Home and at School*. Boston: T.H. Webb, 1842, and New York: Harper and Brothers, 1845. Project Gutenberg EBook #21829, Released June 14, 2007, http://www.gutenberg.org/files/21829/21829-h/21829-h.htm (accessed 23 Feb. 2011).

_____, and Harriet Beecher Stowe. *American Woman's Home: or Principles of Domestic Science*. New York: J.B. Ford, 1869. Project Gutenberg EBook #6598, Released September, 2004, http://www.gutenberg.org/cache/epub/6598/pg6598.html (accessed 23 Feb. 2011).

Bel Geddes, Norman. *Horizons in Industrial Design*. Boston: Little, Brown, 1932.

Bunch, Bryan, and Alexander Hellemans, editors. *The Timetables of Technology*. New York: A Touchstone Book by Simon & Schuster, 1993.

Burke, James. *Connections*. Boston: Little, Brown, 1978.

_____. *The Day the Universe Changed*. Boston: Little, Brown, 1985.

_____. *The Pinball Effect*. Boston: Little, Brown, 1997.

Cheney, Margaret. *Tesla: Man Out of Time*. New York: Simon & Schuster, 2001.

Cowan, Ruth Schwartz. *More Work for Mother: The Ironies of Household Technology from the Open Hearth to the Microwave*. New York: Basic Books, a Division of HarperCollins, 1983.

Derry, T.K., and Trevor I. Williams. *A Short History of Technology from the Earliest Times to A.D. 1900*. New York: Oxford University Press, 1960.

Dreyfuss, Henry. *Designing for People*. New York: Simon & Schuster, 1955.

Frederick, Christine. *The New Housekeeping: Efficiency Studies in Home Management*. Garden City, NY: Curtis Publishing, 1912, and Doubleday, Page, 1913. Google eBook, http://books.google.com/books?id=QiALAAAAIAAJ&printsec=frontcover&dq=the+new+housekeeping+christine+frederick&source=bl&ots=twDJ4wZABf&sig=SQfZBU0ncZ5X3zd86HH7H_Yc6k&hl=en&ei=W8iPTbzVFJGXtweRtKiICQ&sa=X&oi=book_result&c

t=result&resnum=3&ved=0CCQQ6AEwAg#v=onepage&q&f=false (accessed Feb. 23, 2011).
———. *Household Engineering: Scientific Management in the Home.* Chicago: American School of Home Economics, 1921. Household open library, edited April 13, 2010, http://openlibrary.org/books/OL13517487M/Household_engineering (accessed Feb. 23, 2011).
Gantz, Carroll. *The Industrialization of Design.* Jefferson, NC: McFarland, 2010.
Gartman, David. *Auto Opium.* New York: Routledge, 1994.
Giedion, Siegfried. *Mechanization Takes Command: A Contribution to Anonymous History.* New York: Oxford University Press, 1948.
Gorman, Carma R. *The Industrial Design Reader.* New York: Allworth, 2003.
Ikenson, Ben. *Patents, Ingenious Inventions.* New York: Black Dog & Leventhal, 2004.
Israel, Fred L. *The State of the Union Messages of the Presidents.* New York: Chelsea House, Robert Hector, 1966.
Hoover, Frank. *Fabulous Dustpan.* Cleveland: World Publishing, 1955.
Hughes, Thomas P. *American Genesis.* New York: Penguin Books, 1989.
Larson, Erik. *Thunderstruck.* New York: Crown Publishers, a division of Random House, 2006.
Lemay, J.A., and P.M. Zall, editors, *Benjamin Franklin's Autobiography.* New York and London: W.W. Norton, 1986.
Leslie, Frank. *Illustrated Historical Register of the Centennial Exhibition.* New York: Paddington, 1876.
Lifshey, Earl. *The Housewares Story.* Chicago: National Housewares Manufacturers Association, 1973.
Loewy, Raymond. *Industrial Design.* Woodstock, NY: Overlook Press, 1979.
Lynd, Robert S., and Helen Merrell Lynd. *Middletown, a Study in American Culture.* New York: Harcourt, Brace, 1929.
Matranga, Victoria Kasuba. *America at Home, A Celebration of Twentieth Century Housewares.* Chicago: National Housewares Manufacturing Association, 1997.
Messadié, Gerald. *Great Modern Inventions.* Edinburgh: W & R Chambers, 1988.
Pulos, Arthur. *American Design Ethic.* Cambridge, MA: MIT Press, 1983.
Readers Digest. *Stories Behind Everyday Things.* Pleasantville, NY: Reader's Digest Association, 1980.
Smith, Jean Edward. *FDR.* New York: Random House, 2007.
Tambini, Michael, in association with the Cooper-Hewitt National Design Museum of the Smithsonian Institution. *The Look of the Century.* New York: DK Publishing, 1996.
Toledo Museum of Art. *The Alliance of Art and Industry: Toledo Designs for a Modern America.* New York: Hudson Hills Press, 2002.
Trager, James. *The People's Chronology.* New York: Henry Holt, 1992.
Van Doren, Harold. *Industrial Design, A Practical Guide.* New York: McGraw-Hill, 1940.
Van Dulkin, Stephen. *Inventing the 20th Century.* New York: New York University Press, 2000.
Wallechinsky, David, and I. Wallace. *The People's Almanac.* Garden City, NY: Doubleday, 1975.

Corporate Documents

Acheson, L.K. *History of Carpet Cleaning Development.* Hoover Historical Center, 1961. Revised Jan. 1998 by Ann Haines.
The Apex Rotorex Story. Frantz Industries, ca. 1971. Lifshey Collection.
Chronology of Hoover Vacuums 1908–1940. Hoover Historical Center.
Chronology of Sweeper Patents. Hoover Historical Center, pp. 1–12.
Cleaner Sales by Years, 1908–1939. The Hoover Company, Hoover Historical Center.
"The Consumer's Best Friend," *Hoover Worldwide,* Spring 1967.
The Engineering and Manufacturing Facilities of The Hoover Company (promotional brochure). The Hoover Company, 1956.
Evolution of Vacuum Cleaners. General Electric, 1971, pp. 1–19. Lifshey Collection.
The Exciting World of Hoover (promotional brochure). The Hoover Company, ca. 1966.
From Brooms to Vacuum Cleaners. The Lewyt Corporation, ca. 1971. Lifshey Collection.
Gantz, Carroll. Personal Hoover files and records, 1956–1972.
———. Personal Black & Decker files and records, 1972–1986.
Haines, Ann. "Henry Dreyfuss Designs for Hoover." Hoover Historical Center, August 2003.
———. *History of the Broom.* Hoover Historical Center. Revised Jan. 1998.
———. *Hoover's First Suction Cleaner Weighed Nearly 40 Pounds.* Hoover Historical Center. Revised 2010.
———. *Parade of Progress—A Review of the Hoover Company's Innovative Vacuum Cleaner Technology.* Hoover Historical Center. Revised January 2010.
A Heritage of Good Business. Bissell Inc. Corporate Headquarters.
History and Background Information on Eureka Williams Company, Eureka Williams Company, Bloomington, Illinois, 1971. Lifshey Collection.
A History of the Hoover Company (promotional brochure). The Hoover Company, ca. 1970.
The Hoover Company Consolidated Annual Report, 1968.

Hoover Firsts, Floor Care—Vacuum Cleaners. Hoover Historical Center.
Hoover News, Volume 44, No. 3 (March 12, 1965).
How the Vacuum Cleaner Began. The Hoover Company, 1967.
Installed Cleaning Systems Using Suction Only. Hoover Historical Center.
Lewyt, Alex M. *Developing a Successful Merchandising Program.* American Management Association, 1950. Lifshey Collection.
Lifshey, Earl. Papers, articles, and photographs related to his 1973 book, *The Housewares Story,* loaned to the author by Victoria Matranga of the International Housewares Association, Rosemont, Illinois. Lifshey Collection.
Manual Cleaners. Hoover Historical Center.
Simplified History of the Evolution of the Vacuum Cleaner. Hoover Historical Center.
Sutherland, R.H. "Music to a Collector's Ear." *Hoover Worldwide* No. 15, Winter 1969.
Thomas, Lowell. *The Man who Revolutionized the American Home.* Reprinted by Scott & Fetzer Company, ca. 1960s. Lifshey Collection.

Federal Documents

Executive Documents. The Senate of the United States, Second Session, 1858–1859, Vol. 986, XVII, *Household Furniture, etc.* Google eBook, http://books.google.com/books?id=QGpHAQAAIAAJ&pg=PA443&lpg=PA443&dq=shaler+carpet+sweeper&source=bl&ots=zTHOSS-Gvs7&sig=gfumyP3QXCknuFvV4S9_bHmyC4A&hl=en&ei=ps1PTs3lMcmBgAfptonhBg&sa=X&oi=book_result&ct=result&resnum=4&sqi=2&ved=0CFwQ6AEwAw#v=onepage&q=shaler%20carpet%20sweeper&f=false (accessed August 20, 2011).
Report of the Federal Trade Commission on the House Furnishings Industry, Volume III, Kitchen Furnishings and Domestic Appliances, October 6, 1924. Lifshey Collection.
U.S. Patent Office website: http://www.uspto.gov/patft/index.html.

Trade Publications and Articles

Abu-Shalback, Linda. "A Flurry in Floor Care." *Appliance Magazine,* February 2001, pp. 37–39. Matranga files.
Alberts, C.E. "Where Vacuum Cleaners Were 15 Years Ago." *Electrical Merchandising,* June 1931, pp. 64–65. Matranga files.
Blackburn, Tom F. "Evolution of the Vacuum, a Series of Visits with the Pilgrim Founders of the Industry (Part 1)." *Electrical Merchandising,* December 1, 1945, pp. 42, 43, 44, 45, 47. Lifshey Collection.
_____. "Evolution of the Vacuum, a Series of Visits with the Pilgrim Founders of the Industry (Part 2)." *Electrical Merchandising,* February 1, 1946, pp. 46, 47, 48. Lifshey Collection.
_____. "Evolution of the Vacuum, a Series of Visits with the Pilgrim Founders of the Industry (Part 3)." *Electrical Merchandising,* March 1, 1946, pp. 44, 45, 46. Lifshey Collection.
_____. "Evolution of the Vacuum, a Series of Visits with the Pilgrim Founders of the Industry (Part 4)." *Electrical Merchandising,* April 1, 1946, pp. 50, 51, 52. Lifshey Collection.
_____. "Evolution of the Vacuum, a Series of Visits with the Pilgrim Founders of the Industry (Part 5)." *Electrical Merchandising,* May 1, 1946, pp. 52, 53, 54. Lifshey Collection.
"Changes Sweeping," *Home Furnishings Daily,* Jan. 1, 1986, pp. 74–77.
Electrical Merchandising, Vol. 18, No. 4 (April 1918). Matranga files.
Electrical Merchandising, June 1930, pp. 56–57. Matranga files.
Electrical Merchandising, May 1931, p. 72. Matranga files.
Electrical Merchandising, August 1932, p. 46. Matranga files.
"Electrolux Overview," *Appliance Magazine,* February 2001, pp. E-4 to E-54. Matranga files.
"Encyclopedia of Vacuum Cleaners." *Electrical Record* (published by the Gage Publishing Company) Vol. XLI, No. 4 (April 1927), pp. 551–562. Hoover Historical Center.
"Forty Uses for Suction Cleaners." *House Furnishings Journal* (Chicago), April 1918. Lifshey Collection.
"From Sweeping to Suction." *Home Furnishings Daily,* November 25, 1991, pp. 71–83. Hoover Historical Center.
Hinchliff, E.L. "Always Important." *Electrical Merchandising,* June 1930, pp. 56–57.
Hoffman, Manny. "Electric Floor Care Products Making a Clean Sweep." *Home Furnishings Daily,* May 5, 1971. Lifshey Collection.
Hopper, A.R. "Increase Your Sales by Inducing Your Customers to Use Both Types of Vacuum Cleaners." *House Furnishings Review,* June 1914. Lifshey Collection.
"How the Vacuum Cleaner Began." *Electrical Merchandising,* January 1957, pp. 122, 123, 228, 232. Lifshey Collection.
IHA (International Housewares Association). *MarketWatch,* Summer 2004, p. 5. Matranga files.
McClure, Walter. "Whether in Park Avenue or the Bronx, They Buy Cleaners." *Electrical Merchandising,* June 1931, pp. 74–76. Matranga files.
Taggart, Edward. "Evolution of the Carpet Sweeper." *House Furnishings Review,* 1900, pp. 344–350. Hoover Historical Center.
The Vacuum Cleaner, Vacuum Cleaner Manufacturers Association, 1945. Lifshey Collection.

"Vacuum Cleaning and Floor Polisher Market Studies," *Electrical Merchandising*, January, 1956. Lifshey Collection.

"Vacuum Cleaners and Floor Polishers — Market Studies." *Electrical Merchandising*, January, 1959. Lifshey Collection.

Vacuum Companies Incorporation Details. From the 25th Anniversary Vacuum Cleaners Manufacturers Association, October 21–22, 1938, Hot Springs, Virginia. Hoover Historical Center.

Whitehorne, Earl E. "To Harness the Whole Industry for You." *Electrical Merchandising*, Vol. 17, No. 1 (January 1917), pp. 4–6. Matranga files.

Consumer Reports

"Vacuum Cleaners." *Consumer Reports*, July 1951, pp. 293–298. http://www.vacuumland.org/cgi-bin/00ShowCollectionGETD.cgi?photoshow=CR-7-51-002.jpg&dir=/CONSUMER/CU%201951/

"Vacuum Cleaners." *Consumer Reports*, 1954, pp. 1–8. http://www.vacuumland.org/cgi-bin/00ShowCollectionGETD.cgi?photoshow=01@1954%20Consumer%20Reports%20Vacuum%20Report-Page%201.jpg&dir=/CONSUMER/CU1954/

"Vacuum Cleaners." *Consumer Reports*, 1959, pp. 1–12. http://www.vacuumland.org/cgi-bin/00ShowCollectionGETD.cgi?dir=/CONSUMER/CU1959/

Wall Street Journal, October 22, 2004, p. W11C. Matranga files.

Articles

"Bissell's Big Sweep." *Fortune*, March 1960.

"Both Fish and Fowl." *Fortune*, February 1934, pp. 40–43, 88–90, 94, 97–98.

Chicago Tribune, January 8, 1998.

Fenster, J.M. *"Seam Stresses." Great Inventions that Changed the World, A Supplement to Inventions & Technology*, 1997, pp. 12–24.

Gasko, Tom. "Collecting Dust, a History of the Vacuum Cleaner." *Floor Care Professional*, December 1997, pp. 39–40.

_____. "The Air-Way Sanitary System; Looking back at models from 1920–1941, Part 1." *Floor Care Professional*, November 2009. http://www.vdta.com/Magazines/NOV09/fc-GaskoNov09.html.

_____. "The Continuation of Air-Way, Innovative Changes through the Years, Part 2." *Floor Care Professional*, December 2009. http://www.vdta.com/Magazines/DEC09/fc-GaskoDec09.html.

_____. "A Closer Look at Rexair, Great Invention in Vacuum Cleaner History." *Floor Care Professional*, January 2010. http://www.vdta.com/Magazines/JAN10/fc-GaskoJan10.html.

_____. "The American Electrolux, Part 1." *Floor Care Professional*, February 2010. http://www.vdta.com/Magazines/FEB10/fc-GaskoFeb10.html.

_____. "The Cleaner You Never Have to Empty, Part 2." *Floor Care Professional*, June 2010. http://www.vdta.com/Magazines/JUN10/fc-GaskoJun10.html.

_____. "The Modern Electrolux, 1956–2003, Part 3." *Floor Care Professional*, April 2010. http://www.vdta.com/Magazines/APR10/fc-gaskoApr10.html.

_____. "The Women Behind the Vacuum Cleaner! Part 1." *Floor Care Professional*, May 2010. http://www.vdta.com/Magazines/MAY10/fc-GaskoMay10.htm.

_____. "Hoover — It Beats — As it Sweeps — As it Cleans, 1908–1926, Part 2." *Floor Care Professional*, August 2010. http://www.vdta.com/Magazines/AUG10/fc-GaskoAug10.html.

_____. "The Greater Hoover — Positive Agitation, Part 3." *Floor Care Professional*, September, 2010. http://www.vdta.com/Magazines/SEP10/fc-GaskoSep10.html.

_____. "Selling Out the Hoover Salesmen, 1946–1958, Part 4." *Floor Care Professional*, October 2010. http://www.vdta.com/Magazines/OCT10/fc-GaskoOct10.html.

_____. "Hoover 1958–2010: The End of an Era — and a New Beginning, Part 5." *Floor Care Professional*, November 2010. http://www.vdta.com/Magazines/NOV10/fc-GaskoNov10.html.

Graham, Margaret Partlow. "From Broom to Button." *What's New in Home Economics*, April 1945. Hoover Historical Center.

Harris, Sullivan D., editor. "Shaler Carpet Sweeper," in *Ohio Cultivator*, Volume XIV (1858), p. 350. Google eBook, http://books.google.com/books?id=K3d9Djq2Q4AC&pg=PA308&lpg=PA308&dq=ohio+cultivator,+shaler+carpet+sweeper&source=bl&ots=ZdKhC01jh8&sig=poMCvZF6LfarCMUbmZhxYHtMJcI&hl=en&ei=JfJPTq7-J6no0QHf3vjBg&sa=X&oi=book_result&ct=result&resnum=1&sqi=2&ved=0CEsQ6AEwAA#v=onepage&q=ohio%20cultivator%2C%20shaler%20carpet%20sweeper&f=false.

"A History of the Vacuum Cleaner." *Floor Care Professional*, magazine of the Vacuum Dealers Trade Association (VDTA), December, 1997, pp. 33–40. Hoover Historical Center.

"How a Smart Newcomer Cracked a Tight Market," *Kiplinger Magazine*, March 1949. Lifshey Collection.

Jailer-Chamberlain, Mildred. "This is the Way We Cleaned Our Floors: A History of the Vacuum Cleaner." *Floor Care Professional*, December 1997, pp. 33–40. Hoover Historical Center.

Kratch, Edmund. "Stalking Down the Ages With the Broom." *Brooms Brushes & Handles*, July 1923, pp. 41–42. Hoover Historical Center.

Lesko, Jim. "Industrial Design at Carnegie Institute of Technology, 1934–1967." *Journal of Design History* Vol. 10, No. 3 (1997), The Design History Society.

Lotz. "Richard K. Kirby, The Man Behind Less Work." *Machine Design*, July 1948, pp. 106–113. Lifshey Collection.

Siekman, Philip. "Hoover's Well-Vacuumed World." *Fortune*, June 1964, pp. 143–213.

Smith, Abbe. "This Old House: A New Chapter for a Historic Madison Dwelling." *LexisNexis*, October 14, 2007. http://www.allbusiness.com/society-social-assistance-lifestyle/religion-spirituality/13353291-1.html.

Tabor, Robert. "The Whirlwind, the Oldest Vacuum Cleaner In the World, A History of the Vacuum Cleaner." *Floor Care Professional*, December 1997, p. 38.

Wohleber, Curt. "The Vacuum Cleaner." *Invention & Technology*, Spring 2006, pp. 4–5.

Magazine Advertisements

House Beautiful, May, 1909. Lifshey Collection.
Ladies Home Journal, June, 1909. Vicki Matranga files.
Saturday Evening Post, October 9, 1909. Lifshey Collection.

Letters*

Clarence Frantz of Vacuum Cleaner Manufacturers Association (VCMA) to Earl Lifshey, July 28, 1970.

Clarence Frantz of VCMA to Earl Lifshey, January 29, 1971.

Clarence Frantz of VCMA to Earl Lifshey, August 16, 1971.

Clarence Frantz of VCMA to Earl Lifshey, September 24, 1971.

Clarence Frantz of VCMA to Earl Lifshey, December 10, 1971.

Clarence Frantz of VCMA to Earl Lifshey, December 28, 1971.

Clarence Frantz of VCMA to Earl Lifshey, March 31, 1972.

"History and Background Information on Eureka Williams Company," letter to Earl Lifshey from Eureka Williams Company, August 18, 1971.

Personal Consultations

Tom Gasko, Curator, Vacuum Cleaner Museum, St. James, Missouri.

Ann Haines, Operations Coordinator, Hoover Historical Center/Walsh University.

Fred Stachnik, Webmaster, Vacuum Cleaner Collector's Club.

*All letters are from the Lifshey Collection.

Index

Numbers in *bold italics* indicate pages with photographs.

Accrington (UK) 33
Addis, William 23
A.E.G. 81
Aerus LLC 190
Africa 7; North 10
Against All Odds 182
Agan, Frank W. 56; vacuum cleaner 56
Agan, Hiram C. 37–38; sweeper 37, *58*
Air: compressed 39–40, 44; dirty/clean systems 68; impure 18; movement 34
Air Cycling Technology 203
Air Storm Company 186, 197
Air-Tec 194
Air-Way Electric Appliance Company 87, 90, 96, 103–104, 110, 126, 138, 152, 186, 191, 194–195; Centurian 2000 86; DirtMastR 105, 115; GermMaster 138; Model 35 104; Model 55A Sanitizor *115*, 191; Model 66 Sanitizor 125; Model 77 Sanitizer 125; Model 88 Sanitizor 125; RugMaster 152; Sanitary System (disposable) Bag 87–*88*, 110, 125, 128, 182, 203; Select-A-Flow 186; Super Chief 105, 139; Tandem Air 147; Zephyr 115
Airider 193
Akron (OH) 144
Albany (NY) 8
Albee, T.F. 65
Albers, Josef 116
Allegheny International 164; see also Sunbeam
Allen Company 56; vacuum *56*
Altman, B. 101
Altran 196
American Carpet Cleaning Company 36
American Designers Institute 122; *The American Woman's Home: or Principles of Domestic Science* 17–22
American Home Economics Association 43
American Household Inc. 164
American Management Association 101–102, 124
American Motors 163
American Radiator Company 75
American Safety Razor Company 128
American School of Home Economics 78, 80
American Society for Testing and Materials 186
American Society of Industrial Designers 145
American System 14
American Union of Decorative Artists 100
American Vacuum Cleaner Company 68
Ametec 138; Lamb Electric Division 138
Amsterdam (NY) 9; Broom Co. 9; Brush Co. 9
Amway 176–177; Clear Trak 177
Andrews (TX) 151
Angle, Colin 189
Anglo-American Exhibition 78
Apex Electric Manufacturing Company 75, 92, 105–106, 136; Model A3 75; Rotorex 89; tank 114; Which Way 106
Apex, Inc. 173, 176; G-force 173
Apollo 15, 16, and 17 156
Apple 164
Applecroft Home Experiment Station 77
Applica Consumer Products, Inc. 184
Archibald, Nolan D. 170
Arens, Egmont 100–102
Aristotle 51
Armory Show 77

Armstrong, R. Company 56
Arnold Electric Works 69
Arrowsmith, George 21
Art Deco 106, 114; Exhibition 95, 99–100
Art Nouveau 81
Artiebogaget Elektromekanistka AB 83
Arts and Crafts 81
Ashtabula (OH) 32
Asia 10
Aslett, Don 197; Museum of Clean 197
Associated Industries of Massachusetts 102
Association of Home Appliance Manufacturers 190
Atlanta (GA) 174
Atlantic Monthly 100
Aurora (IL) 31; sweeper 31
automobile 3, 47, 93
Avanti 197
Axminster (UK) 11, 12; carpet factory 12

Babbage, Charles 12; Difference Engine 12
"Baby Daisy" *39*, 54
Baekeland, Dr. Leo Hendrik 108
Bakelite 108
Baker, J.B. 28, 37
Balch, John 169
Balmer, James G. 128
Baltimore (MD) 20, 156
Bank of the United States 13
Baseball Hall of Fame 167
Bates, Theodore 8
Bauhaus 81–82, 95, 116
Bavaria 43
Beach, Chester A. 69, 97
Beall, J.R. 164
Beam Industries 185, 196
Beamco 152

222

Index

Beecher, Catherine **15**, 16, 42, 77, 155; *The American Woman's Home: or Principles of Domestic Science* 17–22; *A Treatise on Domestic Economy for the Use of Young Ladies at Home and at School* 16–17
Beecher, Charles 15
Beecher, Edward 15
Beecher, Harriet 17
Beecher, Henry Ward 15
Beecher, Lyman 15
Behrens, Peter 81, 95
Bel Geddes, Norman 100, 105, 116
Belgium 11
Bell, Alexander Graham 29, 38, 40
Bell Telephone 100
bellows 35, 54–55
Benz, Karl 40, 44
Berger, L.D. 70
Berkshire Hathaway 174; *see also* Scott & Fetzer
Berlin (Germany) 83
besom **6**, 7
Beverly Hills (CA) 89
Bewitched 143
Bicoastal Corporation 167
Biévre River 10
Bigelow, Erastus B. 12
Bigelow, Lucius 25
Bigelow Company 12
Bill Cosby Show 148
Bilzarian, Paul 167, 174
Bionaire 171
Birren, Faber 134
Birtman, L.F. 69
Birtman Electrical Company 69, 104, 114, 137, 152; Bee vac **69**, 105; Commander Model 116.722.1 **124**–125; *see also* Kenmore
Bissell, Anna 4, **32**, 99
Bissell, Fred 75
Bissell, John M. 151, 177
Bissell, Mark 177, 182
Bissell, Melville R., Jr. 99
Bissell, Melville R., Sr. 30, 32, 99
Bissell, Melville R., III 128
Bissell Carpet Sweeper Company 32, 99, 113, 144, 177, 182–183, 189, 194, 197; Big Green Clean Machine 177; Carpet Machine 163; Carpet Magic 163; Electro-Foam Shampooer 151, 163; ForEverGreen 194; Little Green Clean Machine 177, **179**; patent **31**; Power Steamer 182; Powerlifter 177; Promax 177; Shampoomaster 128, 146, 151; Spot Lifter 182; Steam n' Clean 182; Steam-Mate 177; stick vac 142; sweepers **30**–33, 61; Sweepmaster 128
Bissell Graphics 151
Bissell Motor Company 69, 75
Black, Robert D. 156
Black, S. Duncan 156
Black & Decker Manufacturing Company 76, 152, 156, 184–185, 197; Collector 163; Dustbuster 160–**161**-**162**–163, 166–170, 185; Household Products Group 170; Kitchen vac **159**, 160; Lunar drill 156; Mini-Buster 171; Mod 4 **157**–158; patent infringement 168–169; Spot-vac 157–160
Black Pool (UK) 49
Black Tuesday 102
Blackberry 3
Blaisdell, George G. 54
Blaisdell Machinery Company 54, 55, 75
Bloomington (IL) 195
Bojack 128
Bonner, P.J. 43
Booth, Hubert Cecil 49–51, 179, 205–206; *The Origin of the Vacuum Cleaner* 50; Puffing Billy **52**
Booth, Stone 52
Bosch 198
Bosse and Smith 24
Boston 22, 28, 36, 52, 146; fire 36; "Welcome" sweeper 30
Bourke, Robert E. 125
Bowers, Maud 106
Brachhausen, Gustav 55
Bradford (PA) 54, 75
Brandon (VT) 40
The Brave Little Toaster 174
Breuer, Marcel 116
Breuer Electric Manufacturing Company 99
Bridgeport (CT) 27, 170
British Design Council 182
British Empire 47
British National Design Museum 182
British Vacuum Cleaner Company 51
Brock, Clarence 103, 110; patent **111**
Brooks, Rodney 189
broom 3, 6–9; birch 7; corn **9**; Gold Bond 9; Indian **7**; "jumping the" 7; mechanical 23; sorghum **8**; splint 7; splinter 7; split 7; twig **6**, 7
broomcorn 8
brush(es) 22, 25–27
Brussels (Belgium) 11
Buchard & Case 66
Buckingham Palace 51
Bucks County (PA) 120
Budapest 41
Burger, Franz 43, 49, 205
Burke Electric Company 54
Burlington (NJ) 12
Burritt, Henry 121
Business Week 177

California 53, 59, 148
Callahan, Martin J. 102
Cambridge (OH) 138
Cambridge Consultants (UK) 196; Stem **196**
Cambridge Fen (UK) 196
Canton (OH) 61, 75, 144
Carey, Augustus 28
Carnegie Institute of Technology 106, 134
carpet(s) 9–13; Axminster 11, 13; beaters **10**; beating machines 29; broadloom 12; Brussels 11, 12; cleaning 16, 29; Oriental 10; Royal Axminster 12; sweeper(s) 22–23, 25, 30; Turkish 11; weaver **11**; weaves **11**; Wilton 11, 12, 13
Carpet Pro 198
Case, Francis Mills 66
Catalyst Award 170
Catholic Church 51
Caulkins, Ernest Elmo 100
Caulkins-Holden Advertising Agency 100
Celtics 7
Centennial Exhibition 29–30, 38
central vacs 202
Centre Georges Pompidou 192
Century of Progress Exposition 105
Cézanne, Paul 78
Chapman, Alonzo E. 53, 205; sweeper 53
Chapman, Dave 106, 114, 136
Charlotte (NC) 195
Charlotte Peters Show 148
Cheaper by the Dozen 76
The Chemistry of Cooking and Cleaning 43
Chicago 22, 47, 55–56, 59, 69, 81, 99, 102, 105, 146; fire 36; World's Fair 41, 43
Chicago Pacific Corporation 170, 175
Chicago, Rock Island and Pacific Railroad 170
China 10, 164
Christiana (PA) 55
Chrysler 163
Church, Melville 75
Church of England 20
Churchill, Winston 119
Cincinnati (OH) 15, 17
Cirrus 198
Civil War 3, 17, 20, 28, 38, 40, 76, 204; veterans 76
Clark, Annette 124
Clark, Don 115
Clarke Company 176, 198
Clean Air Act 206
Clements Manufacturing Company 70, 75, 105; Cadillac 105
Cleveland (OH) 60, 66, 76, 94–95; Institute of Art 106
Clinton Company 12
closet(s): earth 20; water 20
Colonial Fan and Motor Company 71
colonists 7
Colorado 117
Colt, Samuel 20; revolver 25
Columbian Exhibition 41
Columbus (OH) 27
Commonwealth Edison Company 59
Communist Party 95
Compañia Techno Industrial S.A. (Chile) 196
Computer Aided Design 164, 176; Alias 176; AutoCAD 164; Catia 164; Pro/Engineer 164
Computer Numerical Control 164
Conair 171
Congress 13, 14, 20, 145, 150; Senate Resolution No. 127 91
Connecticut 15, 27, 53

INDEX

Connersville (IN) 53; Blower Company 53
Consolidated Foods Corporation 147
Constructivism 95
Consumer Reports 113, 126, 140; 1951 126; 1954 129; 1959 141–142, 184, 195, 203
Consumers Union 113, 126, 129, 141
Cook stove (by Beecher) *19*
Cooking and Cleaning: A Manual for Housekeepers 43
Cooperstown (NY) 204
le Corbusier 95; *L'Esprit Nouveau* 95; *Vers une Architecture* 95
Corliss steam engine 29
Cornell, C.L. Manufacturing Company 55
Corporate Identification Manuals 165
Cor/Vac 152
Cosmo 171
The Cost of Cleanness, Sanitation in Daily Life 43
Cowan, Ruth Schwartz 14, 155, 203; *More Work for Mother* 14, 155, 204
Craftool 146
Crystal Palace 51
Cummings, G.L. 39
Custer, George Armstrong 38
cyclonic separation 110, 117, 171–173, 179

Daewoo Group 198
Daiger, G.P. 130, 135, 139, 144
Danville (KY) 191
Dartmouth 117
Dave Chapman, Inc. 137
Davenport, Thomas 40
Davis, Henry 28
Dayton (OH) 56
Dazey 171; Vac-Man 171
Decker, Alonzo G., Jr. 156, 158
Decker, Alonzo G., Sr. 156
Declaration of Independence 29
Degas, Edgar 78
Delanco County (NJ) 120
Delaware River 12
Denmark 83
Denny, Carl B. 128
Designing for People 131
Dessau (Germany) 81
de Stijl art movement 182
Detroit (MI) 54, 70
Deutsches Werkbund 81
Devonshire (UK) 11
Dewar, Sir James 51
Diamond, Freda 140
Dickenson, Levi 8
Diehl, Arne 146
Diehl, Philip H. 34, 174; electric fan 34
Diehl & Company 174
Diehl Manufacturing Company 34, 99, 174; Motor Products Corporation 174
Diller, Phyllis 148
Dinah Shore Show 148
Dirt Devil 189, 198; *see also* Royal
Dirt Tamer 198
Dohner, Donald 104, 106

Domestic Electric Company 76
Domestic Vacuum Cleaner Company 61, 94
Doty Manufacturing Company 56; cleaner *57*
Doubleday, Abner 204
Douglas 171, 174, 198; Power Broom 174; Quikut 176; Redivac 174
Dover (OH) 60
Dreyfuss, Henry 100, 105, 107, 116, 118–119, 123, 125–126, 131–*132*, 134, 136, 145, 160, 192; *Designing for People* 131
Druids 7
Duchamp, Marcel 78
Duff, Jack 136
Dufour, Corrine 45, 205; electric sweeper 45–47; patent *46*
Dufy, Raoul 78
Duncan, Chris 166
Dunham, Charles M. 164
Dunlap, Albert J. 164
Dunn-Locke Company 54
Duntley Pneumatic Sweeper Company 55, 75; cleaner *57*
DuPont, Pierre 10
Duralumin 104
Durham (NC) 178
Duryea, Charles 40, 44
Duryea, Frank 40, 44
dust 5, 18, 23
Dyson 103, 106, 110, 117, 179–182, 189–190, 198; Ballbarrow 171, 191; DC01 179–*180*, 181–182, 192, 196; DC02 Absolute 179–*181*, 182; DC03 *180*; DC04 182; DC06 189; DC07 189; DC08 189; DC08T *188*; DC14 192; DC15 Ball 191–*192*; DC16 192–*193*; DC41 Ball *198*; Digital motor 191; Eye for Why 192; G-force 173, 179; patents *172*, 185; R&D Center 193; Root Cyclone 189, 192; Trolleyball 171 Wheelboat 171
Dyson, James 171–173, 176, 180–182, 185, 204–206; *Against All Odds* 182; Award 192; Foundation 192; knighted 193

Earl, Harley 100, 128
Earlex Ltd. 33
Earl's Court (France) 49
Eastman, George 40
Eastman Kodak 100, 116
Easton (MD) 163, 170
Eaton-Shore, J. 43
Eco Products 125
Edinburgh (Scotland) 11
Edison, Thomas 40–41, 51, 55
Edson, Jacob 28
Edy Brush Co. 9
Egypt 35, 196
Eighteenth Amendment 86
Eiloart, Tim 196
electric motor 40–42; Universal 69, 97
Electric Renovator Manufacturing Company 53
Electric Suction Sweeper Company 62, 156; assembly line *66*; Model O *63*–65, 156, 203; reorganization 65; *see also* The Hoover Company
Electric Vacuum Cleaner Company 76; *see also* General Electric; Frantz Premier Vacuum Cleaner Company
Electrical Age 176
Electrical Fair 41
Electrical Record 99
electricity 59–60
electrification 108
Electrolux, A.B. 83, 94, 99, 112, 119, 125–126, 140, 147, 154, 158, 166–167, 186, 188, 190, 195–196, 203; Automatic Model E 136, Automatic Model F 136; Backsaver handle 188; Diamond Jubilee 167; Model 12 99, 112, 120; Model 12A 112; Model XXX *112*, 123; Model LX *127*–128; Model LX1 127, 136; Model 1205 *147*, 154; Model G 140; Model L 154; Model PN-1 140; Model PN-2 154; Model PN-4 154; Model Z 70 126; Olympia One 154, 167; Oxygen canister 188; Silverado 167; Super J 154; Trilobite robot 189; *see also* Eureka
Electrolux/Eureka 198–199
Electrolux Group 195
Electrolux Home Products Division 191; *see also* Eureka
Electrolux Small Appliances, N.A. 195
Electrolux USA 190
Elektrolux 83, 93; Model 1 83–*84*; Model V 92–94; *see also* Electrolux
Elin, John 23
Elizabethtown (NJ) 99
Elkins Park (PA) 153
El Segundo (CA) 123
Emer, Inc. 193, 199
Emer USA 193; Designo Italiano 193
Emerson Electric 64–65
Emiletti, Gianni 193
Empire Music Hall (London) 49
engineering 42–43; 101
England 10, 12, 14, 23, 25, 29, 33–34, 39, 43, 45, 47, 81, 116, 126, 166
Entwisle & Kenyon 33
Erber, Stan 130
Erie (PA) 54
L'Esprit Nouveau 95
Eubank Company 33
The Eureka Company 154
Eureka/Electrolux 189
Eureka Vacuum Cleaner Company 70, 75, 78, 90, 92, 104, 119, 121, 139, 143, 158, 166, 171, 178, 191; Boss 166; Challenger 114; Corvette Vac 178; Dream Machine 177; Mighty Mite *166*; Mobile-Aire 140; Model 9 92–93, 96–97, 109; Model 10 103; Model 570 Boss bagless 187–188; Model D-171 118; Model G 104; Model H 104–*105*; 1400-series 147; Model 4440 Powerline Plus Victory 184–*185*; Quickup 171; sales 108; Victory EnviroVac 184; *see also* Electrolux; The Eureka Company; Stecher

Index

Electric & Machine Company; Williams Oil-O-Matic
Eureka-Williams 121, 154
Euripides 22
Euro Pro 199
Europe 22, 41, 51, 77, 81, 92, 96, 99
Evans, Oliver 13

Fairaizi, Max 117, 124
Fairfax 125, 152, 199; Fax-o-Matic *126*
Faraday, Michael 40
Farley, Laurence J. 170
Farnsworth, Philo 98
Fayetteville (NY) 38
Feldman, C. Russell 121
Fels Naptha 29
Fenton (MO) 183
Fessenden, Reginald 40
Fetzer, Carl S. 94
Field Artillery March 90
Filter Queen 199; *see also* Health Mor
Finberg, Raymond 153
fireplace(s) 7, 18
Fisker, P.A. 71
Fisker and Nielson 71
Fistek, Ted 117, 124
Fitch, John 13
Florian, Gordon 125
Florida 110, 184
Flower City Purchasing Company 55
Flushing Meadows (NY) 116
Folwell, F.G. 62; Building 61, 63
Folwell, W.H. 62; Building 61, 63
Forbes 132
Ford, Henry 76, 100; Fordism 76, 81
Ford Motor Company 116, 163; Lincoln Continental 143; Model T 76, 92, 100, 143; Sable 165; Taurus 165
Fortune 105–107, 154, 158
Fort Worth (TX) 183
Fox Theater 148
France 21, 40, 53, 83–84
Franklin, Benjamin 23
Franklin Institute 14
Franklin Simon 99
Frantz, Clarence G. 70–*71*, 75, 88, 92, 106, 153
Frantz, Edward L. 70–*71*, 153
Frantz, Walter A. 70, 75, 153
Frantz Premier Vacuum Cleaner Company 70, 75, 88; Handy-Vac 88; Model C 70; Model D 70; Model K-14-B 76; *see also* Premier Vacuum Cleaner Company
Frease, Harry 75
Frederick, Christine 76, 85; *Household Engineering, Scientific Management in the Home* 78–80; *The New Housekeeping: Efficiency Studies in Home Management* 76
Frick, Henry Clay 52
Fugitive Slave Act 17
Fuller, Alfred C. 60
Fuller Brush Company 89
Fullerton (CA) 174
Fuqua Industries 155

Gasko, Tom 183, 186
Gates, G.W. 31
Gehrig, Lou 98
Geier, Philip A. 72, 92
Geiner, Helen 189
General Compressed Air Company 44
General Electric Company 41, 59, 70, 76, 81, 104, 116, 119–120; 134; 137, 156; Appearance Design Division 113; Housewares Division 153, 170; Model 807 120; Model 811 120; refrigerator 107; washing machine 107
General Electric Vacuum Cleaner Company 88, 130, 142; Model 111 105, 114; Model 815 140; Models C-1, C-2, C-3, *C-4*, C-6, C-7, C-8, C-9, C-10, and C-11 canisters 140; Model FP-1 131; Model MV-1 146; Model R-1 Roll-Easy 140; Model SV-1 146; Model U-1 upright 140; Model U-4 upright 140; *see also* Frantz Premier Vacuum Cleaner Company; Premier Vacuum Cleaner Company
General Motors 93, 100, 103, 106, 163; Cadillac 125; Chevrolet 100
General Signal Corporation 97, 173
Genie Products 171
George Washington University 148
Gerber, Dale C. 130, 135
Germany 40, 81, 83, 89, 95, 101, 116, 167; Olympics 89
Gettysburg (PA) 76
Ghostbusters 160
G.I. Bill 122
Gilbert, A.C. 104
Gilbreth, Frank, Jr. 76; *Cheaper by the Dozen* 76
Gilbreth, Frank, Sr. 76
Gilbreth, Lillian 76
Gillette, King Camp 128
Glasgow 11
Glendale (WI) 164
Glenwillow (OH) 194
global positioning device 3
Gobelin 10
Godfrey, Arthur 139
Goeser, Edwin W. 70
GoldStar Company, Ltd. 191
"Good Design" 131, 165; 177
Good Housekeeping 68; Seal of Approval 103
Goodrich, B.F. Company 108
Goodwill Industries 148
Gore & Edgecomb 30
Goshen (IN) 30
Graham-Paige Motor Company 100, 106
Gramohone 55
Grand Rapids (MI) 30; Brush Company 31–32; Carpet Sweeper Company 31
Grant, Ulysses S. 29
Great Britain 40, 83, 144
Great Depression 102–103, 114, 117, 120, 125, 177
"Great Exhibition" 20, 25
Great Sioux War 38
Greece 119

Greeley, Horace 21
Green, Leslie H. 103, 110, 131
Greenlawn (NY) 77
Gris, Charles Edouard Jenneret 95; *see also* le Corbusier
Gropius, Walter 81, 116
Gross Domestic Product 40
Group SEB 194
Guarantee Sales Company 56
Gue, A.F. 43
Guericke, Otto von 51
Guild, Lurelle 105, 112
Gypsies 7
Gypsy Rose Lee Show 148

Haan Corporation 199
Hadley (MA) 8
Hamilton, Louis H. 69
Hamilton Beach Carpet Washer Company 89, 145
Hamilton Beach Manufacturing Company 69, 153; Hand-held 104; Model AV-1 96; Model 35 Hatbox 145
Hamilton Beach/Proctor-Silex, Inc. 145
Han, Sam 145
Hand, Augustus 45
Hand Stitch Broom Sewing Machine Co. 8
Harding, Warren 86
Hardware Show 161
Harington, Sir John 20
Harley Earl Associates 128
Harned, C. Ray 62, 64
Harris, Daniel 28
Hartford (CT) 15, 53, 75; Female Seminary 15
Harvard University 89
Harvey, James J. 43; machine 43
Hatch, Tracy B. 70
Hatlinger, Joseph J. 37; sweeper 37
Hattersley, John Frank 132
Health-Mor Industries 150; Filter Queen 150, 189; *see also* P.A. Geier Company
Health-Mor Sanitation Systems, Inc. 103, 116–117, 124–125, 131, 142, 171; *see also* P.A. Geier Company
Heming, Edmund 23
Henney Motor Company 121; Henney Kilowatt 121; Wooden Lungs 121
Henry Dreyfuss Associates 132, 144
HEPA filters 128, 150, 180, 184, 186, 190–191
Herrick, Hiram H. 26–27; sweeper *26*, 28
Hertzler, Frederick W. 128
Hess, Daniel 35, 205; patent *36*; sweeper 35
Hiawatha (KS) 114
Hill, T. Russ 110, 131
Hitler, Adolf 89, 116
Hocus Pocus (film) 178
Hoffmann, Heinrich 89
Hohulin, Samuel E. 166
Home Furnishing Network 183; *Weekly Newspaper* 183
Homer 22

Honda 164
Hong Kong 170, 185
Hoover, Dan, Jr. 64
Hoover, Daniel, Sr. 64
Hoover, Frank 64, 132
Hoover, Herbert (U.S. president) 95
Hoover, Herbert W., Jr. 132, *133*, 142, 145
Hoover, Herbert W., Sr. 64, 73, *90*, 92, 119, 132, 135
Hoover, William H. 63, *64*, 156; Company 64, 80
Hoover, Mrs. William H. 63–64
The Hoover Company 87, 92–93, 96, 104–105, 107, 114, 119, 121, 125, 128–129, 132–136, 139, 142, 144–147, 149, 153–155, 158, 170–171, 173–175, 177, 182–183, 189–191, 193–194, 196, 199; 205; Conquest 168; Dimension 1000 196; Dimension series 171; Dirt Finder 104; DirtFINDER 184, 186; Europe 175; factory *134*; Floor MATE 188; Friction Drive Baby 75; Handivac 2; Help-Mate 167; Holiday 125; Hooverette 73; Hygienisac 103; Junior 75, 173; 12" Junior 76; Lark; Model O improved 67, 73; Model 010 125; Model 011 125; Model 1 68, 73; Model 21 131; Model 25 108; Model 26 114; Model 28 *123*, 131; Model 42 126; Model 50 123; Model 55 126; Model 60 118; Model 61 123; Model 63 123, 128; Model 65 *137*, 173; Model 82 Constellation 129–*130*, 153, 193; Model 84 130; Model 100 Dustette 103; Model 102 *82*; Model 103 93; Model 105 79; Model 115 123; Model 125 Dustette 107; Model 150 107–*108*, 114; Model 200 Duster 103; Model 305 114; Model 404 Swingette 2, 148; Model 425 109; Model 475 107, 109; Model 509 149; Model 541 90, 96; Model 543 97; Models 0510 and 0512 143; Model 700 *96*–98; Model 725 103; Model 750 104; Model 800 104; Model 825 107; Model 900 104; Model 1010 145; Model 1060 145; Model 1100 Dial-a-Matic 2, *143*, 166, 177, 182; Model 1151 Dial-a-Matic 148; Model 1170 Dial-a-Matic 148; Model 1176 Dial-a-Matic 148, 184; Model 1340 145; Model 2000 Slim-Line 143; Model 2100 Portable 2, 142–143; Model 2800 Pixie 145, 157; Model 2900 145; Model 2940 Lark 2, 142; Model S3001 Celebrity *152*–153; Model U3301 Concept I 155, *167*; Concept II 177; Model 3500 Floor Washer 2, 139; Model 3600 Floor-a Matic 2, 147; Model U3745–910 PowerMAX 177–*178*; Models U4471 and U4455 174; Model U4473 Elite 600 173–*174*; Elite II 174; Models U5280–900, U5288–900, U5494–900

and U5750–900 Bagless 186; Wind Tunnel bagless 190; Model 5465–900 Wind Tunnel Deluxe *183*–184; Model U6329–930 Power Drive Supreme 177; Norca Model 1; Park 135; Platinum series 194; Portable; sales 86, 108, 114; Self-propelled Wind Tunnel 184; Senior Model 2 *67*; service truck *91*; Side Outlet Model 3 67; song 90; Steam Vac 177, 194, SteamVac Widepath 188; trademark 126; Twin Chamber 186; Worldwide 142, 144–145, 153; *see also* Electric Suction Sweeper Company
Hoover Historical Center 37, 53, 207
Hoover Ltd. (UK) 146, 185–186; Triple Vortex 185–186
Hoover Suction Sweeper Company *72*, 75, 78, 80–*82*, 83, 92; slogan 83; *see also* Hoover Company
Hose, Robert 144
Hotel Secor 75
Hotpoint Electric Heating Company 59
House & Garden 134, 135
House Beautiful 68
House Furnishings Review 39, 45, 77
Household Engineering, Scientific Management in the Home 78–80
"housewifery" 6
Howard, J.E. 43
Howard Hughes Corporation 123
Howe, Elias, Jr. 21
Howe, Jacob 22
Hume, James 25, 27
Hunt, Walter 21
husbandry 6

IBM 143, 164
ID Two 146
IDEO 146
Illinois 31, 69
In Search of Excellence (book) 165
Independence Hall 12
India 10
Indiana 30, 53
industrial artists 82
industrial design 77, 105–106, 122, 131; definition 101, 180–181; typeform 163, 181–182
Industrial Design, a Practical Guide (book) 122
Industrial Design Excellence Awards 177
Industrial Designers Education Association 145
Industrial Designers Institute 143, 145
Industrial Designers Society of America 145, 170, 177
industrial revolution 13, 29
industrialization 14, 20; impact 14
Information Age 176
Innovation Electric Company 45, 52
International Design Fair 173
International Exhibition of Art in Industry 101
International Exhibition of Modern Art 77

International Exposition of Modern Decorative and Industrial Arts 95; *see also* Art Deco
International Team Pump Company 52
Interstate Engineering Corporation 123; Compact Model C1 123; Models C2 through C9 186; Model C-6 *187*; Revelation 123
Invincible Vacuum Cleaner Manufacturing Company 60, 75
Iona Appliance (Canada) 176–177; Fantom 177; Fantom Technologies, Inc. 186
Iowa 35
Ipswitch (MA) 28
Ireland 11
iRobot 189; Ariel Underwater 189; Packbot 189; Roomba 189
IS Robotics 189
Italy 165

Jacquard, Joseph Marie 12; loom 12
Japan 118, 163, 173, 176, 179
Jarden Corporation 164
The Jazz Singer 98
Jedlik, Ányos 40
Jefferson, Thomas 13
Jénatzy, Camille 42
Jobs, Steve 193
John Deere 185; Homelite 185
Johnny Carson Show 148
Johnson, Samuel Curtis 97
Joint Chiefs of Staff 119
Jolson, Al 98; *The Jazz Singer* 98
Jones, Owen 11
Juarez (MEX) 191

Kann, Stan 147–148, 178; *Bill Cosby Show* 148; *Charlotte Peters Show* 148; *Dinah Shore Show* 148; *Gypsy Rose Lee Show* 148; *Johnny Carson Show* 148; *Merv Griffin Show* 148; *Mike Douglas Show* 148
Kansas City (MO) 161
Kansas University 114
Keller Manufacturing Company 55, 60, 83; Keller-Santo vac 60, 75, 83
Kelley, David 146
Kelley, Martin V. 89
Kelvinator 137
Kenmore 104, 125, 137, 143, 152, 189, 191, 199; Commander *124*; Intuition 195; Progressive 195; *see also* Birtman Electric Company; Whirlpool Corporation
Kennedy, John 149
Kennedy, Robert 149
Kenney, David T. 45, 49, 52, 152, 205–206; licenses 66, 89, 92; patents *48*, 55, *74*–75; Vacuum Cleaner Company 45
Kent (OH) 138
Kent, William 23
Kentucky 15
Kenwood 199
Kenyon, Richard Walton 33
Khrushchev, Nikita 137
King, Martin Luther 149

Index

King Edward VII 51
Kiplinger 124
Kirby, James B. 60, 70, 88, 94–95, 109–*110*, 151, 205–206; cleaners 60–61, 94–95; Sani Em-tor 109; *see also* Scott & Fetzer
Kirby Company 145–146, 150, 176, 190, 199; cartoon character 174; Classic 151; Classic Omega 151; Classic III 151; Dual Sanitronic 50 *151*; Dual Sanitronic 80 151; Generation G 174; G series 190; Heritage 174; Heritage II 174; Heritage II Legend 174; Legend II 174; Model 505 151; Model 510 151; Model 516 151; Sentria 190; "Tech Drive" 174; *see also* Scott & Fetzer
Kirby West 151
Kitchen (by Beecher) *18*
KitchenAid 102; Model K 102
Kmart 169
Knapp, Andrew S. 144
Knapp, Robert S. 144
Knapp-Monarch Company 144
Knickerbocker Trust Company 65
Knoll, Florence 138
Knoll Associates 138
Koblenz 199
Koch, Heinrich 42
Kogan, Belle 109
Koons, Jeff 165; installation art 165–166
Kotten, Herman 55; vacuum *57–58*
Krammes, Don 139

labor, division of 6
Ladies' Home Journal 76, 89
Lamb, Edward 138
Lamb Electric Motor Company 76, 137–138, 147
Lambert, Raoul 129
Lamme, B.G. 47
Landers, George 88
Landers, Frary and Clark 88
Lane Theological Seminary 15
Lanning-Stone Sales Company 55
Lappin, Robert 141
Lay, L.B. & Company 32; sweeper "Minneapolis" 32
Leathers, Ward 121
Leaycraft, E.S. 38; patent 110
Leger, Fernand 78
Lehman and Son, Inc. 55, cleaner *58*
Leipzig University 89
Lenin, Vladimir 95
Leno, Jay 178
Levittown (NY) 120
Lewyt, Alex 123–124, 131, 141
Lewyt Corporation 123–124, 140–141; Market Place 124; Model 44 124; Model 77 131, 141; Model 88 131; Model 97 141; Model 105 141; Model 107 140; Model 111 141
LG Corporation 191, 199–200
light bulb 3, 40, 51
Lindbergh, Charles 98
linoleum 22
Lipe, C.E. 8

Loewy, Raymond 100, 105, 116, 119, 122, 125, 131–132, 143, 192
London 20, 23, 34, 49–50, 51, 78, 83, 145–146
London Science Museum 192
Look to Germany—The Heart of Europe 89
Louis XIII 10
Louis XIV 10
Louvre 143
Lowell (MA) 12
Lucia, John 167
Lucier, Francis P. 158, 170
Ludlow (VT) 56
Lux AB 83
Lynd, Helen 93; *Middletown, a Study in American Culture* 93
Lynd, Robert 93; *Middletown, a Study in American Culture* 93
Lysol 188

Machine Age 95
Macy's 99, 101
Madison (CT) 27
Magdeburg (Ger.) 51
Malmesbury (UK) 179
Manhattan Project 119, 128
Mansager, Felix 142, 145, 153
Mansager, Oscar 142, 145
Mansfield (OH) 99
Mansfield Products Company 153
manual suction cleaners 54–58
Marconi, Guglielmo 40
market (vacuum cleaner) 1937 costs 114; 1927 cleaners 98–99; performance 84; sales 86, 118, 120, 131, 138, 150, 175, 185; scale 203; sealed suction 84; 2011 brand options 198–203
Marshall Field 99
Martin, Steve 76
Martinec, Gene 117, 124
Marzano, Stefano 196
Mason, Timothy 23
Massachusetts 13, 28, 36, 43, 60
Massachusetts Institute of Technology 42
Matrix 146
Matushita Electric 191
Maxi Vac, Inc. 177
Maytag Corporation 175, 191
Maytag Floorcare U.K. 193; Satellite 193
McClatchie, Stanley 88–89; *Look to Germany—The Heart of Europe* 89
McCombs 9
McCormick, Cyrus 20, 25; reaper 20, 25
McCreery Manufacturing Company 52
McCrum-Howell Company 52; *see also* Arnold Electric Company
McEnery, Daniel B. 37; patent *38*
McGaffey, Ives W. 35, 205; patent *37*; "Whirlwind" sweeper 35–36, *58*
McLaughlin, Keith 195–196
mechanic arts 14
Memphis furniture 165

"men's work" 5, 15
Merv Griffin Show 148
Metropolitan Museum of Art 77, 99, 192
Metropolitan Vacuum 171, 175
Mexico 194
Michigan 31, 54
Michigan Carpet Sweeper Company 31
Middle West Utilities Company 103
Middletown, a Study in American Culture 93
Miele, Carl 167
Miele Appliances 167, 170–171, 200
Mies van der Rohe, Ludwig 116
Mike Douglas Show 148
Miller, Martin 146
Milwaukee (WI) 115, 178
Minneapolis (MN) 179
Moggridge, Bill 146
Mohawk Valley 8
Moholy-Nagy, László 116
Mondrian, Piet 182
Mont Senis 39
Montana 38
Montgomery, Elizabeth 143; *Bewitched* 143
Montgomery Ward 47, 104, 125, 137; Bureau of Design 104, 106; Model MW 106–*107*; Supreme 125
Monty Python 166
Moore, Edward B. 77
Moore, Mr. 11
Moors 10
mop(s) 22
More Work for Mother 14, 155
Morgan, John Richard 114
Morphy Richards 200
Morse, Samuel F.B. 20, 25, 40
Morton, H. & J. 22
Moscow 137; Olympic Games 154
Moulinex/Hamilton Beach 170–171
Mrs. Doubtfire (film) 178
MultiClean Company 130
Muncie (IN) 93
Munich 146
Museum of Clean 197, 207
Museum of Modern Art 101, 116, 131, 192
Muthesius, Hermann 81

Najimy, Kathy 178; *Hocus Pocus* (film) 178
National Cash Register Company 116
National Electric Light Association 82
National Enameling and Stamping Company 144
National Endowment for the Arts 145
National Super Service Company 70
National Union Electric Company 122, 154
New Berlin (OH) 61, 63, 80; *see also* North Canton
New Hampshire 36
The New Housekeeping: Efficiency Studies in Home Management (book) 76
New Jersey 28, 34, 49

New Orleans (LA) 144
New York 17, 28, 38, 45, 47, 52, 54, 75, 94; *Tribune* 21
New York World's Fair 116
New York Yankees 98
Newcombe, John W. 103, 206; Bagless cleaner 103, 110, 171
Newcombe and Green 103
Newcomen, Thomas 51
Newcomen Society Transactions 50
Newsweek 165
Newton, John 38
Niagara Falls 42
Nilfisk 71, 200
1937 costs 114
1927 cleaners 98–99
Nineteenth Amendment 86
Niskayuna (NY) 8
Nixon, Richard 137, 150
Noe, Dr. William 53
Noney, Augustine W. 27
Norelco 171
Normal (IL) 191
North Canton (OH) 61, 80, 90, 97, 104, 119, 132, 135, 142, 145, 153, 175, 193–194; *see also* New Berlin
Northrop, J.H. 13
Noyes, Eliot 143
Numatic International Ltd. 166, 200; Henry 166
Nutone 152
Nuttal, Mike 146

Oakland (CA) 53
Ohio 15, 27, 32, 52, 55, 60–61, 71; *Cultivator* 27–28
Ohio River 15
Old English 188
Old Greenwich (CT) 112
Olympic Group 196
Ontario (CA) Power Company 59
Oreck, David 144–145, 152
Oreck Corporation 144, 145, 151–152, 176, 196–197, 200; Regina RG3100 187
Organ Power Company 53
The Origin of the Vacuum Cleaner 50
Orr, William 75
Osius, Frederick 69
Oster, John, Sr. 164
Oster Manufacturing Company 64
Otto, Nikolas 44
Owen, Ray 102
"oylcloths" 22

P.A. Geier Company 72, 83, 91, 92, 96, 102, 119, 125, 130; Filter Queen 200 116–117, 131, 124, 171; Filter Queen 350 124–125; Golden Monarch Model 55 131; Health-Mor cleaner 102, 124; Royal Model 1 72; Royal Prince 114; self-propelled 166; Super upright 102; *see also* Health-Mor Industries; Royal
P-38 warplane 125
Paddington (UK) 11
Palm Vacuum Cleaner Company 54, 66
Palo Alto (CA) 146

Panama Canal 47
Panama-Pacific International Exhibition 78
Panasonic 170–171, 176, 200; MC-CG902 195
Paris 10, 21, 49, 92, 95, 99, 143
Parker, Kenneth R. 166
Pascal, Blaise 50
Pass Port Ltd. 187
patents 13; design 20
Patten, Ray 113
Pearl Harbor 118
Peerless 106
Penn Champ Company 151
Pennsylvania 54–55
Pennsylvania Railroad 116
Perfect 200
Perivale (UK) 104
Persia 10
Peters, Tom 165; *In Search of Excellence* 165
Pfaff, G.M., AG 167
Phalen, Gustav Robert 83
Philadelphia 12, 14, 29, 55, 70, 83
Philippines 106
Philips, N.V. 165, 191, 196; Roller Radio 165
Philips Electronics, North America 187
phonograph 3, 40
Picasso, Pablo 78
Pickens (SC) 174
Pittsburgh (PA) 9, 52–53, 164
Plainfield (NJ) 49
plastics 108–109
Plowman, E. Grosvenor 102
Plumb, A.D. 31; "Mystic" sweeper 31
Plumb & Lewis Manufacturing Company 31
pneumatic 39
Pneu-Vac Electric Vacuum Company 60, 76, 96
Poesse, Walter H. 70, 76
Polo, Marco 10
Popular Science 3
Potter, Benjamin F. 31
Potter, Harry 7
Powr-Flite 200
Pratt Institute 106
Preco Inc. 140
Premier Electric Company 153
Premier Vacuum Cleaner Company 70, 75, 88, 104; Model B 70; Deluxe 96; Duplex 96; Grand 105, 114; Spic-Span cleaner 104; *see also* Frantz Premier Vacuum Cleaner Company; General Electric Vacuum Cleaner Company
Principles of Scientific Management 76
Pro Team 200
Proctor-Silex 145–146
"Puffing Billy" 51, *52*, 179
Pullman-Holt 200

Quadrex Company 121
Queen Alexandra 51
Queen Elizabeth I 20
Queen Elizabeth II 193

Queen Victoria 32
Quidditch 7
Quist and Blanch 55

Racine (WI) 69, 89, 97, 164
radio 3
Radio Corporation of America 103–104, 120, 144
Rahway (NJ) 55
Rainbow 200; *see also* Rexaire
Randolph (NY) 55
Ranyard, W. 24
Rashid, Karim 195
Rawson, Merle R. 153, 167–168
Ray, George 75
RCA Whirlpool 137
Reckitt Benckiser 188
Regiment Armory (69th Infantry, NY) 77
Regina Company 55, 93, 97, 109, 119–120, 127, 142, 173, 186; "air pulse" nozzle 173; cleaner 76; Electrikbroom 120–*121*, 171; hand vac 145; Housekeeper 173; Steemer 173
Remington typewriter 30, 38
Renovator Incorporated 89
Replogle, Daniel Benson 87, 206
Revolutionary War 12, 13
Rexair, Inc. 103, 110, 117, 125–126, 130–131, 139, 146, 171, 203; Model B 110; Model C 110, *112*; Model D Rainbow 131; Model D2; Model E series 184; Model E2 191; Rainbow Opportunity 139; Rainbow 110, 176; Series A 110; Switched Reluctance Motor 191
Rhode Island 28
Riccar America Company 174, 176, 200; *see also* Tacony
Richards, Ellen Swallow *42*–43, 77; *The Chemistry of Cooking* 43; *Cleaning and the Cost of Cleanness, Sanitation in Daily Life* 43; *Cooking and Cleaning: A Manual for Housekeepers* 43
Richardson, Earl H. 59
Richfield (OH) 60
Richmond Radiator Company 89
Rietveld, Gerrit 182
Roaring Twenties 86
Robbins & Meyers 53, 76
Robinson, Brooks 167
robotic cleaners 189, 202
Rochester (NY) 55
Rochester Vacuum Cleaner Company 55
Rock, Hugh 23
Rock Island (IL) 69
Rodchenko, Aleksandr 95
Roosevelt, Franklin 105, 118–119; New Deal 108
Roosevelt, Theodore 65, 78
Rowenta A.G. 170, 194, 200; Silence Force *194*
Royal Appliance Manufacturing Company 130, 152, 169, 182, 187, 201; Brum *195*; Dirt Devil *169*, 189; Dirt Devil Broom 169; Dirt Devil upright 173; Kone *195*; Kruz

Index

195; Kurv *195*; Kwik *195*; Model 801 130; Princess 104, 169; Purifier 104; Super 104; tank 114; *see also* P.A. Geier Company
Royal College of Art 171
Royal Philips 201
Rubbermaid 201
The Rumford Kitchen 43
Rural Electrification Administration 108
Ruth, Babe 98
Rutter, Mike 173
Ryobi Motor Products Corporation 174, 182, 185

Sahlin, Gustav 93
St. James (MO) 183
St. Louis (MO) 39, 148, 175; *Dispatch* 45
Salon des Arts Ménages 92
Salvation Army 148
Samsung Electronics 195, 201; GTO Hauzen 195; robotics 195
Sanders, J.V. 131, 139
San Francisco (CA) 53, 78, 146, Museum of Modern Art 192
Sanitaire 196
Sanitation Systems, Inc. 102
Santo Vacuum Cleaner Appliance Business 83; Lux 1 cleaner 83
Sanyo 170–171, 189, 201; DirtCompactor System 188; DirtHunter bagless 188; Model SC-150 Transformax 183
Sanyo-Fisher 176
Sason, Sixten 126
Saturday Evening Post 65, 67
Savannah (GA) 45
Schey, Ralph E. 151
Schmitz, George C. 69
Schott, Walter E. Organization 130
Scientific American 23
Scott, George H. 94
Scott & Fetzer Company 57, 94–95, 102, 109, 139, 145, 151, 174–176, 190; Kirby Model C 109, 125; *see also* Kirby Company
Scovill Manufacturing Company 89, 145, 153; *see also* Hamilton Beach
Sears/Kenmore 152; *see also* Kenmore
Sears Roebuck & Company 47, 55, 57, 104, 125, 137–138, 171, 177; 191; Craftsman 174; design department 114; Model 7006 140
Sebo 201
Sebring (OH) 55
Seck, Werner G. 130
Segesman, Louis 136
Selfridge's 83
Semi-Tech Global 167
Separator Company 83
servants 71, 86–87
Sewing Dealers Trade Association 167, 207
sewing machine(s) 3, 14, 15, 21–22
Shaker 8
Shaler, Rueben 27; bullet 28; carpet sweeper *27*–28; Wheel (Parlor) Skate 28

Sharp Electronics 170–171, 176, 201
Sheelan, Don 173, 186–187
Shelton (CT) 170
Shetland Company 141; Floorsmith 141
Shetland-Lewyt 141, 146, 153; Electra Sponge 146; Fashionables 146; stick vac *141*
Shop Vac Corporation 146, 157, 171, 201
Signal Manufacturing Company 141
Silicon Valley 196
Silver King International 117, 201; Silver King 117, 125
Simplicity 201; *see also* Tacony
Singer, Isaac 21
The Singer Company 167, 174
Singer Manufacturing Company 30, 99, 113, 122–123, 145, 166–167, 182, 202; Diehl Division 174; Magic Carpet S-2, S-3 *122*–123; Model R-1 *113*, 122; original Singer *21*
Singer Sewing Machine Company Inc. 167, 174
Skinner, Benjamin J. 53, 205; cleaner *57–58*; Manufacturing Company 53
Skinner, Halycon 12
Smelie, D.G. 97
Smith, Alexander 12; Carpet Company 12
Smith, E.R. 114
Smith, G.L. 92
Smith, John W. 70
Smith-Corona-Marchant 146, 153
Smith-Hughes Act 82
Smith, Scherr and McDermott 144
Smithsonian: American History Museum 4; Cooper-Hewitt National Design Museum 146; Institution 112, 146
Society of Industrial Designers 122, 131, 144
Sommellier, Germain 39
Sophocles 22
sorghum 6, 8
South Korea 164
Southward, David 196
Soviet Union 81, 95
Spain 10
Spalding, A.J. 204
Spangler, James Murray 4, *61*, 72–73, 81, 156, 180, 194, 205–206; models 61–62; patent *62*
Spanish-American War 47
Sparklin, C.H. 114
Spencer, Ira Hobart 53; Turbine Cleaner Company 53, 75
Spielman, Milton H. 70, 76
Sprague, Frank J. 40–41; Electric Railway and Motor Company 40
Sprague, William Peter 12
Sprengel, Hermann P. 51
Sputnik 138
Stachnik, Fred 178–179; *American Journal* 178; *Jon Stewart's Show* 178; *Late Show with David Letterman* 178; *The Maury Povich Show* 178; *Tonight Show* 178

Stackpole, Greenleaf 22
Stains (UK) 22
Stalin, Josef 119
Standard Vacuum Manufacturing Company 57
Stanley Black & Decker 170
Stanley Works 170
Star-King Building 53
Stark's Vacuum Museum 207
steam engines 29
Stecher Electric & Machine Company 70; Model 1 70; *see also* Eureka Vacuum Cleaner Company
Sterling Vacuum Cleaner Company 55
Stewart, Martha 76
Stowe, Calvin 17
Stowe, Harriet Beecher 17; *Uncle Tom's Cabin* 17
Strausser, Charlie 136
Studebaker 93, 125; Avanti 143
Sturdevant, B.F. Company 72, 75, 92
suction 34
Sullivan, Louis 81
Summers, M.J. *73*
Summerton, Arthur 80; *Treatise of the Vacuum Cleaner* 80
Sunbeam 144, 164; Appliance Division 164
Sunbeam/Oster Corporation 164
Sundberg-Ferar 137
Svensa Elektron AB 83
Swann, E. Russell 134
Sweden 83–84, 92–93, 126
Swedish Design Engineering Prize 188
sweeper(s): carpet 22–23; Boston 28; floor 25; street 23, *24*, 25; Union 28; Weed 28; Welcome 28
Sweeper-Vac Company 60
Syracuse, MO 44
Syracuse, NY 8

Tabor, Robert 36, 167
Tacony, Ken 175
Tacony, N.J. "Nick" 175
Tacony Corporation 174–175, 183, 191; Riccar Radiance *201*; Simplicity 175, 183, 191; Simplicity Gusto *202*; Tandem Air System 191; Vacuum Cleaner Museum 183, 207
Taft, William Howard 65
Taisho Exhibition 78
Taiwan 164, 183
Tate, J.C. 43
Taylor, Frederick Winslow 76; *Principles of Scientific Management* 76
Taylor Society 76
Taylorism 76, 81, 95
Teague, Walter Dorwin 100, 116, 119, 131, 192
Team (Jem) Development 124
Techtronic Industries 169–170, 182, 185, 193–194; Floor Care North America 194
telegraph 20, 25, 40
telephone 29, 40
television 3, 98

Templeton, James 11
Tesla, Nikola 41–42; Electric Company 41
textile mill workers *12*
Thermos: bottle 51; GmbH 51
Thompson, Benjamin 43
Thurman, John S. 44, 49, 203, 205; carpet renovator 44–45, 50; patent *44*
Time 132, 164, 169
Times Building 52
Timonnier, Barthélémy 21
Toepler, Joseph 51
Tokyo 78
Toledo (OH) 52, 69, 75, 87
Tonight Show 178
Torricelli, Evangelista 50
Torrington Electric Company 82–83
Tournai (Belgium) 11
Towson (MD) 158
Tracy, Clarence 87
Tracy, Pratt 87
TRC Acquisition 187
Treatise of the Vacuum Cleaner 80
A Treatise on Domestic Economy For the Use of Young Ladies at Home and at School (book) 16–17
Tri-Star Enterprises, LLC 186, 202
Troxler, Susan 63; *see also* Hoover, Mrs. William H.
Troy (MI) 103
Tueter, Julius 76
turbines 34
Twentieth Century Taste 101
Twentieth Century Vacuum Cleaner Company 55
Two-Rivers (WI) 144

Umbach, Steve 177
Uncle Tom's Cabin (book) 17
United Electric Company 75, 83
United Kingdom 83–84
United Manufacturing and Distributing Company 55
United States 12, 22, 47; population 17, 40
United States Bureau of Standards 84
United States Circuit Court of Appeals 45, 152
United States Defense Department 118
United States Department of Commerce 95
United States Environmental Protection Agency 150, 194, 206
United States Federal Trade Commission 91
United States Food and Drug Administration 142
United States Information Agency 145
United States Patent Office 13, 77, 182
US Steel 116

United Vacuum Appliance Company 53, 75
Universal 103; Model E-115 103

VACUFLO 152
vacuum 50–51; flask 51
Vacuum & Sewing Hall of Fame 167
Vacuum Cleaner Collectors Club 167, 194, 203, 207
Vacuum Cleaner Company of New Jersey 52, 66
Vacuum Cleaner Company of New York 52
Vacuum Cleaners Manufacturing Association 75, 84, 91, 190–191
Vacuum Dealers Trade Association 165, 167, 207
Vacuum Engineering Company 54
Vacuum Hydro Company 69
Van Doren, Harold 101, 122; *Industrial Design, a Practical Guide* 122
Van Gogh, Vincent 78
Vanisher 202
Vassos, John 116, 120
Vax 185, 202
Venture Management Support 187
Vermont 40, 56
Vers une Architecture (book) 95
Victor Talking Machine Company 55
Vienna (Austria) 49, 83
Vietnam War 149
Vital Manufacturing Company 57

Waddington, Hiram 33
Wadsworth, H.N. 23
Wagner E.R. Manufacturing Company 113–114
Wales 7
Walker, George W. 106, 114, 125, 143
Wall Street Journal 189; *Catalog Critic* 189
Wal-Mart 169
Walt Disney Pictures 174; *The Brave Little Toaster* 174
Walton, Frederick 22
Wanamaker 99
War of 1812 13
War of the Currents 41
Wardell, Fred 70, 75, 92, 121
Warren (OH) 71
Washington (D.C) 20, 52
Washington, George 13
Wasserman, Julius 9
Water-Magic 152
Waterbury (CT) 89
Webb, Clifton 76
Webster, George 55
Weimar (Germany) 81
Welch, Jack 170
Wenner-Gren, Axel Leonard 83, 94, 190
West Farms (NY) 12
West Union (IA) 35
Western-Electric Company 72, 96
Western Female Institute 15

Western Territories 17
Westinghouse, George 39, 41–42
Westinghouse Electric and Manufacturing Company 47, 59, 65, 99, 104, 106, 116, 129, 144, 153; Converto-Vac II *153*; Rocket tank *129*
Westman, George L. 43
Westminster Abbey 51
Wheatstone, Charles 40
Whirlpool Corporation 137, 144–145, 152, 170–171, 191, 193; Imperial Mark XII 140; Lady Kenmore Whispertone 140; Miracle Kitchen 137; robot cleaners 188, 189
Whirlpool International 196
White City 41
White Consolidated Industries 137, 153
The White House 52
White Sewing Machine Company 112, 136
White Westinghouse 153
Whitney, Eli 13
Whitty, Thomas 12
Whitworth, Joseph 24–25
Whytock, Richard 11
Wild West 38
Willey, Freeman O. 37; patent *38* sweeper 37, *57*
Williams, Frank W. Company 55
Williams, Robin 178; *Mrs. Doubtfire* 178
Williams Oil-O-Matic 121
Williamsport (PA) 146
Wilton (UK) 10
Wilton, George 43
Wiltshire (UK) 10
Windemere Durable Holdings 184; *see also* Applica Consumer Products, Inc. 184
Windsor (Ontario) 90
Windsor Castle 51
Wisconsin 69
"women's work" 5, 15
Wood Shovel and Tool Company 151
Woodbury machine 23
Worcestershire (UK) 185
Worcester (MA) 60
World Series 98
World Trade Center 189
World War I 80, 99
World War II 33, 89, 110, 115, 119, 125, 136, 144
Wright, H.H. 92
Wright, M.S. Company 92
Wright, Russel 116
Wurlitzer 148

Yonkers, Edward H. 116, 124, 206; patent *117*
Young, James Haddon 25

Zanesville (OH) 93
Zinnkann, Reihard 167
Zippo lighter 54
Zollinger, William R. 61

www.ingramcontent.com/pod-product-compliance
Ingram Content Group UK Ltd.
Pitfield, Milton Keynes, MK11 3LW, UK
UKHW050532150426
5217IPUK00026B/1907

9 780786 465521